INDIGENOUS WATER RIGHTS IN LAW AND REGULATION

Indigenous Water Rights in Law and Regulation responds to an unresolved question in legal scholarship: how are (or how might be) indigenous peoples' rights included in contemporary regulatory regimes for water. This book considers that question in the context of two key trajectories of comparative water law and policy. First, the tendency to 'commoditise' the natural environment and use private property rights and market mechanisms in water regulation. Secondly, the tendency of domestic and international courts and legislatures to devise new legal mechanisms for the management and governance of water resources, in particular 'legal person' models. This book adopts a comparative research method to explore opportunities for accommodating indigenous peoples' rights in contemporary water regulation, with country studies in Australia, Aotearoa New Zealand, Chile and Colombia, providing much-needed attention to the role of rights and regulation in determining indigenous access to, and involvement with, water in comparative law.

ELIZABETH JANE MACPHERSON is a Senior Lecturer at the University of Canterbury. She researches comparative Australasian and Latin American natural resources law and indigenous rights. Her legal practice experience includes representing claimants before New Zealand's Waitangi Tribunal and the Victorian State Government on Aboriginal Affairs.

CAMBRIDGE STUDIES IN LAW AND SOCIETY

Founded in 1997, Cambridge Studies in Law and Society is a hub for leading scholarship in socio-legal studies. Located at the intersection of law, the humanities, and the social sciences, it publishes empirically innovative and theoretically sophisticated work on law's manifestations in everyday life: from discourses to practices, and from institutions to cultures. The series editors have longstanding expertise in the interdisciplinary study of law, and welcome contributions that place legal phenomena in national, comparative, or international perspective. Series authors come from a range of disciplines, including anthropology, history, law, literature, political science, and sociology.

Series Editors

Mark Fathi Massoud, *University of California, Santa Cruz*

Jens Meierhenrich, *London School of Economics and Political Science*

Rachel E. Stern, *University of California, Berkeley*

A list of books in the series can be found at the back of this book.

INDIGENOUS WATER RIGHTS IN LAW AND REGULATION

Lessons from Comparative Experience

Elizabeth Jane Macpherson

University of Canterbury, Christchurch, New Zealand

CAMBRIDGE
UNIVERSITY PRESS

CAMBRIDGE
UNIVERSITY PRESS

University Printing House, Cambridge CB2 8BS, United Kingdom

One Liberty Plaza, 20th Floor, New York, NY 10006, USA

477 Williamstown Road, Port Melbourne, VIC 3207, Australia

314-321, 3rd Floor, Plot 3, Splendor Forum, Jasola District Centre, New Delhi - 110025, India

79 Anson Road, #06-04/06, Singapore 079906

Cambridge University Press is part of the University of Cambridge.

It furthers the University's mission by disseminating knowledge in the pursuit of
education, learning and research at the highest international levels of excellence.

www.cambridge.org
Information on this title: www.cambridge.org/9781108460934
DOI: 10.1017/9781108611091

First published 2019
First paperback edition 2021

A catalogue record for this publication is available from the British Library

Library of Congress Cataloging in Publication data
Names: Macpherson, Elizabeth Jane, 1981–
Title: Indigenous water rights in law and regulation : lessons from comparative experience /
Elizabeth Jane Macpherson, University of Canterbury, Christchurch, New Zealand.
Description: New York, NY : Cambridge University Press, 2019. | Series: Cambridge studies
in law and society | Includes bibliographical references and index.
Identifiers: LCCN 2019019488 | ISBN 9781108473064
Subjects: LCSH: Water rights. | Water use – Law and legislation. | Indigenous peoples –
Legal status, laws, etc. | Water – supply – Management.
Classification: LCC K3496 .M33 2019 | DDC 346.04/32–dc23
LC record available at https://lccn.loc.gov/2019019488

ISBN 978-1-108-47306-4 Hardback
ISBN 978-1-108-46093-4 Paperback

For Sam

CONTENTS

ACKNOWLEDGEMENTS

This book is the culmination of eight years of research, the result of which would not have been possible without the contribution of many people from across the globe. Foremost, I thank the editorial team at Cambridge University Press, especially Finola O'Sullivan, for giving me the opportunity to publish this book and supporting me throughout the process. I am also grateful to the anonymous reviewers for their insightful feedback and support for the project.

The book would not have been finalised without the support of the New Zealand Law Foundation, who funded interviews in Aotearoa New Zealand and Australia and research assistance for two chapters, as well as supporting an International Research Workshop on Indigenous Water Rights in Comparative Law. I am also grateful to the Law Schools at the Universities of Melbourne and Canterbury, both of whom provided research support funds for my project at various times. I was very fortunate to receive the Melbourne University Human Rights Scholarship, which funded my PhD research on indigenous water rights in Chile and Australia at Melbourne Law School, and the International Bar Association's Section on Energy, Environment, Natural Resources and Infrastructure, which provided a generous scholarship to enable my year-long research stay at the Pontificia Universidad Católica de Chile in 2013. To the clever and hardworking research students and assistants who helped me with this project at various times – Nathan de Lautour, Claire Thompson, Stephen Campbell, Julia Torres, Francisco Sherriff and Theo van Woerkam – I thank you all for your patience and diligence. Nathan helped me at the final stages of pulling together the book, including providing invaluable guidance around *tikanga* and *te reo* Māori.

Just before this book was finalised, and in order to test its ideas, we held a research workshop on Indigenous Water Rights in Comparative Law at the University of Canterbury on 7 December 2018, with an impressive line-up of some fifty local and international experts. I thank all those who presented and participated, especially the keynote speakers Jacinta Ruru, Lee Godden and Maria del Pilar Garcia Pachón. I also thank those people who met with me to pass on their knowledge and experience throughout this project, especially the forty-five-odd formal interviewees across Aotearoa New Zealand, Colombia, Australia and Chile.

I have been lucky, throughout this project, to have been mentored by a number of knowledgeable and articulate scholars. I am indebted to Maureen Tehan, Kirsty Gover and Lee Godden of the University of Melbourne, for their confidence in my project and for giving me their time, support, inspiration, guidance and criticism when it was needed. At the University of Canterbury School of Law I would like to acknowledge comparative law scholars John Hopkins and Annick Masselot, and the Heads of School Karen Scott and Neil Boister, all of whom have been so supportive of this project and the work required to finish it.

I have been a member of a number of research centres, which provided fertile space for debate, collaboration and dissemination of ideas. I would especially like to acknowledge the Centre for Resources Energy and Environmental Law at the University of Melbourne and the Waterways Research Centre at the University of Canterbury (particularly Ed Challies). I have also been accommodated as a visiting researcher on numerous occasions at the Centro de Derecho y Gestión de Aguas at the Pontificia Universidad Católica de Chile and the Departamento de Derecho de Medio Ambiente at the Universidad Externado de Colombia. The Ngāi Tahu Research Centre kindly provided a letter of support for my research project. I would like to give special thanks to David Espinoza, formerly of the Chilean General Water Directorate, who gave me valuable government data on indigenous water rights in Chile.

I am very grateful to my colleagues in various countries working on indigenous and water rights, who have been a source of knowledge and debate. Thank you especially to Jacinta Ruru, Lee Godden, Sue Jackson, Katie O'Bryan, Lily O'Neill, Erin O'Donnell, Angus Frith, Lauren Butterly, Lisa Carapis, Francisco Correa, Alejandro and Fransisca Vergara, Daniela Rivera, Josey Lang, Francisco Molina, Tyron Love, Ed Challies, Mihnea Tanasescu, Rutgerd Boelens and the Justicia Hidrica Alliance, Jessica Budds, Baden Vertongen, Tom Perrault, Richard Boast and the brilliant Manuel Prieto. To my female colleagues at the University of Canterbury, who I affectionately refer to as the 'fourth floor feminists', your practical advice, champagne and chocolate have pushed me through. Thank you Elisabeth McDonald, Natalie Baird, Rhonda Powell, Annick Masselot, Karen Scott, Toni Collins and Roisin Burke.

I could not have completed this book without the support of friends and extended family members who provided childcare throughout the process, especially my mother Susan Ellanora Everitt, who flew half way around the world several times to make herself available and, in true reflection of the Māori concept of *whanaungatanga*, shares our life and home. Thank you, too, to Jane Connew, Alanna Vivian, Kath Englund, and Brigit and Andrew Henderson. My father, Mike Macpherson, I thank for his unwavering attention and intellectual appetite.

I have felt a commitment to indigenous justice for as long as I can remember. I am truly indebted to the late Henare Hemi Everitt, my *pāpā whakaangi*, and my *kuia* Aunty Jane Woodford for introducing me to the indigenous world and teaching me the true meaning of biculturalism. (*Kua hinga te Totara o te wao nui a Tane* – A Totara in the great forest of Tane has fallen.)

Finally, I would like to thank my family. I have three beautiful children; Santiago, Leonor and Carmela, who constantly surprise me with their capabilities for patience, pragmatism and fun. My partner Sam has been the single biggest source of encouragement, support, relief, understanding and love since the inception of this book project, and I could not have finished it without him. This is for you.

Some of the research for this book has appeared in the following publications:

Elizabeth Macpherson, 'Beyond Recognition: Lessons from Chile for Allocating Indigenous Water Rights in Australia' (2017) 40(3) *University of New South Wales Law Journal* 1130.

Elizabeth Macpherson and Clavijo Ospina, Felipe, 'The Pluralism of River Rights in Aotearoa New Zealand and Colombia' (2018) 25 *Journal of Water Law* 283.

CHAPTER ONE

INTRODUCTION

In all parts of the world, the development of lands and resources and the survival of the people that inhabit them depend on adequate access to water.[1] Yet, in a range of countries, the indigenous peoples,[2] typically the most disadvantaged members of society, continue to be excluded from the right to use, manage, govern or control the water on their lands.[3]

The issue of how to include indigenous peoples in legal frameworks that allocate and regulate rights to water (or, indeed, whether to do so at all) is increasingly pressing worldwide. We see this, for example, in Māori concerns about the management and distribution of water in Aotearoa New Zealand (the Māori word for New Zealand is Aotearoa, meaning 'the land of the long white cloud') in the face of Crown ignorance of treaty protections, which have been the subject of negotiations, court cases, tribunal inquiries, political debates and social protest. We see it in Aboriginal demands for an equitable share of

---

[1] In this book I am concerned with terrestrial water and not coastal or marine water (which is typically managed by a separate legal framework), and focus on surface water in a natural or artificial resource.

[2] I interchangeably use the terms 'indigenous peoples', 'indigenous groups' or 'indigenous communities' to refer to groups of indigenous people regardless of their structure or status in state law, and do not capitalise the word 'indigenous' on the basis that it is a generic term to describe the indigenous people of any country as opposed to the name of a particular culture or nation.

[3] The terms 'water use rights' or 'rights to use water' are variously used to encompass the many different types of water use rights under laws and policies of different times and jurisdictions, sometimes referred to as 'water licences', 'water permits', 'water allocations' or 'water access entitlements'. In contrast, the term 'water rights', including in the context of 'indigenous water rights' and 'commercial water rights', is used in a generic sense to refer to a right to water regardless of whether it is recognised or provided for in a state law of any jurisdiction.

water and the right to be included in water planning processes, amidst growing frustrations about the failures of the native title model. We see it, also, in indigenous resistance to natural resource development in Chile, causing the desecration of sacred sites and the over-extraction of aquifers, and indigenous frustration about the concentration of water rights in the hands of a wealthy few at the expense of vulnerable groups. We see it in indigenous and afrodescendent agitation for the protection of rivers from extreme pollution in Colombia, and in their demands for the right to participate in water management and governance, in the context of a national peace process, an influx of immigration, and an ongoing struggle against the drug trade. Outside those countries, indigenous exclusion from water law frameworks is also well documented, including, but by no means limited to, Peru,[4] Bolivia,[5] Mexico,[6] Ecuador,[7] the United States[8] and beyond.[9] In all of these places, indigenous demands for rights to use and manage water must compete with increasing demand for water resources from agriculture, urbanisation and industry and must contend with decreasing water supply and quality as a result of climate change.

Indigenous rights to land and resources like water, which are usually communal in nature and may have their origins in traditional (pre-sovereignty) laws and customs or historical practices or be a creature of state law, have been altered, affected and encroached upon as a result of colonisation. Indigenous exclusion from legislation and policy regulating the governance, management and use of water (water law frameworks) is often twofold. First, indigenous peoples are typically excluded from laws establishing processes to manage or govern the use of water. Secondly, indigenous peoples are often excluded from legal frameworks

---

[4] Emily J Hogue and Pilar Rau, 'Troubled Water: Ethnodevelopment, Natural Resource Commodification, and Neoliberalism in Andean Peru' [2008] (3/4) *Urban Anthropology & Studies of Cultural Systems & World Economic Development* 283; Rutgerd Boelens, Armando Guevara-Gil and Aldo Panfichi, 'Indigenous Water Rights in the Andes: Struggles Over Resources and Legitimacy' (2010) 20 *Water Law* 268.

[5] Miriam Seeman, *Water Security, Justice and the Politics of Water Rights in Peru and Bolivia* (Palgrave Macmillan, 2016); Tom Perreault, 'Dispossession by Accumulation? Mining, Water and the Nature of Enclosure on the Bolivian Altiplano' (2013) 45(5) *Antipode* 1050.

[6] Lucero Radonic, 'Through the Aqueduct and the Courts: An Analysis of the Human Right to Water and Indigenous Water Rights in Northwestern Mexico' (2017) 84 *Geoforum* 151.

[7] Mihnea Tanasescu, 'The Rights of Nature in Ecuador: The Making of an Idea' (2013) 70(6) *International Journal of Environmental Studies* 846.

[8] Melanie Durette, 'A Comparative Approach to Indigenous Legal Rights to Freshwater: Key Lessons for Australia from the United States, Canada and New Zealand' (2010) 27(4) *Environmental and Planning Law Journal* 296.

[9] Lee Godden, 'Water Law Reform in Australia and South Africa: Sustainability, Efficiency and Social Justice' (2005) 17(2) *Journal of Environmental Law* 181.

that authorise the substantive use of water, which have since colonisation been largely allocated to other users. Calls from within and beyond indigenous communities to address the first type of exclusion have in some cases led to the development of laws and policies to include indigenous peoples in the management and governance of water resources.[10] In other situations, a perceived need to respond to the second type of exclusion has led governments to devise legal and political mechanisms for the recognition, allocation and reallocation of substantive legal rights for indigenous groups to take and use water.

This book responds to an unresolved question in legal scholarship: how are (or how might be) indigenous peoples' rights included in contemporary regulatory regimes for water. I consider that question in the context of two key trajectories of comparative water law and policy: first, the tendency to 'commoditise' the natural environment and use private property rights and market mechanisms in water regulation; secondly, the tendency of domestic and international courts and legislatures to devise new legal mechanisms for the management and governance of water resources, in particular 'legal person' models. In doing so, I am mindful, applying Morgan's approach, of both rights-based concerns (like questions of identity, culture, entitlement and justice) and regulation-based concerns (like efficiency, institutions, the market and the public interest).[11] The book provides much-needed attention to the role of rights and regulation in determining indigenous access to, and involvement with, water in comparative law.

This book adopts a comparative research method to explore opportunities for accommodating indigenous peoples' rights in contemporary regulatory regimes for water. I examine law and policy in four countries that have attempted, to one extent or another, to recognise and provide for indigenous rights to use water, in order to draw out lessons for an international audience. The four countries: Australia, Chile, Aotearoa New Zealand and Colombia exemplify a spectrum of responses to a similar problem of indigenous water injustice, representing approaches from the global north and south and distinct legal traditions in Australasia and Latin America. Two of the countries included in the study, Australia and Chile, are 'paradigm examples' of the use of market

---

[10] I use the term water 'governance' in a broad sense to mean the exercise of authority and control over water resources via formal and informal legal and policy frameworks. Water governance includes, but is broader than, water management.

[11] B Morgan, *The Intersection of Rights and Regulation: New Directions in Sociolegal Scholarship* (Aldershot, Ashgate, 2007) 1.

mechanisms in water regulation, in which highly uniform water access rights may in some situations be traded independent of land in water markets.[12] In both of these countries, there is major concern about the accumulation of water rights by powerful agricultural and industry interests at the expense of indigenous peoples, and the need for some sort of recognition or redistribution.

At the other end of the spectrum, New Zealand and Colombia have, at least until now, resisted the 'commoditisation' of water rights and the introduction of markets. In those countries, water continues to be allocated by central and local governments via administrative concessions and water cannot be 'owned', although concerns remain about the exclusion of indigenous peoples from legal frameworks distributing use rights. New Zealand and Colombia are also leading global examples of governments addressing indigenous water concerns via the recognition of rivers as legal persons. The New Zealand Government declared the Whanganui River to be a legal person in 2017 as part of a political settlement with local Māori, giving the river the right to hold property, enter into contracts, sue and be sued in its own name.[13] The Whanganui River model inspired the recognition of the Río Atrato as a legal person by the Constitutional Court of Colombia under the guardianship of indigenous and afrodescendent communities.[14] A similar proposal has been developed to involve indigenous people and values in the management of the Yarra River in Australia, albeit without explicitly recognising that the river is a person.[15] In all of these countries, governments have attempted to accommodate indigenous water interests, with varying degrees of success, offering important lessons about the place of indigenous water rights in diverse regulatory approaches.

Drawing on the analysis of the four countries studied, I make a key proposition in this book. That is, that if indigenous peoples are to be finally included in water law frameworks they must have both: the *jurisdiction* to manage, govern or control their water resources in

---

[12] See, e.g., Carl J Bauer, *Siren Song: Chilean Water Law as a Model for International Reform* (Resources for the Future, 2004) 1; Lee Godden, 'Governing Common Resources: Environmental Markets and Property in Water' in *Property and the Law in Energy and Natural Resources* (Oxford University Press, 2010) 413, 426.

[13] *Te Awa Tupua (Whanganui River Claims Settlement) Act 2017.*

[14] *Centro de Estudios para la Justicia Social 'Tierra Digna' and Others v. The President of the Republic and Others* [2016] Corte Constitucional [Constitutional Court], Sala Sexta de Revision [Sixth Chamber] (Colombia) No T-622 of 2016 (10 November 2016) ('*Tierra Digna*').

[15] *Yarra River Protection (Wilip-Gin Birrarung Murron) Act 2017.*

culturally appropriate ways; and a substantive *distribution* of the available consumptive pool of water to use for any purpose. Much of the scholarly or policy literature focuses on one or other of these imperatives. For example, the focus may be that indigenous peoples should be consulted about or involved in decision making on the management of water resources, or that indigenous peoples should be given some sort of substantive water allocation or right. However, all of the country studies in this book suggest dual concerns for indigenous water jurisdiction and distribution. For example, in Aotearoa New Zealand debates about Māori water rights and their reception in law and policy are pushing towards increasing, culturally appropriate, involvement in water resource management or governance, perhaps via legal person models, and also the allocation of a substantive water allocation for Māori to use for any purpose. As will be shown in this book, addressing only one or the other of these imperatives may amount to an incomplete response to indigenous water injustice.

1.1 THE COMPARATIVE STUDY

This book is an exercise in comparative law. Comparative law is the study of a foreign legal system with the purpose of understanding one's own system a little better. Legrand calls this 'comparing in circles', and he likens it to the story of Odysseus going off on a long, fraught journey only to come back home and understand things there a little better.[16] This is true both of studies of the laws of other countries, but also of studies of other systems within one's own country, like indigenous systems of law. Traditionally, the functionalist approach to comparative law recommended comparing systems that were functionally equivalent: like common law with common law, English language with English language or global north with global north.[17] But more recent scholarship on comparative law, like the work of Orucu, Van Hoeke, Lasser and Cotterell, encourages the researcher to 'cast the net wider' and consider countries beyond the usual comparators.[18]

---

[16] Pierre Legrand 'Comparing in Circles' in Penelope (Pip) Nicholson and Sarah Biddulph (eds.), *Examining Practice, Interrogating Theory: Comparative Legal Studies in Asia* (Nijhoff, 2008), preface.

[17] Konrad Zweigert and Hein Kötz, *Introduction to Comparative Law* (Oxford University Press, 1998).

[18] A Esin Orucu, 'Methodology of Comparative Law' in JM Smits Elgar (ed.), *Encyclopedia of Comparative Law* (Edward Elgar Publishing, 2006) 442, 445; Mark Van Hoeke, 'Deep Level Comparative Law' in Mark Van Hoeke (ed.), *Epistemology and Methodology of Comparative Law*

This book is an attempt to assist those considering the legal treatment of indigenous rights to water around the world, via the analysis of comparative, foreign experiences. There is a nascent, yet growing body of comparative law literature on indigenous water rights,[19] although further considerations of comparative experience are needed to answer new and challenging questions as they emerge,[20] especially beyond the usual common law comparators. Significantly, this book provides valuable insight for an English-speaking audience into the complex, yet often inaccessible, law and experience of indigenous water rights in Latin America.

Australia, Chile, Aotearoa New Zealand and Colombia all face similar concerns around indigenous access to water and their place in water regulation, and offer important contributions to comparative domestic and international debates. There are clearly legal, political, institutional, cultural and social differences between the countries studied in this book. For example, Chile and Colombia belong to the civil law tradition, modelled on Roman law, where legislated law is supreme and the courts have a minimal interpretive role (although, as we will see in Chapter 6, Colombia may prove an exception to this general proposition).[21] However, the Australasian and Latin American cases at times exhibit strikingly similar problems in terms of environmental conditions and indigenous disadvantage, and in some cases

(Hart Publishing, 2004) 165; Roger Cotterrell, 'Subverting Orthodoxy, Making Law Central: A View of Sociolegal Studies' in Roger Cotterrell (ed.), *Living Law: Studies in Legal and Social Theory* (Ashgate, 2008); Mitchel De S-O.-L'E Lasser, 'The Question of Understanding' in Pierre Legrand and Roderick Munday (eds.), *Comparative Legal Studies: Traditions and Transitions* (Cambridge University Press, 2003).

[19] Recent examples include Seeman, above n. 5; Elizabeth Macpherson and Felipe Clavijo Ospina, 'The Pluralism of River Rights in Aotearoa New Zealand and Colombia' (2018) 25 *Journal of Water Law* 283; Elizabeth Macpherson, 'Beyond Recognition: Lessons from Chile for Allocating Indigenous Water Rights in Australia' (2017) 40(3) *University of New South Wales Law Journal* 1130; Katie O'Bryan, *Indigenous Rights and Water Resource Management: Not Just Another Stakeholder* (Routledge, 2018); Sue Jackson, 'Enduring and Persistent Injustices in Water Access in Australia' in *Natural Resources and Environmental Justice: Australian Perspectives* (CSIRO Publishing, 2017); Rutgerd Boelens et al., 'Contested Territories: Water Rights and the Struggles over Indigenous Livelihoods' (2012) 3(2) *International Indigenous Policy Journal* 1.

[20] See, e.g., Lee Godden, Raymond L Ison and Philip J Wallis, 'Water Governance in a Climate Change World: Appraising Systemic and Adaptive Effectiveness' (2011) 25(15) *Water Resources Management* 3971, 3972; Sue Jackson, 'Background Paper on Indigenous Participation in Water Planning and Access to Water – Report Prepared for the National Water Commission' in CSIRO (ed.) (2009) 17.

[21] See generally, Javier Barrientos Grandón, 'Juan Sala Bañuls (1731–1806) y El "Código Civil" de Chile (1855)' ['Juan Sala Bañuls (1731–1806) and the Chilean "Civil Code" (1855)'] (2009) 31 *Revista de Estudios Historico-Juridicos* 351; 'Civil Law' [2013] *Columbia Electronic Encyclopedia, 6th Edition* 1.

even similar governmental attempts to respond to those problems. For example, Chile has comparable water conditions to Australia and in parts of both countries water scarcity and pollution in the face of climate change and increasing water demands from agriculture, sanitation and industry are serious problems.[22] In both Colombia and New Zealand, water scarcity is less of an issue with high annual rainfall, although due to climate variability and a history of poor water management, in some parts of both countries waters are reaching full or over allocation,[23] and water quality has been compromised.[24] In all countries, indigenous groups make up the most disadvantaged sector of society and governments have committed (at least to some extent) to reducing indigenous disadvantage, including by supporting indigenous involvement in natural resource decision making and the productive use of indigenous territories.

The country studies in this book all demonstrate similar unresolved challenges for governments engaging with both water regulation and indigenous rights. These include epistemological uncertainty about the recognition of indigenous peoples and rights in legal and political theory and the reconciliation of claims by indigenous peoples concerning natural resources. The studies demonstrate the potential for governments to take novel approaches to comparative natural resource management, including, where culturally appropriate, a tendency towards more 'eco-centric' regulation, pursuant to which natural resources hold inherent value in and of themselves, as opposed to being a resource for human exploitation. The studies suggest increasing recourse to constitutional law and human rights protections to further the interests of indigenous peoples in the protection of their particular water interests. And all cases allude to wider policy issues concerning

---

[22] See R Quentin Grafton et al, 'An Integrated Assessment of Water Markets: A Cross-Country Comparison' (2011) 5(2) *Review of Environmental Economics and Policy* 219, 220.

[23] A water resource is considered fully allocated where with full development of water access entitlements in relation to a particular water resource, the total volume of water able to be extracted by entitlement holders at a given time reaches the environmentally sustainable level of extraction for that system. See Commonwealth of Australia and the Governments of New South Wales, Victoria, Queensland, South Australia, the Australian Capital Territory and the Northern Territory, 'Intergovernmental Agreement on a National Water Initiative', Sch B(i), definition of 'overallocation'.

[24] Macpherson and Clavijo Ospina, above n. 19; Erin O'Donnell and Elizabeth Macpherson, 'Voice, Power and Legitimacy: The Role of the Legal Person in River Management in New Zealand, Chile and Australia' [2018] *Australasian Journal of Water Resources* 1; María del Pilar García Pachón, *Régimen Jurídico de Los Vertimientos en Colombia: Análisis Desde el Derecho Ambiental y el Derecho de Aguas* [*The Legal Regime for Wastewater in Colombia: An Environmental and Water Law Analysis*] (Universidad Externado de Colombia, 2017).

the regulation of water, including whether water is a human right to which all are entitled or a commodity capable of 'ownership', and, if the latter, whether water should be able to be traded in markets.

This book proceeds on the basis that there is much to be learned about the ways in which law may provide for indigenous water rights from considering the experiences of other jurisdictions. As De Stefano points out: '[i]n a globalized world, the use of comparisons between countries is a powerful policy tool ... it makes it possible to identify good management practices, positive supranational trends, the potential need for concerted actions at an international level, or gaps in existing supranational initiatives'.[25] There are clearly risks involved in comparative research and it is difficult to understand deeply and do justice to a foreign legal system.[26] In this book I pay close attention to social, cultural, political and historical context in order to avoid misinterpreting other 'legal languages'. Unless stated otherwise, all translations have been provided by the author, and where possible, foreign language terms are retained (and italicised). The book includes a glossary of common foreign language terms at the end although, as a sad reflection on the process of colonisation in Australia and Latin America, the only indigenous language commonly used in legal documents, and therefore referenced in this book, is *Te Reo Māori*.

In undertaking this research I have been 'critically reflective',[27] by acknowledging, reflecting on, and accounting for my own cultural difference, assumptions and perspectives. I do not, for example, identify as indigenous, although I was lucky to be raised in a bicultural and bilingual Māori/*Pākehā* (New Zealand European) family. Nor am I Latin American, and although my children are and I have for many years been linguistically and culturally immersed in Latin American society, I inevitably have my own *Pākehā* assumptions and perspectives.

The critical appreciation of context in this book comes not only from an analysis of law and interdisciplinary scholarship, but from extensive doctrinal and empirical research carried out in all four countries during the course of eight years. The book draws on a rich archive of primary sources collected and interviews undertaken in each country to reveal data and analysis that has until now been uncollected, untranslated and

---

[25] Lucia De Stefano, 'International Initiatives for Water Policy Assessment: A Review' (2010) 24 (11) *Water Resources Management* 2449, 2450.

[26] Mitchel De S.-O.-L'E. Lasser, 'The Question of Understanding' in Pierre Legrand and Roderick Munday (eds.), *Comparative Legal Studies: Traditions and Transitions* (Cambridge University Press, 2003) 212–13.

[27] Lasser, above n. 18, 198.

unpublished. This includes in-depth archival research conducted at government offices, archives and libraries across the countries studied and forty-five semi-structured interviews with government officials, former politicians, community representatives, academics, activists and legal practitioners working in the field of indigenous water rights in Chile, Colombia, Aotearoa New Zealand and Australia. Although the interviewees are all highly experienced in water law and policy and indigenous rights, and have held various senior government, community and practitioner roles in the field, all views expressed by them are interpreted as being their own and are not presented as the view of any organisation in an official capacity.

Of course, indigenous water rights, and regulatory frameworks governing their exercise, exist regardless of (and in spite of) the extent of any formal recognition or mandate by the state. Those rights can be described as 'property', regardless of whether they are recognised as such.[28] However, they are usually understood as being distinct from the propertised (or 'commodified')[29] rights to land and resources typically recognised or allocated by western governments.[30] The task in this book is to understand what governments are doing, or may do, to provide for indigenous water rights. Because the focus of the book is on the inclusion of indigenous water rights in state law and policy, the interviews are primarily with government representatives working in water law and governance or experts on state laws, although some of these representatives and experts were in fact indigenous. The comparative experiences uncovered in the book provide new perspectives on the reasons why indigenous water rights are needed, and the role law might play to provide for them.

1.2   OVERVIEW OF THE BOOK

In Part I of the book, I consider the key conceptual challenges inherent in attempts by governments to provide for indigenous water rights, and

---

[28] See Godden, above n. 12, 413.
[29] See Jonnette Watson Hamilton and Nigel Banks, 'Different Views of the Cathedral: The Literature on Property Law Theory' in *Property and the Law in Energy and Natural Resources* (Oxford University Press, 2010) 19.
[30] Indigenous land and resource rights are sometimes presented as falling within 'common property regimes'. See Carol M Rose, 'Expanding the Choices for the Global Commons: Comparing Newfangled Tradeable Allowance Schemes to Old Fashioned Common Property Regimes' (1999) 10 *Duke Environmental Law & Policy Forum*; Seán Kerins, 'Social Enterprise as a Model for Developing Aboriginal Lands' (Australian National University, 2013).

to accommodate indigenous interests in legal frameworks for the regulation of water.

In Chapter 2 I explore the tensions inherent in debates about indigenous water rights in legal and political theory, setting up the key propositions for this book. I argue that legal and policy mechanisms that seek to recognise cultural relationships with water and involve indigenous peoples in water governance should strive towards recognising indigenous water relationships, but, more importantly, indigenous water jurisdiction. This argument is central to the consideration of the four country studies included in this book, in which law and policy is sometimes able to provide a space for indigenous groups to exercise jurisdiction in planning and governing their water resources. I also contend that the reason states should provide for indigenous water rights is an imperative of distribution. Such rights are needed not only to remedy the historical injustice of non-recognition, but also because indigenous exclusion from water law frameworks is ongoing.

In Chapter 3 I introduce the two key regulatory tendencies relevant to the study of indigenous water rights in comparative law. One of these developments is the idea that governments should 'commoditise' the natural environment and use private property rights and market mechanisms in water regulation and allocation; an approach typically counterposed with the idea of treating access to water as a fundamental human right, to which all are entitled. The other is the tendency to devise new legal mechanisms like 'legal personality' to protect the 'rights of nature' and address social or community concerns around water governance and quality. Both trajectories play out repeatedly in debates about indigenous rights to water in comparative law, and resulting legal and policy frameworks in the country studies considered in this book. I argue that most regulatory frameworks are in fact a combination of public and private interests and transactions, and suggest that both private and public mechanisms may have a place in debates about how best to provide for indigenous water rights.

Part II of the book includes the four country studies, which critically examine legal and policy mechanisms providing for indigenous water rights in Australia, Aotearoa New Zealand, Colombia and Chile, respectively.

In Chapter 4 I consider the limited recognition of traditional, cultural water rights in Australian law. In the Australian model, often put forward as international best practice for water regulation, property rights in water and water markets accompany government oversight and planning. However, while Australian water law has undergone drastic reforms since the early 1990s, and indigenous land rights are now broadly recognised in Australia, little has been done to provide indigenous peoples with the right to use water on their lands for commercial and productive purposes. Native title rights to water have been interpreted narrowly by the courts according to traditional and cultural uses, and are usually accounted for as in-stream cultural and conservation values in water catchments, distinguishing them from the consumptive rights held by other users. Yet indigenous Australians continue to make up the most disadvantaged sector of Australian society and Australian governments have committed to reducing that disadvantage, including by supporting the development of indigenous lands. The Australian experience demonstrates the difficulties inherent in recognising historical indigenous rights to land and resources, as indigenous water practices change over time and conflict with other uses. The study highlights the need for an allocative model, enabling both the reservation of water for indigenous allocation and the redistribution of water rights in fully allocated catchments.

In Chapter 5 I consider water rights for Māori in Aotearoa New Zealand. In the 2017 settlement between local Māori and the Crown the Whanganui River was recognised as '*Te Awa Tupua*' ('an indivisible and living whole, incorporating all its physical and meta-physical elements'), and declared to be a legal person. The arrangement gives certain guardianship and governance rights, but not property rights, to the Māori people that traditionally owned the river. At a national level, however, Māori continue to agitate, both politically and before courts and tribunals, for the right to 'own' their water resources, amid cautious government plans for water law reform. The New Zealand study raises interesting questions about the nature of water in law; as a private right to be held and allocated, or a public interest incapable of ownership. Māori variously seek both recognition of their distinctive water relationships and influence and control over water governance and a substantive share of the consumptive pool of water for any purpose including economic development. The study of indigenous rights to water in Aotearoa New Zealand in this chapter demonstrates the variability of indigenous water demands, and a need for multi-faceted responses to indigenous water exclusion.

In Chapter 6 I examine indigenous water rights in Colombia and, specifically, the declaration by Colombia's highest court that the Atrato River is a 'legal subject' in response to indigenous concerns about water management. This watershed case of November 2016 was an action for protection of constitutional rights brought in the Colombian Constitutional Court on behalf of a number of indigenous and afrodescendent communities, in response to serious environmental and humanitarian damage caused by illegal mining in the region of Chocó. I show in this chapter how the legal person model for the Atrato was adopted to reflect the 'biocultural rights' of indigenous and tribal communities, but the approach is clearly not a complete answer to indigenous water exclusion. Indigenous peoples also need substantive water allocations, in order to have a sufficiently strong voice in lobbying for water access and influencing decision making about river management. Yet, because the Atrato is a subject it has representatives from the community, or guardians, and they have a voice on behalf of the River, where previously they had none. The Colombian study is highly significant, in that it underscores the potential for legal person models to create new jurisdictions for indigenous peoples in which to participate in river sharing, governance and use.

The final country study, Chapter 7, examines indigenous water rights recognition and distribution in Chile. In this chapter, I consider the recognition of the *ancestral* water rights of indigenous peoples in article 64 of the *Indigenous Law* and the creation of an Indigenous Land and Water Fund for the acquisition of rights in the market. I argue that the recognition of *ancestral* water rights is an incomplete response to the ongoing exclusion indigenous peoples experience from rights allocated within water law frameworks, because it continues to exclude groups who have lost water access to other users. The Fund, by contrast, specifically responds to the situation where indigenous peoples have been unable to continue to exercise their water relationships. In the case of already fully allocated water resources, the Fund finances the purchase of water use rights in markets for redistribution to indigenous landholders. An interesting lesson from the Chilean experience is that market mechanisms may, in some situations, enable 'creative' responses to the injustice in water rights distribution. However, by setting aside a share of water use rights before water resources are already fully allocated, governments reduce the cost of buying back water use rights for allocation to indigenous peoples in the future.

In the third and final part of this book, I bring together the findings from the comparative country studies to generalise some observations about the current state of, and potential for, indigenous water rights in state law. I argue that governments must finally address historical water injustice, and respond to the exclusion indigenous peoples have experienced, and continue to experience, from water law frameworks. I argue that this cannot be done, if indigenous peoples lack either the jurisdiction to exercise authority and influence over water management and governance in their territories, or a fair distribution of substantive rights to use water under legal and policy frameworks. I conclude the book with a reflection on how a more complete response to indigenous water injustice might look.

PART I

CONCEPTUALISING INDIGENOUS WATER RIGHTS

CHAPTER TWO

JUSTIFYING INDIGENOUS WATER RIGHTS:
Jurisdiction and Distribution

2.1 INTRODUCTION

Demands for indigenous rights to manage, access and use water have been grounded in ideas of a distinct indigenous culture, and have been received as such in scholarly and policy debates. Like indigenous rights to land and other resources, indigenous water rights are usually presented as being somehow different from the way in which we think of water rights in western law; something warranting 'recognition'. The literature abounds with descriptions of the cultural relationships with water held by indigenous peoples, often involving a deep spiritual and ancestral connection between people and resource, and an obligation on people to govern and care for the resource for present and future generations, rather than just exploit it for their own gain.[1]

   In the first part of this chapter, I consider the framing of indigenous rights to water as distinctive 'cultural' rights. The distinct indigenous culture in which indigenous water rights are located is routinely associated with pre-contact, subsistence or environmental resource uses as opposed to modern, commercial or market-based uses. However, this portrayal has, for the most part, been at odds with indigenous demands for water rights, which, although they emphasise indigenous difference and an holistic view of the natural environment, do not restrict land or

---

[1] See, e.g., Brendan Tobin, *Indigenous Peoples, Customary Law and Human Rights: Why Living Law Matters* (Routledge, Taylor & Francis Group, 2014) 141.

resource relationships to particular sorts of use. On the contrary, indigenous water claims often emphasise a desire for access to, or ownership of, water for a range of purposes, certainly including the management and use of water for cultural, spiritual and environmental ends, but not at the expense of a substantive allocation for commercial use.

I then set up the first of two key themes for the consideration of indigenous water rights in comparative law in this book. That is, that legal and policy mechanisms that seek to recognise cultural relationships with water and involve indigenous peoples in water governance should strive towards recognising indigenous water relationships but, more importantly, indigenous water jurisdiction. This argument is central to the consideration of the four country studies included in this book, in which law and policy is sometimes able to provide a space for indigenous groups to exercise jurisdiction in planning and governing their water resources.

In the second part of this chapter, I consider the justifications typically put forward to support claims for indigenous water rights, both cultural and commercial. This consideration is important because the rationale underlying legal and policy models places parameters around what the model can, and also cannot, do. For present purposes, I group the justifications commonly made for indigenous water rights into two sets.[2]

The theory most typically put forward to support indigenous water rights is the idea of 'reparative justice', which focuses on a need to respond to historical injustice inflicted through colonisation via recognising rights that have their origins in pre-sovereignty use. However, reparative justifications for indigenous water rights are problematic for two reasons. If the object of indigenous water rights is to recognise pre-sovereignty water interests, must the rights be constrained by pre-sovereignty notions of resource use? What happens when rights to the resources traditionally used by that indigenous group have (post-sovereignty) been allocated to other users? Reparative justifications have been responsible for recognition mechanisms like the Australian common law doctrine of native title, which recognises interests arising out of traditional laws and customs from before British sovereignty. As will be shown in Chapter 4, this model has enabled only very limited recognition of indigenous rights to water for traditional, cultural purposes. Reparative justifications also informed the recognition of ancestral

---

[2] The distinction here is for heuristic purposes and is just one way of organising the diverse scholarship on indigenous land and resource rights.

water rights in Chile which, as will be shown in Chapter 7, have offered similarly little potential for resolving indigenous water injustice.

Increasingly, demands for recognition or allocation of indigenous water rights are supported by ideas of 'equality', 'equity' or 'distributive justice'. Instead of focusing on the recognition of traditional, historical or ancestral rights, focus has turned to the fact that indigenous peoples *continue* to be excluded from water law frameworks. Such justifications direct reform to the *ongoing* injustice in the distribution of water use rights. I finish this chapter by introducing the second of the two key themes for the consideration of indigenous water rights in comparative law in this book: that the need for indigenous water rights is in fact a distributive imperative. By shifting the focus of indigenous water laws and policies towards redressing water injustice now, there is less risk of pigeon-holing indigenous water rights through a narrow 'cultural' lens. It is logical that indigenous water rights should encompass the full spectrum from cultural to commercial purposes, and be allocated (or, if necessary, redistributed) as such.

2.2 CULTURAL RELATIONSHIPS WITH WATER AND INDIGENOUS WATER 'JURISDICTION'

The distinctly cultural relationships of indigenous peoples with the natural world have been well-traversed in the theoretical literature. Hendrix, for example, argues that the basis of indigenous land claims is a complex set of arguments about the cultural relationship between indigenous people and the natural world.[3] Tobin argues for indigenous rights to natural resources as, 'vital for protection of their cultural integrity and their survival as distinct peoples'.[4]

The cultural relationships of indigenous peoples and their water resources, too, are well-documented. For example, O'Bryan argues: '[f]or Indigenous people around Australia, and indeed the world, water is an essential part of country, culture and identity'.[5] Jackson has also explained in the Australian context that:[6]

---

[3] See, e.g., Burke A Hendrix, 'Context, Equality, and Aboriginal Compensation Claims' (2011) 50 *Dialogue: Canadian Philosophical Review* 669, 672.

[4] Tobin, above n. 1, 141.

[5] Katie O'Bryan, 'The National Water Initiative and Victoria's Legislative Implementation of Indigenous Water Rights' (2012) 7(29) *Indigenous Law Bulletin* 24.

[6] Sue Jackson, 'Recognition of Indigenous Interests in Australian Water Resource Management, with Particular Reference to Environmental Flow Assessment' (2008) 2(3) *Geography Compass* 874, 876–7.

Indigenous Australians hold distinct cultural perspectives on water relating to identity and attachment to place, environmental knowledge, resource security and the exercise of custodial responsibilities to manage inter-related parts of customary estates.

In Chile, too, indigenous peoples are recognised in scholarly and policy circles as being sociologically and culturally distinct from the dominant society, which warrants preservation.[7] During the Parliamentary Debates on Chile's *Indigenous Law*, Octavio Jara explained:[8]

> ... these cultures have systems of life, ways of living, customs, traditions, work methods, language, religion, technical knowledge, institutions, artistic expressions and values that distinguish them from the global culture.

In Colombia, the peoples of Chocó who live alongside the Atrato River depend on the river for their physical and spiritual sustenance,[9] and have distinct relationships with the river not just as their ancestral territory, but as a 'space to reproduce life and recreate culture'.[10] For the Atrato communities the river is a social, economic, logistical, spiritual and territorial space, forming the core of their distinct cultural identity.[11]

---

[7] See, e.g., Jose Aylwin, *Pueblos Indígenas de Chile: Antecedentes Históricos y Situación Actual* [*Indigenous Communities of Chile: History and Current Situation*] (Instituto de Estudios Indigenas Universidad de la Frontera, 1994) Vol. 1; Lila Barrera-Hernández, 'Indigenous Peoples, Human Rights and Natural Resource Development: Chile's Mapuche Peoples and the Right to Water' (2005) 11(1) *Annual Survey of International & Comparative Law*; Lila Barrera-Hernandez, 'Got Title Will Sell: Indigenous Rights to Land in Chile and Argentina' in Aileen McHarg et al. (eds.), *Property and the Law in Energy and Natural Resources* (Oxford University Press, 2010); Nancy Yañez and Raul Molina, *Las Aguas Indigenas en Chile* [*Indigenous Waters in Chile*] (LOM Ediciones, 2011); Milka Castro Lucic et al., *El Derecho Consuetudinario en La Gestión del Riego en Chiapa. Las Aguas Del 'Tata Jachura'* [*Customary Rights in Irrigation Management in Chiapa. The Waters of 'Tata Jachura'*] (Konrad Adenauer Stiftung, 2017).

[8] Chile, 'Discusion Sala' [Parliamentary Debates], Camera De Diputados [House of Representatives], 21 January 1993 (Octavio Jara) in Biblioteca del Congreso [Library of National Congress of Chile] Nacional de Chile, 'Historia de la Ley No 19.253 Establece Normas Sobre Protección, Fomento y Desarrollo de los Indígenas, y Crea la Corporación Nacional de Desarrollo Indígena [History of Law No. 19.253 to Establish Norms for the Protection, Creation and Development of the Indigenous, and to Create the National Corporation of Indigenous Development]' (5 October 1993) 150–1.

[9] Camilo Antonio Hernandez, *Ideas y Practicas Ambientales del Pueblo Embera del Chocó* (Cerec, 1995) 12.

[10] *Tierra Digna* [2016] Corte Constitucional [Constitutional Court], Sala Sexta de Revision [Sixth Chamber] (Colombia) No T-622 of 2016 (10 November 2016) 165.

[11] Acta Final de Inspección Judicial, Appendix 1, Part B and [6.2] in *Tierra Digna* [2016] Corte Constitucional [Constitutional Court], Sala Sexta de Revision [Sixth Chamber] (Colombia) No T-622 of 2016 (10 November 2016); interview with Pilar Garcia (Bogotá, 30 August 2017).

In Aotearoa New Zealand, Māori relationships with natural resources are based on distinct *tikanga* Māori (Māori laws and customs), which lay out intricate rules for their use and management.[12] Those rules are different from western laws around resource use, with reciprocal obligations between interdependent people and living natural resources based on culturally distinct values of *whanaungatanga* (kinship) and *whakapapa* (genealogies).[13]

'Culture' is a term used in many different contexts and is capable of many different interpretations. Kymlicka uses culture in the sense of a 'societal culture', 'which provides its members with meaningful ways of life across the full range of human activities, including social, educational, religious, recreational, and economic life, encompassing both public and private spheres. These cultures tend to be territorially concentrated, and based on a shared language.'[14] Coulthard defines culture as 'the interconnected social totality of distinct mode of life encompassing the economic, political, spiritual and social'.[15]

The idea that indigenous rights derive from a distinct indigenous 'culture' can be traced back to the indigenous rights movement of the 1960s and 70s. The movement sought recognition of indigenous identity and land rights in response to policies that aimed to assimilate indigenous cultures into western culture.[16] A number of political theorists have since considered what this 'culture' imperative really means. Taylor argued that the recognition of cultural identity and difference is 'a vital human need'.[17] Kymlicka argued that group-specific rights for ethnic and national minorities are needed to ensure that all citizens are treated with genuine equality (with some limits).[18] Tully argued that states must accommodate cultural diversity,

---

[12] Nin Tomas, 'Maori Concepts of Rangatiratanga, Kaitiakitanga, The Environment, and Property Rights' in David Grinlinton and Prue Taylor (eds.), *Property Rights and Sustainability* (BRILL, Martinus Nijhoff Publishers, 2011) Vol. 11, 228.

[13] Joseph Williams, 'Lex Aotearoa: An Heroic Attempt to Map the Māori Dimension in Modern New Zealand Law' (2013) 21 *Waikato Law Review* 1, 3.

[14] See, e.g., Will Kymlicka, *Multicultural Citizenship: A Liberal Theory of Minority Rights* (Oxford University Press, 1996) 2–3.

[15] Glen Sean Coulthard, *Red Skin, White Masks: Rejecting the Colonial Politics of Recognition* (University of Minnesota Press, 2014) 65–6.

[16] Karen Engle, *The Elusive Promise of Indigenous Development: Rights, Culture, Strategy* (Duke University Press, 2010) 55.

[17] See, e.g., Charles Taylor, 'Multiculturalism: Examining the Politics of Recognition' in Amy Gutmann (ed.), *The Politics of Recognition* (Princeton University Press, 1994) 25–33.

[18] Kymlicka, above n. 14, 2.

including allowing indigenous peoples and institutions to 'speak in their own languages and customary ways'.[19]

Yet where indigenous rights to land and resources are based upon an idea of indigenous cultural difference there is a tendency to restrict the purposes for which the rights can be exercised. The practice of indigenous land and resource rights for commercial ends, for example, may no longer be considered 'cultural' if it means borrowing methods or technology from modern, western cultures or traditions.[20] Commercial or capitalist resource use is sometimes presented as being inconsistent with 'authentic' indigenous culture.[21] As an example, in Australia's Murray Darling Basin, indigenous groups have campaigned for 'cultural flows', the title of which accentuates the distinct cultural nature of indigenous water interests.[22] However, the policy has thus far been implemented by leveraging off existing 'in-stream' environmental flow protections, without the allocation of a substantive water use right. Such protections do not authorise indigenous groups to take and use water for commercial purposes, prompting Jackson and Langton to label them 'essentialist'.[23] Indigenous groups in other parts of Australia have distanced themselves from the cultural flows concept and emphasise a desire for commercial as well as cultural rights.[24] In New Zealand, Maori water rights claimants before the Waitangi Tribunal have expressed a desire to 'walk in two worlds: to resist assimilation and protect their matauranga Maori and tikanga (knowledge and law) but also to benefit commercially from development'.[25]

If indigenous peoples seek recognition of their distinct cultural relationships with water, what they appear to seek is the space to practice or 'recreate' that culture with respect to their water resources. Strelein and Tran argue, '[t]he decolonization strategy is to create

---

[19] See, e.g., James Tully, *Strange Multiplicity* (Cambridge University Press, 1995) 184.

[20] Duncan Ivison, 'The Logic of Aboriginal Rights' (2003) 3(3) *Ethnicities* 321, 325.

[21] See generally Engle, above n. 16, 186–7.

[22] Murray Lower Darling Rivers Indigenous Nations and Northern Murray–Darling Basin Aboriginal Nations, *Agreed Definition of Cultural Flows* www.mdba.gov.au/explore-the-basin/ communities/indigenous-communities.

[23] Sue Jackson and Marcia Langton, 'Trends in the Recognition of Indigenous Water Needs in Australian Water Reform: The Limitations of "Cultural" Entitlements in Achieving Water Equity' (2012) 22(2/3) *Journal of Water Law* 110, 117.

[24] See, e.g., Northern Australian Land and Sea Management Alliance, 'Knowledge Series: Indigenous People's Right to the Commercial Use and Management of Water on Their Traditional Territories' (2013) www.nailsma.org.au.

[25] The Waitangi Tribunal, 'The Interim Report on The National Freshwater and Geothermal Resources Claim' (The Waitangi Tribunal, 2012) 38, 44.

a space for indigenous governance to continue to "breathe".[26] Boelens et al. explain such an approach as 'establishing the necessary conditions under the law (access to water and autonomy for management), in order to stay out of the way of the law'.[27] Webber refers to recognising 'broad interests', but avoiding overly prescribing the content of the right or its conditions,[28] leaving as much of the internal administration of the right as practicable with the indigenous people themselves.[29] The way in which indigenous peoples choose to 'use' natural resources, may not in fact coincide with western notions of authentic indigenous culture. Tobin argues that one of the defining aspects of self-determination is the right of indigenous peoples, sovereign over their natural resources, to freely dispose of their natural wealth, should they wish to do so.[30] These ideas resonate with how lawyers perceive 'jurisdiction', as legal power to make decisions or rules. Dorsett and McVeigh describe jurisdiction as authority to perceive, pronounce and inaugurate law.[31]

Thus, states must allow indigenous groups to practice their cultural ways over, and exercise their cultural relationships with respect to, their water bodies, applying their own laws and customs in water management and governance. This is the imperative of indigenous water jurisdiction.

2.3 SUBSTANTIVE INDIGENOUS WATER RIGHTS AND THE IMPERATIVE OF 'DISTRIBUTION'

According to liberal political theory, 'rights' protect 'fundamental human interests'.[32] The idea of indigenous rights has arisen in reaction to the event of colonisation and its related injustices because, as Webber explains, 'the need to frame interests as rights only arises

---

[26] Lisa Strelein and Tran Tran, 'Building Indigenous Governance from Native Title: Moving Away from "Fitting in" to Creating a Decolonized Space' (2013) 18(1) *Review of Constitutional Studies* 19, 47.

[27] Rutgerd Boelens et al., 'Special Law' in Dik Roth, Rutgerd Boelens and Margreet Zwarteveen (eds.), *Liquid Relations: Contested Water Rights and Legal Complexity* (Rutgers University Press, 2005) 167.

[28] Jeremy Webber, 'Beyond Regret: Mabo's Implications for Australian Constitutionalism' in Duncan Ivison, Paul Patton and Will Sanders (eds.), *Political Theory and the Rights of Indigenous Peoples* (Cambridge University Press, 2000) 60, 70, 84, 86.

[29] See also Daniel Fitzpatrick, '"Best Practice" Options for the Legal Recognition of Customary Tenure' (2005) 36(3) *Development and Change* 450.

[30] Tobin, above n. 1, 121.

[31] Shaunnagh Dorsett and Shaun McVeigh, *Jurisdiction* (Routledge, 2012) 4.

[32] See Ivison, above n. 20, 322–4.

when they are subjected to the threats posed by colonisation and one is forced to find a means of protection that is comprehensible to a non-indigenous system of law'.[33] As pointed out by Tanasescu, '[r]ights have an essential connection to the making of claims'.[34]

It is widely accepted that demands for indigenous rights to land and resources need justification in order for the state to pass laws providing for them. Ivison argues that 'well-developed reasons drawing on individual societal interests are needed to compel the state to take action to provide for indigenous land and resource rights'[35] and Waldron points out that 'indigenous claims are not self-justifying'.[36] However, there remains no consensus in the literature on whether indigenous groups should be entitled to specific land and resource rights; nor is there consensus on the reason why indigenous land and resource rights might be needed. One ongoing debate, for example, is the inconsistency of indigenous group rights with the liberal objective of individual equality.[37] In the following section, I consider the justifications typically put forward to support indigenous water claims.

2.3.1 Indigenous Water Rights and Reparative Justice

The idea that states must 'recognise' indigenous groups, and their ongoing rights to land and resources, has become the central claim of the international indigenous rights movement.[38] Such claims are underpinned by ideas of 'legal pluralism', on the basis that indigenous rights and law exist independently of state law and should be recognised by the state.[39] Demands for indigenous water rights in Australia, for

---

[33] See Webber, above n. 28, 60, 64; see also Henry Reynolds, 'New Frontiers' in Paul Havemann (ed.), *Indigenous Peoples' Rights: In Australia, Canada & New Zealand* (Oxford University Press, 1999) 7.

[34] Mihnea Tanasescu, *Environment, Political Representation, and the Challenge of Rights: Speaking for Nature* (Houndmills, Basingstoke, Hampshire: Palgrave Macmillan, 2016) 34.

[35] Ivison, above n. 20, 321, 329.

[36] Jeremy Waldron, *Why Is Indigeneity Important?* (American Political Science Association, 2005) 1, 81–2.

[37] See, e.g., Brian M Barry, *Culture and Equality: An Egalitarian Critique of Multiculturalism* (Cambridge, Polity, 2001) 317; Chandran Kukathas, 'Are There Any Cultural Rights?' [1992] (1) *Political Theory* 105, 107.

[38] Coulthard, above n. 15, 1–2; see also Asebe Regassa Debelo, 'Contrast in the Politics of Recognition and Indigenous People's Rights' 7(3) *AlterNative: An International Journal of Indigenous Peoples* [258]; Nancy Fraser, 'From Redistribution to Recognition? Dilemmas of Justice in a "Post-Socialist" Age' (1995) 21(2) *New Left Review* 68 Fraser argues that, 'The "struggle for recognition" is fast becoming the paradigmatic form of political conflict in the late twentieth century'.

[39] See, e.g., Sebastien Grammond, 'The Reception of Indigenous Legal Systems in Canada' in Albert Breton et al. (eds.), *Multijuralism: Manifestations, Causes, and Consequences* (Ashgate Publishing Limited, 2009); see generally Kirsty Gover, 'Legal Pluralism and State-Indigenous

example, are often made with reference to the historical failure of colonial states to recognise indigenous water rights at the acquisition of sovereignty, as in:[40]

> In exploring the entitlement of Aboriginal people to rights and interests in water, it is undeniable that prior to the assertion of British sovereignty Aboriginal people had rights under traditional law and custom to the use, management and regulation of their waterways.

Those who argue for the recognition of indigenous rights to land and resources sometimes point to the prior status of pre-sovereignty indigenous water rights and indigenous 'firstness'.[41] They draw on theories of 'reparative' or 'transitional' justice,[42] which may be described as 'a responsibility to amend for past wrongs'.[43] Williams uses the term 'transitional justice' to refer to 'a process by which the new order agrees either to uphold pre-existing rights recognised in the old usurped order or to make good on those that were unfairly taken away'.[44] Dodds argues that the state is under an obligation to provide for indigenous land rights in response to indigenous dispossession and historical injustice.[45] Kuppe similarly conceptualises the need for recognition as follows:[46]

> Their land and resource rights, can be defined as collective rights vested in the Aboriginal nation that holds it and are justified by the fact that these rights were exercised independently of and historically before the emergence of the modern state (and/or also before emergence of its legal predecessor in title). Thus, as much as possible, the recognition of these

Relations in Western Settler Societies' (International Council on Human Rights Policy, 2009).

[40] Tony McAvoy, 'The Human Right to Water and Aboriginal Water Rights in New South Wales' (2008) 17(1) *Human Rights Defender* 6.

[41] See, e.g., Jason Behrendt and Peter Thompson, 'The Recognition and Protection of Aboriginal Interests in NSW Rivers' (2004) 3 *Journal of Indigenous Policy* 37, 42; Sue Jackson, 'Aboriginal Access to Water in Australia: Opportunities and Constraints' in R. Quentin Grafton and Karen Hussey (eds.), *Water Resources Planning and Management* (Cambridge University Press, 2011) 601–2; Bardy McFarlane, 'The National Water Initiative and Acknowledging Indigenous Interests in Planning' (2004); Patricia Lane, 'Native Title and Inland Waters' (2000) 4(29) *Indigenous Law Bulletin*.

[42] Webber, above n. 28.

[43] Margaret Urban Walker, *What Is Reparative Justice?* (Marquette University Press, 2010) 9.

[44] Joe Williams, *Confessions of a Native Judge: Reflections on the Role of Transitional Justice in the Transformation of Indigeneity* (Australian Institute of Aboriginal and Torres Strait Islander Studies, 2008) 3 http://lryb.aiatsis.gov.au.

[45] Susan Dodds, 'Justice and Indigenous Land Rights' (1998) 41(2) *Inquiry. An Interdisciplinary Journal of Philosophy* 187, 199.

[46] René Kuppe, 'The Three Dimensions of the Rights of Indigenous Peoples' (2009) 11(1) *International Community Law Review* 103, 107.

rights is to counter-balance the negative effects of the historic colonization.

Johnson argues that indigenous land and resource rights seek to repair the deteriorated relationship between indigenous groups and the state by returning lost things (or compensating for their loss), explaining:[47]

> reconciliation, at least in post-colonial New Zealand, Australia and Canada denotes an acknowledgement by the state that indigenous people have suffered historical injustices (including land and resource dispossession) and that they deserve redress in the form of symbolic acts and the redistribution of social and economic goods.

In repairing the relationship between the state and indigenous groups caused by historical injustice, the state also improves its own legitimacy as ruler, by obtaining indigenous consent to be governed.[48] However, if indigenous land and resource rights are justified solely with reference to a historical lack of recognition, there is a risk that rights may only be recognised today if they continue to exhibit some connection to rights that existed before colonisation (referred to in this book as a 'problem of continuity').[49] The use of water for modern purposes today may be quite different from the use of water by indigenous groups pursuant to pre-sovereignty traditional laws and customs. This may raise a question of whether the use of water for contemporary, including commercial, purposes derives from pre-sovereignty traditional laws and customs at all. The idea of restoring indigenous water rights not recognised at the acquisition of sovereignty also produces inevitable tensions between indigenous rights and water use rights held by others (referred to in this book as a 'problem of priority'). Both of these problems, we

---

[47] See Miranda Johnson, 'Reconciliation, Indigeneity, and Postcolonial Nationhood in Settler States' (6) 14(2) *Postcolonial Studies* 187, 189; see also Damien Short, *Reconciliation and Colonial Power: Indigenous Rights in Australia* (Ashgate, 2008) 15; Kirsty Gover, *Tribal Constitutionalism: States, Tribes, and the Governance of Membership* (Oxford University Press, 2010) 15–16; Dodds, above n. 45, 188.

[48] See Paul Keal, 'Indigenous Self-Determination and the Legitimacy of Sovereign States' (2007) 44(2/3) *International Politics* 287, 288; Short, above n. 47, 177.

[49] See also Duncan Ivison, Paul Patton and Will Sanders, 'Introduction' in Duncan Ivison, Paul Patton and Will Sanders (eds.), *Political Theory and the Rights of Indigenous Peoples* (Cambridge University Press, 2000) 3; Gover, *Tribal Constitutionalism: States, Tribes, and the Governance of Membership*, above n. 47, 12; Manuhuia Barcham, 'The Limits of Recognition' in Benjamin Richard Smith and Frances Morphy (eds.), *The Social Effects of Native Title: Recognition, Translation, Coexistence* (ANU E Press, 2007); Arif Bulkan, 'Disentangling the Sources and Nature of Indigenous Rights: A Critical Examination of Common Law Jurisprudence' (2012) 61(4) *International & Comparative Law Quarterly* 823, 835.

will see in the countries studied in this book, undermine the potential for recognition of substantive rights to water for indigenous peoples. As Strelein puts it in the context of Australian native title, such recognition is:[50]

> ... a practical compromise by the courts in light of the lateness of the recognition and the impact on titles and interests granted over Indigenous peoples' land over the past 200 years'.

2.3.2 Indigenous Water Rights and Distributive Justice

Reparative justice 'looks backward to the scene of wrongdoing and its impact on the victim of wrong'.[51] However, indigenous exclusion from water law frameworks is not merely something that happened at the acquisition of sovereignty, but is ongoing. Waldron has argued that, 'compensation for lost land and resource rights is not reparation for injustice that started and finished in the past, but as a way to put a stop to ongoing or persisting injustice and restore resources to those who have continued to be entitled to them'.[52] In this sense, indigenous rights are needed in order to stop perpetuating injustice.[53]

Increasingly, arguments for recognition, restitution or allocation of indigenous rights are backed up by theories of 'distributive justice',[54] often relying on the work of egalitarian legal and political philosophers like Rawls and Dworkin.[55] In such accounts, indigenous rights may be attached to claims that all people should be entitled to a 'fair share' of resources or wealth.[56] Distributive claims typically allude to the unfair distribution of access or property rights, and the dominance of powerful interests, accumulating access and use rights at the expense of the vulnerable. For example, the encroachment on indigenous territories and exclusion of indigenous groups from allocative frameworks has been compared to the enclosure of landholdings in Britain from the

---

[50] See Lisa Strelein, 'Conceptualising Native Title' (2001) 23 *The Sydney Law Review* 95, 99, 123.
[51] Walker, above n. 43, 44.
[52] Jeremy Waldron, 'Superseding Historic Injustice' (1992) 103(1) *Ethics The University of Chicago Press* 4, 14.
[53] Barcham, above n. 49, 203.
[54] See, e.g., Jackson, 'Enduring and Persistent Injustices in Water Access in Australia', Introduction, above n. 19.
[55] John Rawls, A *Theory of Justice (Revised Edition)* (Oxford University Press, 1999); and Ronald Dworkin, *Law's Empire* (Cambridge: Belknap Press, 1986).
[56] See, e.g., A John Simmons, 'Historical Rights and Fair Shares' (1995) 14(2) *Law and Philosophy* 149, 184; relying on Robert Nozick, *Anarchy, State, and Utopia* (New York: Basic Books, 1974); see also Fraser, above n. 38.

sixteenth century, which benefited large landowners at the expense of peasant farmers.[57]

Distributive demands for indigenous rights to land and resources also often allude to broader debates around the need to support indigenous economic development and socio-economic well-being.[58] Because of this imperative of equity, a minority group may require differentiated rights in order to enjoy substantive equality with the majority. In this way, Ivison explains that indigenous rights are, 'a necessary (although hardly sufficient) condition of aboriginal peoples being treated equally, given the legacy of colonialism and the challenges they face today in living decent lives according to their own lights'.[59]

A common criticism of indigenous rights based on distributive justice is that indigenous demands for a fair share of resources cannot be differentiated from those of other marginalised groups. As an example of this, the indigenous rights movement has at times shown ambivalence around whether the disadvantage of marginalised indigenous groups is related to their ethnicity or their class.[60] In this regard, Kymlicka suggests that 'group-differentiated rights' are only warranted to the extent necessary to rectify the disadvantage experienced by the minority,[61] and achieve substantive equality.

In the context of indigenous land and resource rights, Hendrix, Hewitt and Barrera-Hernández have argued that property rights or money must be redistributed to indigenous groups to correct the ongoing disadvantage.[62] Pearson has argued that indigenous groups need formal title to land and resources as a way to accumulate capital and raise themselves out of poverty,[63] referring to the controversial

---

[57] See, e.g., David Harvey, *The New Imperialism* (Oxford University Press, 2003)148; Perreault, 'Dispossession by Accumulation? Mining, Water and the Nature of Enclosure on the Bolivian Altiplano', Introduction above n. 5; Michael De Alessi, 'The Political Economy of Fishing Rights and Claims: The Maori Experience in New Zealand' (2012) 12(2/3) *Journal of Agrarian Change* 390; MPK Sorrenson, 'Folkland to Bookland: F.D. Fenton and the Enclosure of the Māori "Commons"' (2011) 45 *New Zealand Journal of History* 149.

[58] See generally Maureen Tehan, 'Customary Land Tenure, Communal Titles and Sustainability: The Allure of Individual Title and Property Rights in Australia', in *Comparative Perspectives on Communal Lands and Individual Ownership: Sustainable Futures* (Routledge, 2010) 362–3; Lee Godden et al., 'Accommodating Interests in Resource Extraction: Indigenous Peoples, Local Communities and the Role of Law in Economic and Social Sustainability' (2008) 26(1) *Journal of Energy & Natural Resources Law* 1, 25.

[59] Ivison, above n. 20, 338.

[60] See Engle, above n. 16, 189.

[61] Kymlicka, above n. 14, 4.

[62] See, e.g., Hendrix, above n. 3, 676; Anne Hewitt, 'Commercial Exploitation of Native Title Rights – a Possible Tool in the Quest for Substantive Equality for Indigenous Australians?' (February 12) 32(2) *Adelaide Law Review* 227, 230–3; Barrera-Hernández, above n. 7.

[63] Noel Pearson, 'Properties of Integration' *The Australian* (Australia), 14 October 2006.

work of Peruvian neoliberal economist Hernando de Soto.[64] Fraser has argued that the remedy for economic injustice is 'redistribution', under which heading she groups various types of political-economic restructuring including redistributing income, reorganising the division of labour, subjecting investment to democratic decision making and transforming basic economic structures.[65]

Increasingly, those who advocate for better protection or recognition of indigenous rights to water, particularly substantive or commercial water rights, point to a need for such rights to address indigenous economic disadvantage, and enable social equity. Jackson and Altman have argued:[66]

> Provision needs to be made by the Australian state for relatively impoverished Aboriginal land owners to have adequate access to water for their interrelated consumptive and non-consumptive, or commercial and enviro-cultural, purposes.

Morgan, Strelein and Weir write that indigenous peoples have been marginalised from their economic resource base through a failure to recognise their water rights and that indigenous groups have a right to be included in the economic benefits derived from water resource management.[67] Altman, Jackson and Nikolakis have all variously argued in the Australian context for substantive water rights for indigenous groups in the interests of equity, distributive justice, and indigenous economic development.[68]

Where reparative and distributive theories are used together, they appear to provide a more robust explanation for the need for indigenous-differentiated rights.[69] As Kuppe explains:[70]

---

[64] Hernando de Soto, *The Mystery of Capital: Why Capitalism Triumphs in the West and Fails Everywhere Else* (London: Bantam, 2000).

[65] See Fraser, above n. 38, 73.

[66] Sue Jackson and Jon Altman, 'Indigenous Rights and Water Policy: Perspectives from Tropical Northern Australia' (2009) 13(1) *Australian Indigenous Law Review* 27, 43.

[67] Monica Morgan, Lisa Strelein and Jessica Weir, *Indigenous Rights to Water in the Murray Darling Basin: In Support of the Indigenous Final Report to the Living Murray Initiative* (Native Title Research Unit, Australian Institute of Aboriginal and Torres Strait Islander Studies, 2004) 8.

[68] See Sue Jackson and Marcus Barber, 'Recognition of Indigenous Water Values in Australia's Northern Territory: Current Progress and Ongoing Challenges for Social Justice in Water Planning' (2013) 14(4) *Planning Theory & Practice* 435; William Nikolakis and R Quentin Grafton, 'Fairness and Justice in Indigenous Water Allocations: Insights from Northern Australia' (2014) 16 *Water Policy* 19; see, e.g., Jackson and Altman, above n. 66, 41.

[69] See also Fraser, above n. 38, 73. Fraser argues that '[re]distributive remedies generally presuppose an underlying conception of recognition'.

[70] Kuppe, above n. 46, 103.

Indigenous peoples currently experience three levels of injustice: They are the trans-generational victims of historic colonization; they are politically disenfranchised; and their cultural diversity is not officially recognized. In turn, indigenous peoples struggle for the recognition of their specific rights, in order to overcome the injustice they are currently experiencing.

However, as Fraser suggests, according reparative and distributive justice to indigenous peoples will be conceptually challenging:[71]

> People who are subject to both cultural injustice and economic injustice need both recognition and redistribution. They need both to claim and to deny their specificity. How, if at all, is this possible?

Indigenous groups certainly suffered an historical injustice at the acquisition of sovereignty, when their land and resource rights were not recognised. This historical injustice of non-recognition sets indigenous experiences aside from other disadvantaged groups. It began a chain of events leading to the current unfairness in the distribution of water rights. As Ivison warns, 'it does not follow that the "aboriginal" in aboriginal rights is redundant'.[72] However, instead of focusing only on culture and historical injustice, states must correct the ongoing inequity in the distribution of water rights by which other water users, instead of indigenous groups, hold almost all the rights to use water.

If indigenous water rights are needed to respond to injustice in the distribution of water rights, the answer is finally to include indigenous peoples in water law frameworks by providing them the right to use water for both cultural and commercial purposes. Where water use rights have now been allocated to other users, and as demand for water continues to grow for irrigation, sanitation and industry as well as ecological or environmental needs, there may need to be some sort of redistribution of those rights.

2.4 CONCLUSION

In this chapter, I have discussed the conceptual tensions involved in discussions about indigenous water rights. These discussions, until now, have focused on the cultural nature of indigenous water interests, which has limited the legal entitlements that flow from their

---

[71] Fraser, above n. 38, 74.
[72] See Ivison, above n. 20, 338.

recognition. I have argued that legal and policy mechanisms that seek to recognise cultural relationships with water and involve indigenous peoples in water governance should strive towards recognising indigenous water relationships but, more importantly, indigenous water jurisdiction. This argument is central to the consideration of the four country studies included in this book, where law and policy are sometimes able to provide a space for indigenous groups to exercise jurisdiction in planning and governing their water resources.

I have found in this chapter that indigenous water rights are often treated as being of a particular traditional, cultural nature, because they are framed as an imperative of 'reparative justice'. That is, they are necessary because indigenous water rights were not recognised at the acquisition of sovereignty and indigenous peoples were therefore wrongfully dispossessed of those rights. The remedy for the historical injustice of non-recognition is recognition (or, at least, compensation). But this approach deals inadequately with situations where indigenous groups no longer exercise relationships with particular water resources in the way they did in pre-sovereignty times, and other users now hold rights to the same resources.

I have suggested instead that the reason states should provide for indigenous water rights is an imperative of distribution. Such rights are needed not only to remedy the historical injustice of non-recognition but because indigenous exclusion from water law frameworks is ongoing. This distributive injustice is just another type of disadvantage experienced by indigenous peoples, who in many contexts remain the most disadvantaged sector of society. I contend here that the focus must shift from recognising pre-sovereignty rights that have continued in a substantially uninterrupted manner to attending to the current injustice in the distribution of water use rights. Such a redistributive mechanism could finally include indigenous peoples in water law frameworks, in the way that might have occurred had indigenous land and water rights been recognised at the acquisition of sovereignty.

CHAPTER THREE

REGULATING INDIGENOUS WATER RIGHTS:

Nature, Humans and Markets

3.1 INTRODUCTION

In the previous chapter, I discussed the conceptualisation of indigenous rights to water in legal and political theory. However, any in-depth consideration of indigenous peoples and water necessarily requires an appreciation, not only of the legal rules determining entitlements to access and use water, but of the regulatory frameworks that manage the way in which such entitlements are exercised. Research on indigenous rights to water often focuses on one or the other of these perspectives. Yet, increasingly, socio-legal scholars like Morgan argue that socio-legal research must combine rights-based (naming, claiming and blaming) and regulation-based (rule making, monitoring and enforcement) perspectives.[1] With that directive in mind, in this chapter I introduce two regulatory tendencies relevant to the study of indigenous water rights in comparative law.

One of these developments is the idea that governments should 'commoditise' the natural environment and use private property rights and market mechanisms in water regulation and allocation, as opposed to treating access to water as a fundamental human right, entitled to all. The other is the tendency to devise creative legal mechanisms, like recent laws recognising rivers as legal persons, intended to protect the

---

[1] Bronwen Morgan (ed.), *The Intersection of Rights and Regulation: New Directions in Sociolegal Scholarship* (Aldershot, Ashgate, 2007).

rights of nature and address social or community concerns around water governance and quality. The first of these tendencies has typically provoked substantive demands for some sort of indigenous water right and corresponding water allocation. As discussed in the previous chapter, this is a claim to a 'fair share'[2] of available resources as an imperative of distribution. The second tendency has provided fertile ground for indigenous demands for greater collaboration in, and control of, water management and governance on the basis of their particular cultural and spiritual water relationships. In the previous chapter, I called this an imperative of jurisdiction. Both trajectories play out repeatedly in debates about indigenous rights to water in comparative law, and resulting legal and policy frameworks, in the country studies considered in this book.

I begin this chapter by introducing the 'rights of nature movement', and highlight its co-option by indigenous groups to agitate for increased autonomy in, or control of, water resource management or governance via legal person models. I discuss the interdependent, yet uneasy, relationship between the rights of nature and international human rights norms. Later, I consider the use of private property and markets in the regulation of water, and the opposition to such approaches from those who consider access to water to be a basic human right. I argue that most regulatory frameworks for water in fact involve a combination of public and private interests, mechanisms, protections and transactions, and I suggest that both private and public tools may have a place in debates about how best to provide for indigenous water rights.

3.2 HUMAN RIGHTS AND NATURE'S RIGHTS

Western legal frameworks that regulate the management and use of natural resources, like water, have traditionally focused on the utility such resources provide to humans, both socially and economically. Such an approach is often described as a 'propertised' or 'commodified' one, in which property law concepts are used to distribute and govern resource use by humans.[3] This stance positions water, like other natural resources, as a resource to be exploited to extract

---

[2] See A John Simmons, 'Historical Rights and Fair Shares' [1995] (2) *Law and Philosophy* 149.
[3] See David R Boyd, *The Rights of Nature: A Legal Revolution That Could Save the World* (ECW Press, 2017) xxvi; Malcolm Langford and Anna FS Russell, *The Human Right to Water: Theory, Practice and Prospects* (Cambridge University Press, 2017) 11.

utility for human endeavours via legal and policy frameworks that allocate property and use rights. It is often described as 'anthropo-centric', because humans are the central concern in the regulatory model.[4]

The anthropocentric approach to the regulation of natural resources is well-embedded in international legal frameworks. As an example, the *Rio Declaration* of 1992 adopted the banner of 'sustainable develop-ment', which places humans 'at the centre' of concern.[5] Since then, the international community has many times turned its attention to envir-onmental concerns, particularly in response to the impact of climate change.[6] However, international law continues to emphasise the need for human development, and to pursue environmental protections within the human rights framework as necessary conditions for the exercise of a number of other fundamental human rights. The right to a clean environment, for example, now recognised in the constitutions of more than eighty countries worldwide,[7] is characterised as a human right. So too is the human right to water and sanitation, protected since 2010 by a United Nations General Assembly resolution.[8] The connec-tion between environmental law and human rights is evident in the characterisation of international environmental protections like the right to safe and clean drinking water and sanitation as being 'a human right that is essential for the full enjoyment of life and all human rights', as part of a 'third generation of human rights'.[9] The human rights basis for international environmental law has also been co-opted for the recognition and protection of indigenous rights.[10] International envir-onmental law invariably acknowledges the interests of indigenous

---

[4] Elisa Morgera and Kati Kulovesi, *Research Handbook on International Law and Natural Resources* (Edward Elgar Publishing, 2016) 5; B Donnelly and P Bishop, 'Natural Law and Ecocentrism' (2006) 19(1) *Journal of Environmental Law* 89, 90; Catherine J Iorns Magallanes, 'Maori Cultural Rights in Aotearoa New Zealand: Protecting the Cosmology that Protects the Environment' (2015) 21(2) *Widener Law Review* 273, 275.

[5] *Report of the United Nations Conference on Environment and Development* (UN Doc A/CONF.151/26 (Vol. I) (12 August 1992) annex I ('Rio Declaration on Environment and Development') principle 1.

[6] Conference of the Parties, United Nations Framework Convention on Climate Change, Adoption of the Paris Agreement, UN Doc FCCC/CP/2015/L.9/Rev.1 (12 December 2015) Annex ('Paris Agreement').

[7] David R Boyd, 'The Constitutional Right to a Healthy Environment' (2012) 54(4) *Environment: Science and Policy for Sustainable Development* 3, 3.

[8] *UN General Assembly, The Human Right to Water and Sanitation* (Sixty-Fourth Session, 2010) UN Doc A/64/L.63/Rev.1 2010.

[9] See generally Langford and Russell, above n. 3, 2–3 for a discussion of the evolution of the human right to water in international law.

[10] See Ronald Niezen, *The Origins of Indigenism: Human Rights and the Politics of Identity* (University of California Press, 2002) 220.

peoples,[11] on the assumption that indigenous peoples share benefits with environmental outcomes. The Parties to the 2015 *Paris Agreement* acknowledged that actions taken to address climate change, 'should . . . respect, promote and consider their respective obligations on human rights' and 'the rights of indigenous peoples'.[12] The preamble to the 2010 *Cancún Agreement* emphasises that parties should undertake action to address climate change that 'fully respects human rights' and 'recognises' the need for 'effective' indigenous participation.[13]

Human rights norms are increasingly relied upon to protect indigenous communities and their collective claims to land and natural resources, although indigenous rights have occupied an uneasy space in human rights law. Indigenous demands for land and resources often rest on ideas of collective self-determination, nationhood and sovereignty, in contrast to the western liberal idea of universally recognised individual human rights within the nation-state.[14] Indigenous understandings of and attitudes towards the environment may depart from conceptions of universal human rights. For example, Radonic's case study on the indigenous Yaqui Tribe in the Mexican State of Sonora shows how, from an indigenous perspective, a river, in encompassing the multivalent associations of indigenous practices and identities, is more important that an abstract human right to water.[15] Notwithstanding, a range of international and domestic human rights instruments now protects indigenous rights,[16] and indigenous communities typically invoke these protections when advocating for recognition and protection of their distinct rights to land and resources.

[11] See, e.g., 'Declaration of the United Nations Conference on the Human Environment, Report of the United Nations Conference on the Human Environment, Stockholm, 5–17 June 1972' (1972) A/CONF.48/14/Rev.1 (1972) www.un-documents.net/aconf48-14r1.pdf.

[12] Conference of the Parties, United Nations Framework Convention on Climate Change, Adoption of the Paris Agreement, UN Doc FCCC/CP/2015/L.9/Rev.1 (12 December 2015) Annex ('Paris Agreement') preamble.

[13] Conference of the Parties, United Nations Framework Convention on Climate Change, *Report of the Conference of the Parties on Its Sixteenth Session, Held in Cancun from 29 November to 10 December 2010 – Addendum – Part Two: Action Taken by the Conference of the Parties at Its Sixteenth Session* (UN Doc FCCC/CP/2010/7/Add.1 (15 March 2011) Decision 1/CP.16 ('The Cancun Agreements: Outcome of the Work of the Ad Hoc Working Group on Long-Term Cooperative Action under the Convention') preamble.

[14] Kirsty Gover, *Indigenous Rights and Governance in Canada, Australia, and New Zealand* (Electronic Resource, Oxford University Press, 2012). Chapter 2, 'Indigeneity', intro.

[15] Radonic, Introduction above n. 6, 157.

[16] See, e.g., *United Nations Declaration on the Rights of Indigenous Peoples*, GA Res 61/295, UN GAOR, 61st Sess, 107th Plen Mtg, Agenda Item 68, Supp No 49, UN Doc A/RES/61/295 (2 October 2007, Adopted 13 September 2007) ('UNDRIP'); *Convention Concerning Indigenous and Tribal Peoples in Independent Countries (No. 169) [1989] 28 ILM 1382 (Entered into Force 5 September 1991)* ('Convention 169').

In recent years, a number of governments at local and national level, have taken a new approach in their laws and policies around the regulation of natural resources like water. Pursuant to this so-called 'ecocentric' shift, nature is conceived of as holding its own rights, rather than merely being an entitlement for humans to use.[17] In fact, this approach provides that humans owe entitlements to nature, to protect nature's rights and interests.[18] The origin of the ecocentric movement is often credited to Christopher Stone, who in the 1970s argued in his seminal article *Should Trees have Standing* that the legal rights of trees, oceans, animals and the environment as a whole should be provided for in order to give them a 'voice' and thereby ensure their protection,[19] although Thomas Berry's work was also instrumental.[20] As Tanasescu explains, the idea of human representatives providing a voice for nature is central to the idea of nature's rights.[21]

Ecocentrism inspired the development of 'wild law'[22] or 'earth jurisprudence', which sought revolutionary change in law and regulation,[23] based on a belief that 'humanity and the non-human world belong to the same moral order' as members of 'Earth's Community'.[24] Proponents of earth jurisprudence often contrast human rights to property in natural resources with the inherent, inalienable rights of nature.[25] However, those arguing for ecocentric laws and policies still use the concept of 'rights', and often specifically 'human rights', to secure nature's protection.[26]

---

[17] See also Meg Good, 'The River as a Legal Person: Evaluating Nature Rights-Based Approaches to Environmental Protection in Australia' [2013] (1) *National Environmental Law Review* 34, 34.

[18] See Michelle Maloney and Peter Burdon, *Wild Law – In Practice* (Taylor and Francis, 2014).

[19] Christopher Stone, 'Should Trees Have Standing? Towards Legal Rights for Natural Objects' (1972) 45 *Southern California Law Review* 450; Christopher Stone, *Should Trees Have Standing? Law, Morality, and the Environment* (Oxford University Press, 2010).

[20] Thomas Berry, *The Great Work: Our Way into the Future* (Bell Tower, 1999).

[21] Tanasescu, Chapter 2 above n. 34.

[22] See Nicole Rogers and Michelle M Maloney, *Law as If Earth Really Mattered: The Wild Law Judgement Project* (Routledge, 2017).

[23] Michelle Maloney, 'Building an Alternative Jurisprudence for the Earth: The International Rights of Nature Tribunal' [2016] (1) *Vermont Law Review* 129, 132; Sophia Imran, Khorshed Alam and Narelle Beaumont, 'Reinterpreting the Definition of Sustainable Development for a More Ecocentric Reorientation: Reinterpreting the Definition of Sustainable Development' (2014) 22(2) *Sustainable Development* 134.

[24] Vito De Lucia, 'Towards an Ecological Philosophy of Law: A Comparative Discussion' (2013) 4 (2) *Journal of Human Rights and the Environment* 167, 188; Berry, above n. 20, 4.

[25] See Abigail Hutchison, 'Whanganui River as a Legal Person' [2014] (3) *Alternative Law Journal (Gaunt)* 179, 179; Mari Margil, 'The Standing of Trees: Why Nature Needs Legal Rights' (2017) 34(2) *World Policy Journal* 8.

[26] See Macpherson and Clavijo Ospina, Felipe, Introduction above n. 19.

International movements for the environment continue to push towards greater protection of nature, sometimes adopting an ecocentric perspective, in which nature is conceived of as holding its own rights and interests. This approach is evident, for example, in the agenda of the United Nations Harmony with Nature Programme, which pro-motes 'the balanced integration of the economic, social and environ-mental dimensions of sustainable development through harmony with nature'.[27] There is also a draft 'International Declaration on the Rights of Mother Earth' spearheaded by the Global Alliance for the Rights of Nature, which declares a number of human rights-based protections for all natural entities.[28] More recently the Earth Trusteeship initiative, has proposed the 'Hague Principles for a Universal Declaration on Responsibilities for Human Rights and Earth Trusteeship', which emphasise human responsibilities for nature as part of an earth community.[29]

At the domestic level, a number of countries have passed laws recognising the rights of nature to one degree or another. The first major law protecting the rights of nature was the 2008 Ecuadorian Constitution, which recognises the rights of 'Pachamama' (Mother Earth).[30] In 2009, Bolivia also recognised the rights of Mother Nature in its Constitution and subsequently passed a *Law of the Rights of Mother Earth*.[31] The development of rights of nature laws in Ecuador and Bolivia has applied the concept of *Buen Vivir*, the idea of living a good life, inspired by indigenous communal societal goals.[32]

Rights of nature protections, although they vary much in scope and content, have since been recognised by national and local governments in other places, including Mexico and the United States.[33] The first US rights of nature ordinance was passed by the Tamaqua Borough,

---

[27] *Report to the United Nations General Assembly on Harmony with Nature* (UN Doc A/C.2/73/L.39/ Rev.1) (21 November 2018).

[28] 'Universal Declaration of Rights of Mother Earth' (World People's Conference on Climate Change and the Rights of Mother Earth Cochabamba, Bolivia, 22 April 2010).

[29] *The Hague Principles for a Universal Declaration on Human Responsibilities and Earth Trusteeship |* *Earth Trusteeship* www.earthtrusteeship.world/the-hague-principles-for-a-universal-declaration-on-human-responsibilities-and-earth-trusteeship/.

[30] *Constitución de la República del Ecuador 2008* (Ecuador).

[31] *Constitución del Estado Plurinacional de Bolivia 2009* (Bolivia); *Ley de Derechos de La Madre Tierra (Ley 071) 2010* (Bolivia).

[32] See Tom Perreault, 'Tendencies in Tension: Resource Governance and Social Contradictions in Contemporary Bolivia' in *Governing Resource Extraction* (2017); Craig M Kauffman and Pamela L Martin, 'Can Rights of Nature Make Development More Sustainable? Why Some Ecuadorian Lawsuits Succeed and Others Fail' (2017) 92 *World Development* 130.

[33] *Constitución Política de la Ciudad de México 2016* (Mexico City).

Pennsylvania, in 2006. Since then, over 100 local authorities have passed similar ordinances, including several larger cities such as Barrington, New Hampshire, Athens Ohio, Pittsburgh and Santa Monica.[34] In 2016, the General Council of the Ho-Chunk Nation was the first indigenous Tribal Nation to pass a rights of nature amendment to its tribal constitution providing that '[e]cosystems and natural communities within the Ho-Chunk territory possess an inherent, fundamental, and inalienable right to exist and thrive'.[35] In other places there are proposals before courts and governments for rights of nature protections.[36]

3.3   RIVER RIGHTS AND INDIGENOUS RIGHTS

Early proponents of the rights of nature, like Stone, focused on the need for natural resources to have 'standing' before the courts, in order to protect their own rights. The same year that Stone first published his article on standing, Douglas J argued for standing for natural resources as legal persons in his dissent in the Supreme Court decision in *Sierra Club* v. *Morton*, in the interests of allowing public interest litigation to prevent environmental harms.[37] The idea that natural resources should have the rights of legal persons has since been proposed, and sometimes implemented, in a growing body of jurisprudence.

In most western legal systems, at least those evolving out of the Roman law tradition, 'legal persons' are generally restricted to natural persons (individuals), corporations, and sometimes ships.[38] Giving rivers legal personality is intended to provide them with the standing humans and corporations have to protect their interests against others, where necessary in the courts.[39] At a more philosophical level, Naffine argues that the conferment of legal personality represents what values

---

[34] Cameron La Follette and Chris Maser, *Sustainability and the Rights of Nature: An Introduction* (CRC Press, 2017) 25.

[35] Community Environmental Legal Defence Fund, *Press Release: Ho-Chunk Nation General Council Approves Rights of Nature Constitutional Amendment* (18 September 2016) https://celdf .org/2016/09/press-release-ho-chunk-nation-general-council-approves-rights-nature-constitutional-amendment.

[36] Erin O'Donnell, *Legal Rights for Rivers: Competition, Collaboration and Water Governance* (Routledge, 2018) 159.

[37] *Sierra Club* v. *Morton* (1972) 405 USSC 727.

[38] M Davies and N Naffine, *Are Persons Property? Legal Debates about Property and Personality* (Ashgate Publishing, 2001).

[39] Erin O'Donnell, *Constructing the Aquatic Environment as a Legal Subject: Legal Rights, Market Participation, and the Power of Narrative* (PhD Thesis, University of Melbourne, 2017).

matter to a society.[40] A number of recent legal and political develop-ments concerning the rights of nature have involved declarations that rivers, forests or mountains are legal persons, which hold legal rights and have the necessary standing to sue and be sued, enter into con-tracts, and hold property in their own name.[41]

Most commonly, these declarations or grants of legal personality have related to rivers. The first legislative recognition of a river as a legal person occurred in New Zealand in March 2017, when the Whanganui River (Te Awa Tupua) was declared to be a legal person under a political settlement between the New Zealand Government and Māori of the Whanganui Iwi (tribe), considered in detail in Chapter 5.[42] The *Te Awa Tupua (Whanganui River Claims Settlement) Act 2017 (Te Awa Tupua Act)* declares that the Whanganui River is a legal person, with 'all the rights, powers, duties, and liabilities of a legal person'[43] and devises a complicated collaborative governance regime, presided over by a 'guardian' called 'Te Pou Tupua', to act in the interests of the river and enforce its rights.[44]

A similar approach was adopted by a regional court for rivers in India in April 2017,[45] albeit in recognition of the spiritual relationships of Hindu communities with the river, rather than a specific indigenous or tribal relationship, although the decisions were appealed and their status is unclear.[46] In May of the same year the Constitutional Court of Colombia released its decision recognising the Río Atrato as a legal person in recognition and protection of the human rights of indigenous and afrodescendent communities in the face of serious environmental contamination and social deprivation, considered in depth in Chapter 6.[47] Lawsuits seeking declarations that the Colorado River is a legal person were filed in October 2017 in the United States,[48] although the

---

[40] Ngaire Naffine, *Law's Meaning of Life: Philosophy, Religion, Darwin and the Legal Person* (Hart Publishing, 2009) 11.

[41] SM Solaiman, 'Legal Personality of Robots, Corporations, Idols and Chimpanzees: A Quest for Legitimacy' [2017] (2) *Artificial Intelligence and Law* 155, 157.

[42] *Te Awa Tupua (Whanganui River Claims Settlement) Act 2017* (NZ).

[43] Ibid. s 14.

[44] Ibid. s 18.

[45] *Mohd Salim con State of Uttarakhand & others*, 20 March 2017, High Court of Uttarakhand (WPPIL 126/2014) (India).

[46] See Erin O'Donnell, 'At the Intersection of the Sacred and the Legal: Rights for Nature in Uttarakhand, India' [2017] *Journal of Environmental Law* 1.

[47] *Centro de Estudios para la Justicia Social 'Tierra Digna' and Others v. The President of the Republic and Others*, No T-622 of 2016, Corte Constitucional [Constitutional Court], Sala Sexta de Revision [Sixth Chamber] (Colombia) (10 November 2016).

[48] *Colorado River System v. State of Colorado*, Case No. 1:17-cv-02316-RPM (D Colo).

case was eventually withdrawn.[49] Activism for recognition of the Río Magdalena in Mexico as a legal person, relying on rights of nature protections in Mexican state constitutions,[50] is well underway.[51] Most recently, in February 2019, the High Court of Bangladesh has declared its intention to recognise the Turag river as a 'living being' and 'legal person', although the judgement is yet to be released.[52] In other situations, the inherent value of rivers is recognised, without the grant of legal personality, such as the arrangement for the Yarra River in Victoria, Australia, discussed in Chapter 4.[53]

The fact that rivers may be treated by legal frameworks as persons is, as O'Donnell notes, a 'transformative idea'.[54] Laws that recognise legal rights for rivers have been seized by proponents of rights for nature as the embodiment of an ecocentric approach to natural resource regulation.[55] Proponents of nature's rights believe that protecting the rights of rivers reflects their inherent value and will enhance the protection of water resources from human exploitation. However, as the study of New Zealand and Colombian examples in this book shows, regulatory models that protect the rights of rivers have been largely driven, not by environmentalists, but by indigenous and tribal communities, who claim distinct relationships with water based on their cosmovision of guardianship, symbiosis and respect, as opposed to western liberal utilitarianism.[56]

Indigenous relationships with natural resources are often cited as evidence of diverse societal attitudes or regulatory approaches to natural resource use.[57] As Māori legal scholar Linda Te Aho explains, 'we see ourselves as direct descendants of our earth

[49] See Marianne Goodland, *Lawsuit Seeking 'personhood' for Colorado River Dismissed* (5 December 2017) Colorado Springs Gazette http://gazette.com/lawsuit-seeking-personhood-for-colorado-river-dismissed/article/1616604.

[50] See, e.g., *Constitución Política de la Ciudad de Mexico 2017* art 13A(2).

[51] See Earth Law Center, *Mexico on the Vanguard for Rights of Nature* (21 November 2017) Earth Law Centre www.earthlawcenter.org/blog-entries/2017/11/mexico-on-the-vanguard-for-rights-of-nature.

[52] 'HC Starts Delivering Verdict on Illegal Structures on Turag River's Bank' *The Daily Star*, 31 January 2019 www.thedailystar.net/city/illegal-structures-on-turag-rivers-bank-high-court-start-delivering-verdict-1695025.

[53] Yarra River Protection (Wilip-Gin Birrarung Murron) Act 2017.

[54] Erin O'Donnell, above n. 36, 158.

[55] See, e.g., Laura Villa, *The Importance of the Atrato River in Colombia Gaining Legal Rights* (5 May 2017) Earth Law Center www.earthlawcenter.org/new-blog-1/2017/5/the-importance-of-the-atrato-river-in-colombia-gaining-legal-rights.

[56] See also Macpherson and Clavijo Ospina, Felipe, Introduction above n. 19.

[57] See, e.g., Joanne Clapcott et al., 'Mātauranga Māori: Shaping Marine and Freshwater Futures' (2018) 52(4) *New Zealand Journal of Marine and Freshwater Research* 457.

mother and sky father and consequently not only "of the land" but "as the land"'.[58] Legal rights for rivers appear to have gained most traction where pursued by indigenous groups. The strongest protections of a river's rights as a legal person have been in Aotearoa New Zealand and Colombia, in both cases against a history of strong indigenous activism for the river. However, the rights of nature, and specifically legal rights for rivers, are not synonymous with, and may even at times run contra to, indigenous interests or claims.[59]

The rights of nature movement has grown out of, and is still driven principally by, a non-indigenous perspective as a 'western legal construct'.[60] O'Donnell has argued that, 'constructing the environment as a legal subject, although it brings a suite of new legal powers and is often necessary to support significant improvement in environmental outcomes in the context of water rights and water markets, can undermine the cultural narrative that considers the environment worthy of protection at all'.[61] O'Donnell and I have together on other occasions considered the ability of regulatory models that involve legal rights for rivers to deliver improved environmental outcomes, given the influence of the broader regulatory and institutional setting and community and policy attitudes.[62] Such approaches are part of a broader 'collaborative turn' in comparative natural resource regulation towards greater community involvement in resource governance.[63]

Legal rights for rivers, at least where they directly involve indigenous peoples, are an attempt to resolve historical and contemporary

---

[58] Linda Te Aho, 'Indigenous Challenges to Enhance Freshwater Governance and Management in Aotearoa New Zealand – The Waikato River Settlement' (2009) 20(5–6) *Journal of Water Law* 285, 285.

[59] Virginia Marshall, *Overturning Aqua Nullius: Securing Aboriginal Water Rights* (Aboriginal Studies Press, 2017).

[60] O'Bryan, Introduction above n. 19, 208.

[61] Erin O'Donnell, above n. 36, 183–4.

[62] Erin O'Donnell, 'Competition or Collaboration? Using Legal Persons to Manage Water for the Environment in Australia and the United States' (2017) 34 *Environmental and Planning Law Journal* 503, 503, 504, 515; O'Donnell and Macpherson, above n. 24; Elizabeth Macpherson and Erin O'Donnell, '¿Necesitan Derechos los Ríos? Comparando Estructuras Legales Para Regulación de Ríos en Nueva Zelanda, Australia y Chile [Do Rivers Need Rights? Comparing Legal Structures for River Regulation in New Zealand, Australia and Chile]' (Pontificia Universidad Católica de Chile, 2017).

[63] See Suzanne von der Porten, Rob E de Loë and Deb McGregor, 'Incorporating Indigenous Knowledge Systems into Collaborative Governance for Water: Challenges and Opportunities' (2016) 50(1) *Journal of Canadian Studies/Revue d'études canadiennes* 214, 215; P Memon and N Kirk, 'Role of Indigenous Māori People in Collaborative Water Governance in Aotearoa/ New Zealand' (2012) 55(7) *Journal of Environmental Planning and Management* 941, 941.

grievances about resource use and reconstitute governance arrangements.[64] They are inherently tied up with a process of addressing the rights, concerns and relationships of indigenous peoples with respect to their water resources and providing a space for indigenous peoples to be involved in river management and governance. These laws recognise cultural water relationships as reparative responses to historical injustice involving indigenous and tribal groups, while the outcome of negotiated agreements for indigenous rights always depends on, and is affected by, power imbalances between indigenous peoples and states.[65] I argue throughout this book that legal person models may offer an opportunity to recognise indigenous jurisdictions for the exercise of indigenous management obligations towards the natural world as guardians. Yet I am mindful that legal rights for rivers are not tantamount to indigenous rights and, against huge variance between and within indigenous communities, may not in every case represent what indigenous peoples want.

Legal pluralists like Webber argue that recognition mechanisms do not preserve indigenous law, or indigenous jurisdictions, in their pristine, pre-contact state,[66] and there is an inevitable process of 'accommodation' (or 'approximation') in laws that recognise indigenous interests.[67] Indigenous rights are, as Webber explains, 'mediated rights'.[68] As a consequence, state laws that provide for indigenous rights are often accused of translating, transforming and 'essentialising' indigenous interests in order to accommodate them within state law, and failing to account for the continually changing state of indigenous law.[69] Where states declare rivers to be legal persons, they may be

---

[64] See Katherine Sanders, '"Beyond Human Ownership?" Property, Power and Legal Personality for Nature in Aotearoa New Zealand' (2018) 30(2) *Journal of Environmental Law* 1; Macpherson and Clavijo Ospina, Felipe, Introduction above n. 19.

[65] Lily Maire O'Neill, *A Tale of Two Agreements: Negotiating Aboriginal Land Access Agreements in Australia's Natural Gas Industry* (PhD Thesis, University of Melbourne, 2016); Lily O'Neill, 'The Role of State Governments in Native Title Negotiations: A Tale of Two Agreements' (2014) 18(2) *Australian Indigenous Law Review* 29.

[66] Jeremy Webber, 'Beyond Regret: Mabo's Implications for Australian Constitutionalism', in Duncan Ivison, Paul Patton and Will Sanders (eds.), *Political Theory and the Rights of Indigenous Peoples* (Cambridge University Press, 2000) 60.

[67] See Arif Bulkan, 'Disentangling the Sources and Nature of Indigenous Rights: A Critical Examination of Common Law Jurisprudence' (2012) 61(4) *International & Comparative Law Quarterly* 823, 835.

[68] Webber, above n. 66, 63–70.

[69] Glen Coulthard, 'Subjects of Empire: Indigenous Peoples and the "Politics of Recognition" in Canada' (2007) 6(4) *Contemporary Political Theory* 437, 437; CA Zorzi, 'The "Irrecognition" of Aboriginal Customary Law' (2000) 5; Webber, above n. 66, 64–6; Rutgerd Boelens, 'The Politics of Disciplining Water Rights' (2009) 40(2) *Development and Change* 307, 316–20.

recognising and accommodating indigenous relationships, all the while mediating and limiting indigenous jurisdictions.[70]

3.4 PRIVATE RIGHTS AND THE PUBLIC INTEREST

Perhaps the oldest debate concerning the regulation of water is whether it should be treated by regulatory frameworks as a public or a private good.[71] This comes down to the following questions: is access to water a basic human right, which should be provided by governments to all people, free of charge? Or is access to water an economic commodity, capable of being allocated and administered applying commercial principles like private bargaining in markets?[72]

Those who believe water should to be treated as a private or economic good argue that governments are poorly equipped to efficiently regulate and distribute water, and the market will do a better job.[73] A water market is an arrangement that allows for commercial transactions (or trade) in water rights. Fisher describes a market as 'a set of circumstantial arrangements – whether locational, temporal, institutional, cultural or otherwise – according to which commercial transactions relating to assets take place'.[74] Water markets are typically taken to depend on the existence of '[c]lear and nationally-compatible characteristics for secure water access entitlements'.[75] Those who support the use of markets in water regulation consider that they encourage the more efficient, and productive, use of water through increased 'competition', thus ensuring a sufficient number of users in a market so that one, or a few, cannot dominate the market.[76] The economic philosophy underlying the market-based water reforms in Australia, for example,

---

[70] Macpherson and Clavijo Ospina, Felipe, see Introduction above n. 19.
[71] See Langford and Russell, above n. 3, 11–12; Bronwen Morgan, *Water on Tap: Rights and Regulation in the Transnational Governance of Urban Water Services* (Cambridge University Press, 2011, 2011) 2.
[72] Maude Barlow, 'The World's Water: A Human Right or a Corporate Good? Whose Water Is It?' in B McDonald and D Jehl (eds.), *The Unquenchable Thirst of a Water-Hungry World* (National Geographic Society, 1st ed, 2003).
[73] See J Braithwaite, *Regulatory Capitalism: How It Works, Ideas for Making It Work Better* (Edward Elgar Publishing, 2008).
[74] DE Fisher, 'Markets, Water Rights and Sustainable Development' (4) 23(2) *Environmental and Planning Law Journal* 100, 103.
[75] Commonwealth of Australia and the Governments of New South Wales, Victoria, Queensland, South Australia, the Australian Capital Territory and the Northern Territory, Introduction above n. 23 cl. 23.
[76] Fisher, above n. 74, 104–5.

was that water markets would facilitate the transfer of water use rights to where they produce the highest economic value.[77]

Those who see water as a public good typically rely on human rights protections to argue for centralised government control and distribution of water, according to social priorities protecting vulnerable interests like the economically and socially disadvantaged, indigenous peoples or cultural minorities and the environment (sometimes called a 'command and control' approach).[78] Those who argue that water is a human right rather than an economic good do not accept that markets are able to value and provide for social and environmental outcomes effectively.[79] They typically argue for public ownership of water resources and public guarantees of access. Such human rights-based claims have often been made by or on behalf of indigenous groups in countries where water markets exist or are proposed, like Latin American countries and Australia, on the basis that water markets allow other water users to accumulate water use rights to the detriment of indigenous groups.

There are limitations, however, in taking a purely human rights-based approach to the place of indigenous water interests in water regulation, given the difficulty of domestic implementation of human rights norms, the limited drinking water and sanitation focus of the human right to water, and the 'remarkable gap' between global consensus around access to water as a human right and the legal frameworks that factually govern water access.[80] Those supportive of indigenous rights to natural resources have frequently declaimed property and formalisation as weapons of colonisation and dispossession. Yet, indigenous peoples find themselves in a situation where they are both dependent on, and threatened by, natural resource development.[81] Some indigenous communities do not want to be shut out of

---

[77] Commonwealth of Australia and the Governments of New South Wales, Victoria, Queensland, South Australia, the Australian Capital Territory and the Northern Territory, Introduction above n. 23. See generally William D Nikolakis, R Quentin Grafton and Hang To, 'Indigenous Values and Water Markets: Survey Insights from Northern Australia' (2013) 500(12) *Journal of Hydrology* 12.

[78] See, e.g., Barrera-Hernández, Chapter 2 above n. 7; Gary Raumati Hook and Lynne Parehaereone Raumati, 'Cultural Perspectives of Fresh Water' (2011) 2 *MAI Review* 1; McAvoy, Chapter 2 above n. 40.

[79] See, e.g., in relation to environmental water protections Laurence H Tribe, 'Ways Not to Think about Plastic Trees: New Foundations for Environmental Law' (1874) 83 *Yale Law Journal* 1315, 1317–22.

[80] Van Koppen, Barbara, 'Water Allocation, Customary Practice and the Right to Water', in *The Human Right to Water: Theory, Practice and Prospects* (Cambridge University Press, 2017) 57.

[81] See Tobin, Chapter 2 above n. 1, 120.

distributive frameworks, and hope rather to benefit from natural resource development.

Market-based water regulation is still exceptional in international practice. Aside from trailblazers Chile and Australia, market mechanisms are used for the regulation of water to one degree or another in west coast states of the United States and South Africa, while limited water trading exists in parts of China, Mexico and Spain, although all countries have different legal and institutional frameworks.[82] There have been calls, at times, for Aotearoa New Zealand to introduce markets in the regulation and distribution of water, although, as will be discussed in Chapter 5, such proposals have not, at the time of writing this book, been adopted.

In reality, most comparative regulatory approaches to the management of water combine a mixture of government regulation for the prioritisation of basic human needs and environmental protections together with market mechanisms or water trading. A clear example of this is Australia, which has 'a hybrid governance system of collaborative planning of water resources together with market mechanisms and statutory regulation that provides a robust model for potential application in many countries facing water over-allocation across multiple jurisdictions, periodic drought and water scarcity'.[83] Such a hybrid approach in many ways makes sense, given that there are multiple competing uses of water, some of which, like drinking water, recreation or environmental health are clearly public and others, like agriculture, industry or tourism, are private interests.[84] This is the context within which indigenous peoples make their substantive water claims.

3.5 CONCLUSION

In this chapter, I have introduced two key tendencies of modern water regulation relevant to the analysis of indigenous water rights in this book. Each of these tendencies has, as one time or another, been the

---

[82] Grafton et al., Introduction above n. 22, 221; K William Easter, Mark W Rosegrant and Ariel Dinar, 'Formal and Informal Markets for Water: Institutions, Performance, and Constraints' (1999) 14(1) *The World Bank Research Observer* 99.

[83] Elizabeth Macpherson et al., 'Lessons from Australian Water Reforms: Indigenous and Environmental Values in Market-Based Water Regulation', in *Reforming Water Law and Governance: From Stagnation to Innovation in Australia* (Springer, 2018) 214.

[84] Erin O'Donnell and Macpherson, Elizabeth, 'Challenges and Opportunities for Environmental Water Management in Chile: An Australian Perspective' (2012) 23(1) *Journal of Water Law* 24.

site for legal and policy debate about how best to provide for indigenous water interests. Are indigenous water rights a basic human right? Are they a right to an economic good; a distribution of available water; applicable to commercial as well as subsistence purposes and potentially transferable in markets? Are indigenous water rights essentially environmental rights, whereby indigenous interests run concurrent with the rights of nature? Or have legal rights for rivers been coincidental and pragmatic solutions to indigenous claims for greater jurisdiction over water management and governance in an attempt to reconstitute state-indigenous relations? What I will show is that all of these conceptual questions play out in the four countries studied in this book, where indigenous peoples struggle for water jurisdiction and distribution via available legal models.

PART II

COMPARATIVE COUNTRY
STUDIES

CHAPTER FOUR

THE LIMITED RECOGNITION OF INDIGENOUS WATER RIGHTS IN AUSTRALIA

4.1 INTRODUCTION

Indigenous Australians have, since the arrival of and settlement by the British Crown,[1] been excluded from regulatory regimes controlling access and use of water. Their cultural interests in and relationships with water were, until relatively recently, completely ignored in water planning and policy. In recent years some efforts have been made to reflect indigenous water values in water planning frameworks and allow limited input from indigenous Australians in decision making about water via consultation and representation on water boards. These efforts are admirable, although certainly falling well short of the imperative of indigenous water jurisdiction, discussed in Chapter 2. An even more pressing challenge in the Australian context is the unfair distribution of water, in which Aboriginal and Torres Strait Islander peoples have been 'locked out' of water law frameworks for anything beyond limited traditional, cultural water rights. The distributive injustice around water access in Australia is the focus of this chapter.

The only generalised legal mechanism that recognises indigenous rights to use water in Australia is 'native title'. Native title is a doctrine

---

[1] The issue of the manner and effect of British acquisition of sovereignty in Australia is contested. See Linda Popic, 'Sovereignty in Law: The Justiciability of Indigenous Sovereignty in Australia, the United States and Canada' (2005) 4 *Indigenous Law Journal* 117; Angela Pratt, 'Treaties vs. Terra Nullius: Reconciliation, Treaty-Making and Indigenous Sovereignty in Australia and Canada' (2004) 3 *Indigenous Law Journal* 43.

of the common law, now enshrined in the *Native Title Act 1993* (Cth), which recognises rights based in traditional laws and customs that have been substantially maintained and observed since the moment of colonisation. In determining rights to access and use water, Australian water law frameworks have been influenced, adversely, by the conception of indigenous water rights under native title. However, native title law has developed narrowly through the decisions of the Australian courts and legislature. In particular, the native title model has provided very little opportunity for the recognition of indigenous rights to use water for commercial purposes.

Until recently, the Australian academic and policy literature suggested hope that commercial water rights might be recognised as part of a native title determination.[2] Some thought that native title jurisprudence would 'evolve' to enable the recognition of 'commercial water rights' in the future, encouraged by a number of recent determinations that have recognised a right to take resources for 'any purpose'.[3] However, the study of Australian law and experience in this chapter shows little potential for any meaningful allocation of water to indigenous communities under native title. Where native title rights to water have been recognised by the Australian courts they have tended to be limited to traditional, cultural water rights that resemble pre-sovereignty water interests, at the expense of commercial use. Further, these traditional, cultural native title rights to water are extinguished or ineffective where other users hold inconsistent rights. In sum, native title is ill disposed to commercial indigenous water claims, because native title recognises traditional, or pre-sovereignty, interests. If indigenous rights to water are to be provided for in Australia for cultural and commercial purposes alike, this is most probably (and effectively) to be achieved alongside, but outside of, the limited native title recognition model.

In this chapter I suggest that an allocation model could be devised, and enshrined in legislation, to distribute (and if necessary redistribute)

---

[2] See, e.g., Alexander Walter Gardner et al., *Water Resources Law* (LexisNexis Butterworths, 2009); Lane, Chapter 2 above n. 41; McFarlane, above n. 41; Michael O'Donnell, 'Briefing Paper for the Water Rights Project by the Lingiari Foundation and ATSIC' in Lingiari Foundation (ed.), *Background Briefing Papers* (2002); Sue Jackson, 'National Indigenous Water Planning Forum – Background Paper on Indigenous Participation in Water Planning and Access to Water' (National Water Commission, CSIRO, 2009).

[3] See, e.g., O'Donnell, 'Briefing Paper for the Water Rights Project by the Lingiari Foundation and ATSIC', above n. 2; see also Samantha Hepburn, 'Native Title Rights in the Territorial Sea and Beyond: Exclusivity and Commerce in the Akiba Decision' (2011) 34(1) *University of New South Wales Law Journal* 159, 159–81 in the context of commercial fishing rights under native title.

water rights to indigenous groups. Statutory allocation models have already been successfully implemented for indigenous land rights in Australia; invented precisely because of the inability to recognise native title. However, indigenous land rights legislation originally overlooked the question of water, and still does not authorise indigenous peoples to use water on their lands for all purposes. In some parts of Australia, particularly the State of Victoria, change is underway to respond to indigenous cultural and commercial water demands via new water policy and statutory mechanisms, in the fresh context of Aboriginal self-determination and treaty making. While these new policy developments are limited in their application, they transcend the assumption that indigenous water rights must be limited to traditional, cultural purposes.

The study of indigenous water rights in Australian law and experience in this chapter tells much about the futility of attempting to recognise indigenous water interests based on pre-sovereignty notions of resource use where indigenous water users are required to compete with others for water rights. The study highlights the need for an allocative model, enabling both the reservation of water for indigenous use and the redistribution of water rights in fully allocated catchments.

4.2   INDIGENOUS EXCLUSION FROM WATER LAW FRAMEWORKS IN AUSTRALIA

4.2.1   Aboriginal and Torres Strait Islander Peoples in Australia

The indigenous peoples of Australia, otherwise known as the Aboriginal and Torres Strait Islander peoples, are ancient and culturally variegated. The human occupation of Australia has been dated back 60,000 years and traditionally included more than 700 culturally and linguistically distinct groups or nations operating under their own traditional systems of law and custom.[4] Australia was colonised by the British Crown from 1788 relying on the legal fiction of *Terra Nullius* to assume full beneficial ownership of the entire territory.[5] The indigenous peoples of Australia now make up 2.8 per cent of the total population.[6]

---

[4] Australian Museum, *Introduction to Indigenous Australia* (15 May 2018) https://australianmuseum.net.au/indigenous-australia-introduction.
[5] Reynolds, Chapter 2 above n. 33, 129.
[6] Australian Bureau of Statistics, *Census: Aboriginal and Torres Strait Islander Population* (27 June 2017) www.abs.gov.au.

Australia was settled without consent or treaty. From the British acquisition of sovereignty until the end of the twentieth century, Australian Governments did not recognise that indigenous Australians held any right or title to their traditional territories, lands and waters. An indigenous land rights movement emerged in the 1960s, in line with international tendencies, demanding equal rights and land redress. In response, certain States and Territories passed indigenous land rights legislation, which allocated land rights to indigenous groups, beginning with the *Aboriginal Land Rights (Northern Territory) Act 1976* (Cth).[7] Then, with the 1992 landmark *Mabo and Others v. The State of Queensland [No. 2]* ('*Mabo [No. 2]*'),[8] the High Court of Australia recognised native title rights to land and waters stemming from pre-sovereignty times, which the Commonwealth later regulated through the *Native Title Act 1993* (Cth).[9] The *Native Title Act* was passed by the Commonwealth Parliament in response to the recognition of the common law doctrine of native title in *Mabo [No. 2]*, partly to provide additional certainty around the circumstances in which native title can be recognised and the impact of recognition on the rights held by other Australians. It provides for the recognition of native title in a court determination, provides for the validation of 'past acts' which could be invalid after the recognition of native title and allows for 'future acts' affecting native title to proceed in certain circumstances. Today, more than 30 per cent of the total land in Australia is held by indigenous groups pursuant to a mixture of native title and titles granted under land rights legislation.[10]

Indigenous groups in Australia have suffered historical injustices and continue to experience economic and social disadvantage. As the

[7] The *Aboriginal Land Rights (Northern Territory) Act 1976* (Cth) was the first significant piece of land rights legislation in Australia; it arose out of the recommendations of the Aboriginal Land Rights Commission of 1973 presided over by Justice Woodward, following an unsuccessful claim for native title in the Northern Territory in *Milirrpum and Others v. Nabalco Pty Ltd and the Commonwealth of Australia* (1971) 17 FLR 141 ('*Milirrpum*') (known as the 'Gove Land Rights Case').

[8] *Mabo and Others v. The State of Queensland [No. 2]* (1992) 175 CLR 1 ('*Mabo*').

[9] The High Court of Australia in *Western Australia v. Ward* (2002) 213 CLR 1, 25 (Gleeson CJ, Gaudron, Gummow and Hayne JJ) held that when determining native title first recourse must be made to the Act and not the decisions of the common law, which are relevant only for whatever light they cast on the Act.

[10] Jon Altman and Francis Markham, 'Value Mapping Indigenous Lands: An Exploration of Development Possibilities' (Centre for Aboriginal Economic Policy Research The Australian National University, 2013). According to Altman and Markham indigenous land held under land rights legislation and exclusive possession native title makes up 22 per cent of the total landmass in Australia. If non-exclusive possession native title is included, the figure is 31 per cent.

preamble to the *Native Title Act* states, 'Aboriginal peoples and Torres Strait Islanders have become, as a group, the most disadvantaged in Australian society'. Among the disadvantages, indigenous land and resource rights in Australia have been affected, altered and encroached upon as a result of colonisation, whereby governments have taken control of the management of water and allocated rights to use water resources to other users instead of indigenous groups.[11] Of course, indigenous groups throughout Australia used land and water prior to the acquisition of sovereignty without holding any written title, recognisable in the western sense. Some indigenous groups, particularly those that live in remote areas least affected by the water demands of third parties, have managed in the years since colonisation to continue to make use of water without holding water use rights for hunting, fishing and foraging as well as other non-subsistence or market uses,[12] relying on open or public access. In these areas, state vesting and control of water is only a 'theoretical proposition'.[13]

In recent years, indigenous nations and other groups in Australia have reasserted their demands for self-determination under the banner of 'Treaty'. Drawing inspiration from countries like Aotearoa New Zealand and Canada, there is growing consensus among Aboriginal Australians and Torres Strait Islanders for a comprehensive treaty or settlement to legitimise the colonisation of Australia and enable redress for historical and contemporary grievances. This new movement has led to the declaration of a statement signed by over 300 indigenous Australians at Uluru in 2017, asserting continued indigenous sovereignty and favouring constitutional reform and agreement making.[14] State-based treaty projects have started in Victoria, South Australia (later discontinued), Tasmania and the Northern Territory, with varying degrees of implementation.[15] However, as Butterly notes, the process is likely to move slowly given the controversy of, and likely public backlash towards, the policy, particularly in northern regions.[16]

Indigenous demands for redress for historical injustices have invariably referred to water. The North Australian Indigenous Land and Sea Management Alliance (NAILSMA), for example, released a *Policy*

---

[11] See generally Marshall, Chapter 3 above n. 59.
[12] Jackson and Altman, Chapter 2 above n. 66, 35.
[13] Ibid., 36.
[14] Referendum Council, *Uluru Statement From The Heart* (26 May 2017) www .referendumcouncil.org.au.
[15] See, e.g., *Advancing the Treaty Process with Aboriginal Victorians Act 2018* (Vic).
[16] Interview with Lauren Butterly (Sydney, 11 July 2018).

Statement on North Australian Indigenous Water Rights,[17] in November 2009, which they position as 'a pragmatic response to the Council of Australian Governments Water Reform Agenda . . . and the sudden pace of development in the north of Australia'. NAILSMA emphasise the need for indigenous water rights in order to support the economic development of land:[18]

> Incentivising Indigenous people to pursue water using activities and water trade would also help increase the productive use of land and water resources, contributing to both Indigenous and broader social and economic outcomes.

The First Peoples' Water Engagement Council, active until 2012, argued that commercial water rights are 'not an alternative to addressing access to cultural and customary water but an additional policy to improve the economic lives of indigenous people'.[19] More recently, the National Cultural Flows Project is specifically seeking an allocation of water for indigenous peoples. This project focuses on the Murray Darling Basin and seeks the allocation of 'Cultural Flows', defined as 'water entitlements that are legally and beneficially owned by Indigenous Nations of a sufficient and adequate quantity and quality, to improve the spiritual, cultural, environmental, social and economic conditions of those Indigenous Nations. *This is our inherent right.*'[20]

4.2.2 Australian Water Law Frameworks and Indigenous Rights
Australia is often referred to as the driest inhabited continent and water demand is fiercely competitive due to pressure from irrigated agriculture, urbanisation, industry and climate change.[21] Because of this, the efficient and effective regulation and management of water in Australia is both technically complex and politically challenging. In response to

---

[17] North Australian Indigenous Land and Sea Management Alliance, *A Policy Statement on North Australian Indigenous Water Rights* (November 2009) www.nailsma.org.au/nailsma/forum/downloads/Water-Policy-Statement-web-view.pdf.

[18] Northern Australian Land and Sea Management Alliance, Chapter above n. 24.

[19] Lawlab, 'Options Paper for the First Peoples' Water Engagement Council (FPWEC): Options for an Indigenous Economic Water Fund (IEWF)' www.nwc.gov.au.

[20] Murray Lower Darling Rivers Indigenous Nations and Northern Murray–Darling Basin Aboriginal Nations, Chapter 2 above n. 22. While the focus of the National Cultural Flows Research Project was on the Murray Darling Basin, the project was established for the benefit of all First Nations across Australia.

[21] Elizabeth Macpherson and Erin O'Donnell, '¿Necesitan Derechos los Ríos? Comparando Estructuras Legales para la Regulación de los Ríos en Nueva Zelanda, Australia y Chile [Do Rivers Need Rights? Comparing Legal Structures for River Regulation in New Zealand, Australia and Chile]' (2017) 25 *Revista de Derecho Administrativo Económico* 103.

the challenges, and like the Chilean system considered in Chapter 7, Australia has adopted an integrated-market system of water regulation; involving centralised and localised water planning, a cap on water extractions,[22] and water rights transfers in markets.[23]

When it acquired sovereignty over the land now known as Australia, the British Crown vested in itself the sovereign title to all land and waters on the continent, and Australia inherited the riparian system of water regulation from the British common law.[24] Riparian rights provided landholders with an incidental right to use water from natural water sources passing through or adjacent to their land in a defined channel, enjoying little restriction other than a requirement to make 'reasonable use' of the water.[25] Riparian rights attached to the ownership of the riparian land giving the landholder access to the watercourse, and the holder of the right was required to use the water for reasonable use connected with the use of the land.[26] Neither the British Crown nor subsequent Australian governments recognised any particular rights to land or resources for indigenous Australians until the late twentieth century. Accordingly, indigenous groups in Australia did not typically hold land title, and did not, therefore, hold riparian water use rights as an incident of landholding.

In the late nineteenth century, the Crown in each State and Territory was vested with the right to the 'use, flow and control' of surface and ground water.[27] The States did so for various reasons including that the riparian rights doctrine was not well suited to Australia's water conditions.[28] At the same time, the States

---

[22] Council of Australian Governments, 'Water Reform Framework (Communiqué)' www .environment.gov.au/water/publications/action/pubs/policyframework.pdf.

[23] Commonwealth of Australia and the Governments of New South Wales, Victoria, Queensland, South Australia, the Australian Capital Territory and the Northern Territory, Introduction above n. 23.

[24] Jennifer McKay, 'The Legal Frameworks of Australian Water: Progression from Common Law Rights to Sustainable Shares' in L Crase (ed.), *Water Policy in Australia: The Impact of Change and Uncertainty* (Resources for the Future, 2008) 46, 46–50; DE Fisher, *Water Law* (LBC Information Services, 2000) 3.

[25] *Chasemore* v. *Richards* (1843) 77 ER 82 (Lord Wensleydale); *John Young and Co* v. *Bankier Distillery and Co* [1893] AC 691 'Reasonable use' was required to not bring substantial diminishment to the flow so as to affect the riparian rights of other water users.

[26] *Jones* v. *Kingborough* (1950) 82 CLR 282, 324.

[27] The first of these was *The Irrigation Act 1886* (Vic) which in its section 4 vested all water in the Crown and substantially abrogated riparian water rights. The vesting clauses are recorded today in the *Water Management Act 2000* (NSW) 392; *Water Act 2000* (Qld) 26; *Water Act 1989* (Vic) 7; *Rights in Water and Irrigation Act 1914* (WA) 5A; *Water Resources Act 2007* (ACT) 7; *Water Act 1992* (NT) 9.

[28] B Evans and P Howsam, 'A Critical Analysis of the Riparian Rights of Water Abstractors in England and Wales' (2005) 16(3) *Journal of Water Law* 90; Poh-Ling Tan, *Legal Issues Relating*

implemented a statutory system of water licences and concessions (water use rights) to authorise the 'consumptive use' of water: now defined in the Australian context as 'the use of water for private benefit consumptive purposes including irrigation, industry, urban and stock and domestic use'.[29]

The vesting of the right to 'use, flow and control' of surface and ground water under Australian water legislation did not vest 'ownership' of water in the Crown in each State and Territory, rather, its purpose was to allow the Crown to control the use of water, which was treated as a public or common resource.[30] In this sense the vesting of the 'use, flow and control' of surface and ground water can be distinguished from the vesting of minerals in the Crown under Australian minerals legislation.[31]

Under the new statutory water law frameworks, a water use right would be required in order to take and use water for consumptive purposes. Otherwise, landholders retained limited water use rights, exercisable for 'domestic and stock' purposes without the need for a licence or concession, which were the remnant of common law riparian rights.[32] The statutory water use rights were also attached to land, and were intended to support the productive use of land through activities such as irrigated agriculture.[33] Indigenous groups, who still did not hold land titles, did not enjoy access to statutory water use rights as an incident of landholding, and could not, therefore, lawfully make use of water on or adjacent to their traditional territories.

Today, the States primarily determine who may and may not take and use water, and regulate the way in which to do so. This is done via water legislation and corresponding water resource plans, which set out the amount of water that can be taken from particular water resources

to *Water Use*, in *Property: Rights and Responsibilities, Current Australian Thinking* (Land and Water Australia, 2002) 15–19.

[29] Commonwealth of Australia and the Governments of New South Wales, Victoria, Queensland, South Australia, the Australian Capital Territory and the Northern Territory, Introduction above n. 23, Sch. B(i).

[30] *ICM Agriculture* v. *the Commonwealth* [2009] 240 CLR 140, 203 (Hayne, Kiefel and Bell JJ) in respect of ground water and 173 (French CJ, Gummow and Crennan JJ) referring to water as *publici juris* and 'common property'.

[31] *Western Australia* v. *Ward* (2002) 213 CLR 1, 186 (Gleeson CJ, Gaudron, Gummow and Hayne JJ) explaining that the vesting of minerals gave the Crown 'full dominium' in the substances.

[32] See, e.g., the limited domestic and stock rights preserved in the *Water Management Act 2000* (NSW) 52, which do not require a water access entitlement. Similar provisions apply in other Australian States and Territories.

[33] See BR Davidson, *Australia: Wet or Dry? The Physical and Economic Limits to the Expansion of Irrigation* (Melbourne University Press, 1969); T Langford-Smith and J Rutherford, *Water and Land: Two Case Studies* (Australia National University Press, 1966).

for a range of purposes. The exception is in Australia's largest water catchment, the Murray Darling Basin, covering more than one million square kilometres in the settled, south east of Australia. There, the Commonwealth has implemented specific water legislation, and a Basin-wide water resource plan, discussed below.[34]

The Commonwealth Government devised a national approach to water planning and regulation in Australia under the National Water Initiative,[35] first agreed to by the Council of Australian Governments in 1994.[36] Implementing the Initiative at State level is intended to produce a 'nationally-compatible, market, regulatory and planning based system of managing surface and groundwater resources for rural and urban use that optimises economic, social and environmental outcomes'.[37] As part of the Initiative, States agreed to cap water rights, improve transparency of water pricing, separate water rights from land titles and establish water trading, and allocate water to the environment. The new approach to water regulation was premised on a mixture of centralised water planning and trade in water use rights that were now defined as 'water access entitlements' and could be transferred separately from landholding. A water access entitlement is defined as 'a perpetual or open-ended share of the consumptive pool of a specified water resource, as determined by the relevant water plan'.[38]

The unbundling of water use rights was intended to encourage the more efficient, and productive, use of water through increased competition.[39] The economic philosophy underlying the reforms provided that water markets would facilitate the transfer of water use rights to where they produce the highest economic value.[40] However, unlike the other market-based regulatory framework considered in this book (Chile, Chapter 7), these 'water access entitlements', held separately from land, are not treated as rights of 'property', because this would be

---

[34] *Water Act 2007* (Cth); *Basin Plan 2012* (Cth).
[35] Commonwealth of Australia and the Governments of New South Wales, Victoria, Queensland, South Australia, the Australian Capital Territory and the Northern Territory, Introduction above n. 23.
[36] Council of Australian Governments, above n. 22.
[37] Ibid. cl. 23.
[38] Commonwealth of Australia and the Governments of New South Wales, Victoria, Queensland, South Australia, the Australian Capital Territory and the Northern Territory, Introduction above n. 23.
[39] Fisher, Chapter 3 above n. 74, 104–5.
[40] Commonwealth of Australia and the Governments of New South Wales, Victoria, Queensland, South Australia, the Australian Capital Territory and the Northern Territory, Introduction above n. 23.

inconsistent with the vesting of all water in the Crown and because the common law does not allow ownership of water in its natural state.

In 2007, the Commonwealth reinforced its responsibility for water resource management within the Murray Darling Basin by passing the *Water Act 2000* (Cth), and later the *Basin Plan 2012* (Cth). The legislation established a sustainable limit on water extractions in the Basin and was accompanied by an ambitious project for water recovery for the environment via water purchase programmes and efficiency projects.[41]

It was not until the end of the twentieth century that Australian Governments began to recognise and allocate land titles for indigenous groups under native title and land titles granted pursuant to land rights legislation. Importantly, the recognition and allocation of land rights to indigenous groups coincided with the unbundling of water use rights from land titles under the National Water Initiative. This detachment of water use rights from landholding reinforced indigenous exclusion from water law frameworks, for two reasons.

First, even if indigenous groups obtained land rights through indigenous land rights regimes they would not (as a matter of water law) acquire the right to use water on those lands in the manner they might have had their land rights been recognised or allocated prior to unbundling. If the riparian rights of indigenous groups had been recognised at the acquisition of sovereignty, the groups would have acquired riparian water use rights. Indigenous groups with riparian rights could later have converted these to state-based water licences and concessions, which also attached to land title. Second, unbundling also reinforced indigenous exclusion from water law frameworks because water use rights became available for purchase in water markets, independent of landholding. This meant that other users could potentially acquire water use rights on or affecting indigenous lands.

Because indigenous landholders do not typically enjoy water use rights as an incident of land title, over and above the limited entitlement to use water for domestic and stock purposes 'as of right', specific laws would be needed to provide for such rights. Yet, indigenous land rights legislation does not address indigenous water rights. As will be discussed below, while native titleholders have had traditional-cultural water rights recognised in certain circumstances, native title rights to water have not, to date, been recognised for commercial purposes and

---

[41] See O'Donnell and Macpherson, Elizabeth, Chapter 3 above n. 84.

most native title determinations explicitly restrict native title rights to water to non-commercial use.

4.3   RECOGNISING INDIGENOUS WATER RIGHTS UNDER NATIVE TITLE

The common law doctrine of native title developed from the decision of the High Court in *Mabo [No. 2]*[42] in response to the British Crown's failure to recognise indigenous land rights at the acquisition of sovereignty in reliance on the legal fiction of *Terra Nullius* (land of no one).[43] Native title is the dominant legal mechanism that recognises indigenous land and resource rights in Australia. It is sometimes described with reference to ideas of legal pluralism, because the origin of native title rights and interests is in traditional laws and customs existing at the time sovereignty was acquired over the territory in question. As Brennan J famously stated in *Mabo* [No. 2], '[n]ative title has its origin in and is given content by the traditional laws acknowledged by and the traditional customs observed by the indigenous inhabitants of a territory'.[44] However, while native title is recognised by the common law, it is not 'an institution of the common law'. Traditional laws and customs are a source of law independent of and pre-dating the sovereignty asserted by the Australian state. Pearson has drawn on ideas of legal pluralism to argue that Australian native title law is a 'recognition space' between state and Aboriginal systems of law.[45]

The 'recognition and protection' of pre-existing native title rights to land and waters is now provided for in the *Native Title Act*. Indigenous water rights, like land, may be recognised in the native title process. Section 223(1) of the *Native Title Act* provides that native title rights and interests comprise:

> . . . the communal, group or individual rights and interests of Aboriginal peoples or Torres Strait Islanders in relation to land or waters, where:

---

[42] *Mabo* (1992) 175 CLR 1.
[43] See generally Reynolds, Chapter 2 above n. 33; Maureen Tehan, 'A Hope Disillusioned, and Opportunity Lost? Reflections on Common Law Native Title and Ten Years of the Native Title Act' (2003) 27(2) *Melbourne University Law Review* 523, 527.
[44] *Mabo* (1992) 175 CLR 1 (Brennan J).
[45] Noel Pearson, 'The Concept of Native Title at Common Law' [1997] (5) *Australian Humanities Review*.

a. the rights and interests are possessed under the traditional laws acknowledged, and the traditional customs observed, by the Aboriginal peoples or Torres Strait Islanders; and
b. the Aboriginal peoples or Torres Strait Islanders, by those laws and customs, have a connection with the land or waters; and
c. the rights and interests are recognised by the common law of Australia.

Native title rights to water are communal or group rights, as the right is vested in a body corporate.[46]

The *Native Title Act* provides for rights to water in two specific ways. First, native title rights to water may be recognised as part of a court determination under section 225 of the Act, discussed below. Native titleholders can also rely on limited statutory water use rights under section 211 of the *Native Title Act* without the need for a water use right, in the enjoyment of their native title rights and interests.

Where native titleholders have obtained a determination of native title rights to land to the exclusion of all others, they have also been recognised as holding an accompanying right to make decisions about access to and use of land and waters, although only to the extent that this is not inconsistent with the rights and interests of others granted under legislation (including the water use rights held by other users) which take priority.[47] In any event, such rights are procedural rather than substantive in nature and do not provide any positive entitlement to access and use water.

Section 211 applies to the situation where state-based water laws prohibit the taking and use of water without a water licence or permit.[48] It follows from section 211 that native titleholders may access waters and hunt, fish, gather, or carry out cultural or spiritual activities on those waters,[49] 'in exercise or enjoyment of their native title rights and interests' without the need for a water use right.[50] However, section 211 only authorises water use without a water use right for satisfying, 'personal, domestic or non-commercial communal needs',[51] and otherwise goes no further than the rights all landholders enjoy to use water

---

[46] *Native Title Act 1993* (Cth) 1993, s. 224 ('*Native Title Act*').
[47] See, e.g., *Northern Territory of Australia v. Alyawarr, Kaytetye, Warumungu, Wakaya Native Title Claim Group* (2005) 145 FCR 442, 504–7.
[48] *Native Title Act 1993*, s. 211(1)(b).
[49] Ibid. s. 211(3).
[50] Ibid. s. 211(2)(b).
[51] Ibid. s. 211(2)(a).

for limited 'domestic and stock' purposes without the need for a water access entitlement.

It is the court, in making a determination of native title, which determines the nature and extent of native title rights and interests. Section 225 of the *Native Title Act* sets out the matters that the court must include in its determination:

> A determination of native title is a determination whether or not native title exists in relation to a particular area (the determination area) of land or waters and, if it does exist, a determination of:
>
> a. who the persons, or each group of persons, holding the common or group rights comprising the native title are; and
> b. the nature and extent of the native title rights and interests in relation to the determination area; and
> c. the nature and extent of any other interests in relation to the determination area; and
> d. the relationship between the rights and interests in paragraphs (b) and (c) (taking into account the effect of this Act); and
> e. to the extent that the land or waters in the determination area are not covered by a non-exclusive agricultural lease or a non-exclusive pastoral lease – whether the native title rights and interests confer possession, occupation, use and enjoyment of that land or waters on the native title holders to the exclusion of all others.

The 'native title rights and interests' to be determined by the court have been characterised in the Australian jurisprudence as a 'bundle of rights',[52] most notably in the decision of the High Court of Australia in *Western Australia* v. *Ward*, which held:[53]

> ... recognising that the rights and interests in relation to land which an Aboriginal community may hold under traditional law and custom are not to be understood as confined to the common lawyer's one-dimensional view of property as control over access reveals that steps taken under the sovereign authority asserted at settlement may not affect every aspect of those rights and interests. The metaphor of 'bundle of rights' which is so often employed in this area is useful in two respects. It draws attention first to the fact that there may be more than one right or

---

[52] See, e.g., *Yanner* v. *Eaton* (1999) 201 CLR 351, 17–31. In *Yanner* v. *Eaton* the High Court explained that while there is no one definition of property, property rights are a 'description of a legal relationship with a thing' and are typically presented as a 'bundle of rights', including rights such as the right to use and enjoyment and the right to exclude or control access.

[53] *Western Australia* v. *Ward* (2002) 213 CLR 1, 95 (Gleeson CJ, Gaudron, Gummow and Hayne JJ).

interest and secondly to the fact that there may be several kinds of rights and interests in relation to land that exist under traditional law and custom. Not all of those rights and interests may be capable of full or accurate expression as rights to control what others may do on or with the land.

The various 'sticks' in the bundle of rights may range from a right of exclusive possession to limited use rights, such as: the right 'to derive sustenance from the land'; 'to hunt and gather food on the land'; 'to hold ceremonies on the land'; 'to care for the land according to environmental requirements, including burning the land'; and 'to regulate access to the land'.[54]

However, rights to 'own' water, understood in Australian law as 'a legal right to have and to dispose of possession and enjoyment of the subject matter',[55] cannot be one of the sticks in the bundle. That is the case, because section 223(1)(c) of the *Native Title Act* provides that native title rights and interests must be 'recognised by the common law of Australia',[56] and the common law does not allow ownership of water in its natural state.[57] Accordingly, native title determinations often include a proviso that, '[n]otwithstanding anything in this determination, the native title rights and interests do not confer on native title holders rights of ownership in respect of flowing water'.[58]

Further, the courts have held that native title rights to water cannot be exclusive, as exclusive rights would be inconsistent with the legislative vesting of the right to the control and use of water in the Crown.[59] The High Court in *Western Australia v. Ward* explained the impact of the statutory vesting of water control and use in Western Australia on native title as follows:[60]

> Part III of the Rights in Water and Irrigation Act provides (in s. 4(1)) that the 'right to the use and flow and to the control of the water' in natural waters 'shall, subject only to the restrictions hereinafter

---

[54] These were the various sticks in the bundle claimed by the native title applicants in ibid. 48 (Gleeson CJ, Gaudron, Gummow and Hayne JJ).

[55] *Yanner v. Eaton* (1999) 201 CLR 351, 25.

[56] See *Commonwealth v. Yarmirr* (2001) 208 CLR 49, 49 (Gleeson CJ, Gaudron, Gummow and Hayne JJ), explaining that 'recognise' in this context means that the common law 'will, by the ordinary processes of law and equity, give remedies in support of the relevant rights and interests to those who hold them'.

[57] *ICM Agriculture v. The Commonwealth* [2009] 240 CLR 140, 173 (Hayne, Kiefel and Bell JJ).

[58] See, e.g., *Kaurareg People v. Queensland* [2001] FCA 657, cl. 8.

[59] *Western Australia v. Ward* (2002) 213 CLR 1, 152 (Gleeson CJ, Gaudron, Gummow and Hayne JJ).

[60] Ibid. (Gleeson CJ, Gaudron, Gummow and Hayne JJ).

provided, and until appropriated under the sanction of this Act, or of some existing or future Act of Parliament, vest in the Crown'. It deals with riparian rights (s. 14) and allows riparian owners to apply for special licences to divert and use water (s. 15). The vesting of waters in the Crown was inconsistent with any native title right to possession of those waters to the exclusion of all others.

Native title determinations similarly usually confirm that native title rights to water are non-exclusive, for example:[61]

> Subject to paragraphs 10, 11 and 12 below the nature and extent of the native title rights and interests in relation to the land and waters described in Part 1 of Schedule 1 are: . . . b) in relation to Water, the non-exclusive rights to: . . . (iii) take and use the Water of the area, for personal, domestic and non-commercial communal purposes.

The impossibility, based on current Australian jurisprudence, for native title rights to water to be rights of 'ownership' or exclusive, does not (in and of itself) limit the potential for native title rights to water to be recognised or exercised for commercial purposes. This is the case because the Australian courts have not been prepared to characterise any water use rights as exclusive rights of ownership,[62] although the water use rights held by other users are often allocated and exercisable for commercial purposes. However, the potential for native title to provide for commercial water rights in Australia is further limited by problems of 'continuity' and 'priority' in the following ways.

4.3.1 The Problem of Continuity

When the High Court of Australia in *Mabo [No. 2]* recognised the potential continuance of native title over land and water, it also introduced the idea that native title rights will only be recognised where a group can prove that it has continued to acknowledge traditional laws or observe traditional customs which form the basis for the rights since prior to the acquisition of sovereignty. Brennan J explained:[63]

> . . . when the tide of history has washed away any real acknowledgement of traditional law and any real observance of traditional customs, the foundation of native title has disappeared.

---

[61] *Brooks on behalf of the Mamu People* v. *State of Queensland (No. 4)* [2013] FCA 1453, cl. 8.
[62] *ICM Agriculture* v. *The Commonwealth* [2009] 240 CLR 140.
[63] *Mabo* (1992) 175 CLR 1, 60 (Brennan J).

This idea of 'continuity' is reflected in section 223 of the *Native Title Act*, which requires that native title rights and interests in land or waters are 'possessed under the traditional laws acknowledged, and the traditional customs observed, by the Aboriginal peoples or Torres Strait Islanders'. Those traditional laws acknowledged and traditional customs observed have been construed by the Australian courts in the present tense. That is, that the rights and interests must currently be possessed pursuant to traditional laws acknowledged and traditional customs observed by the native title holder.[64] Further, section 223 requires that 'the Aboriginal peoples or Torres Strait Islanders, by those laws and customs, have a connection with the land or waters'.

The application of section 223 in the Australian jurisprudence has produced two concerns with respect to continuity.[65] The first is whether the traditional laws and customs must be continuously acknowledged or observed since pre-sovereignty times, or whether some break in the chain of continuity is allowed for. The second is whether the laws and customs can adapt or change since pre-sovereignty times and still be considered to be 'traditional'.

The High Court of Australia in *Members of the Yorta Yorta Aboriginal Community* v. *State of Victoria and Others* ('*Yorta Yorta*')[66] confirmed that section 223 required a native title applicant to particularise the traditional laws and customs establishing their native title right. Those laws and customs must have been 'acknowledged and observed by the ancestors of the claimants at the time of sovereignty'.[67] Further, the acknowledgement and observance of those laws and customs must have continued 'substantially uninterrupted since sovereignty', being 'passed down generation to generation'.[68] Former government native title lawyer, Sophia Angelis, explains how in practice this 'connection' is shown by looking at the traditional practices or customs on the land, songs associated with the land or water and traditional uses of the land such as fishing.[69]

Even in the case of land, it is very difficult for indigenous groups to prove the content of pre-sovereignty traditional laws and customs

[64] *Members of the Yorta Yorta Aboriginal Community* v. *State of Victoria and Others* (2002) 214 CLR 422, 444 ('*Yorta Yorta*') (Gleeson CJ, Gummow and Hayne JJ).
[65] See Australian Law Reform Commission, 'Connection to Country: Review of the Native Title Act 1993 (Cth)' (126, 2015) 19.
[66] *Yorta Yorta* (2002) 214 CLR 422.
[67] Ibid. 456 [87].
[68] Ibid.
[69] Interview with Sophia Angelis (Melbourne, 6 July 2018).

establishing their rights some 200 years after the acquisition of sovereignty.[70] Proving that those traditional laws and customs have been acknowledged and observed, in a substantially uninterrupted manner, since the acquisition of sovereignty, is also difficult.[71] It may be especially difficult to prove the continuity of traditional laws and customs authorising water use because rights to use water were an incident of landholding until the end of the twentieth century. Because indigenous people did not hold title to land adjacent to water resources, they may have been unable to continue to acknowledge traditional laws or observe traditional customs with respect to water where their access was prevented by the rights of adjacent landholders.[72] The courts have stressed that a physical connection is not necessarily required in order to satisfy the requirement in section 223(1)(b) that the native titleholders, by their pre-sovereignty laws and customs, 'have a connection with the land or waters'.[73] Yet, it is certainly harder to prove a connection to particular waters, pursuant to traditional laws and customs authorising water use at the acquisition of sovereignty, where water resources are no longer used by the claimant.[74]

The requirement that laws and customs establishing indigenous water interests be passed down generation to generation in a substantially uninterrupted manner presents particular challenges for claims concerning water use for modern, commercial or economic purposes. The use of water for commercial ends may be quite different from the way indigenous groups used water under pre-sovereignty traditional laws and customs.[75] This may raise a question of whether the use of water for commercial purposes in fact derives from pre-sovereignty traditional laws and customs at all.[76] The Court in *Yorta Yorta* observed:[77]

---

[70] See Hewitt, Chapter 2 above n. 62, 251–56; Australian Law Reform Commission, above n. 65, 75–76.

[71] Interview with Sophia Angelis (Melbourne, 6 July 2018), above n. 69. See generally Australian Law Reform Commission, above n. 65, 243.

[72] See Jackson and Marcia Langton, Chapter 2 above n. 23.

[73] *De Rose* v. *South Australia (No. 2)* (2005) 145 FCR 290, 306 (Wilcox, Sackville and Merkel JJ).

[74] *Yanner* v. *Eaton* (1999) 201 CLR 351, 373 [38] (Gleeson CJ, Gaudron, Kirby and Hayne JJ), '[r]egulating particular aspects of the usufructuary relationship with traditional land does not sever the connection of the Aboriginal peoples concerned with the land (whether or not prohibiting the exercise of that relationship altogether might, or might to some extent)'.

[75] Interview with Sophia Angelis (Melbourne, 6 July 2018), above n. 69.

[76] See Ivison, Chapter 2 above n. 20, 327; Hewitt, Chapter 2 above n. 62, 256.

[77] *Yorta Yorta* (2002) 214 CLR 422, 455 (Gleeson CJ, Gummow and Hayne JJ).

The key question is whether the law and custom can still be seen to be traditional law and traditional custom. Is the change or adaptation of such a kind that it can no longer be said that the rights or interests asserted are possessed under the traditional laws acknowledged and the traditional customs observed by the relevant peoples when that expression is understood in the sense earlier identified?

Aside from the problems of proof, once native title rights to water have been recognised they must be exercised in accordance with traditional laws and customs. Determinations usually include a condition to the effect that:[78]

The native title rights and interests are subject to and exercisable in accordance with: (a) the valid laws of the [relevant State or Territory] of Australia and the Commonwealth of Australia; and (b) the traditional laws acknowledged and traditional customs observed by the native title holders.

The requirement that native title rights to water must be exercised in accordance with traditional laws and customs casts further doubt on the application of native title rights to water to commercial purposes.

4.3.2 The Problem of Priority

Even if native title rights to water can be established pursuant to traditional laws and customs, the *Native Title Act* provides for their extinguishment or ineffectiveness to the extent of any inconsistency with rights granted to other users. The High Court in *Western Australia v. Ward* explained that the 'relevant enquiries' when considering the grant of inconsistent rights are:[79]

... whether rights have been created in others that are rights inconsistent with native title rights and interests, and whether the Crown has asserted rights over the land that are inconsistent with native title rights and interests.

In 1975, the *Racial Discrimination Act 1975* (Cth) commenced. The Act guarantees to all Australians the right to equality before the law regardless of race, colour, or national or ethnic origin.[80] Relying on the *Racial Discrimination Act*, the Australian Courts have held that

---

[78] *Japalyi v. Northern Territory of Australia* [2014] FCA 421, [7].
[79] *Western Australia v. Ward* (2002) 213 CLR 1, 136 (Gleeson CJ, Gaudron, Gummow, and Hayne JJ).
[80] *Racial Discrimination Act 1975* (Cth), s. 10.

indigenous Australians are entitled to 'immunity from legislative inter-
ference with their enjoyment of their human right to own and inherit
property as it clothes other persons in the community', upholding their
right to consultation or compensation for the extinguishment of native
title rights and interests after 1975.[81] As a consequence of the recognition of native title rights and interests by the Australian courts in *Mabo
[No. 2]*, the grant of inconsistent rights to others after 1975 could be
rendered invalid because of its discriminatory impact on native title.

In response to concerns around the uncertainty produced by the
recognition of native title on other titleholders, the *Native Title Act*
created a complicated regime to determine the impact of the grant of
inconsistent rights (called 'acts') on native title rights and interests.[82]
The rules vary depending on whether the inconsistent right is a 'past
act', 'intermediate period act', or 'future act', as defined in the *Native
Title Act*.

Where other users were granted inconsistent water use rights under
water legislation prior to 1975, however, native title rights to water will
have been extinguished, without the need for consultation or
compensation.[83] Inconsistent water legislation, grants of land rights,
Crown reservations[84] and water control and storage works[85] may also
have extinguished native title rights to water before 1975, if native title
rights and interests are inconsistent with those laws, rights, reservations
or works.[86]

The first set of rules in the *Native Title Act* deals with acts that took
place between 1975 and 1 January 1994 ('past acts'), which could have
a discriminatory impact on native title and therefore be invalid under
the *Racial Discrimination Act*.[87] The past act provisions of the *Native
Title Act* provide that the grant of water use rights to others under water
legislation between 1975 and 1 January 1994 is valid to the extent of
any inconsistency with native title rights and interests,[88] although

---

[81] *Mabo v. Queensland [No. 1]* (1988) 166 CLR 186, 218–19 (Brennan, Toohey and Gaudron JJ).
This was because s. 10 of the *Racial Discrimination Act 1975* (Cth) guarantees native titleholders
'security of enjoyment in that property to the same extent as the title holders of other races'.

[82] See generally Gardner et al., above n. 2, 257–65. Gardner et al. discuss in detail the impact of
different 'acts' on native title rights to water in Australia.

[83] *Western Australia v. Ward* (2002) 213 CLR 1, 263–5 (Gleeson CJ, Gaudron, Gummow, Hayne
JJ).

[84] Ibid. 152, 138 (Gleeson CJ, Gaudron, Gummow, Hayne JJ).

[85] *Native Title Act 1993*, ss. 15, 19.

[86] Ibid. s. 227.

[87] Ibid. ss. 13A, 228.

[88] Ibid. ss.19, 228.

compensation may be owed.[89] Crown reservations and public water works may also be valid as past acts.[90]

The second set of rules in the *Native Title Act* deals with acts that took place between 1 January 1994 and 23 December 1996 ('intermediate period acts').[91] The intermediate period act provisions of the *Native Title Act* provide that the grant of water use rights to others under water legislation during the intermediate period is valid to the extent of any inconsistency with native title rights and interests,[92] again with compensation potentially payable.[93] Crown reservations and public water works may also be valid as intermediate period acts.[94]

In addition to the past act and intermediate period act provisions, the *Native Title Act* specifically validates inconsistent rights granted after 1993, in its 'future acts' provisions.[95] The future acts provisions confirm that certain acts that affect native title after 1 July 1993 are valid despite their impact on native title.[96] Additional validation provisions concerning water were inserted into the future acts regime in 1998, in response to concerns about the certainty of the rights and interests held by other titleholders after the decision in *Wik Peoples* v. *Queensland*.[97] Michael O'Donnell characterises the impact of the future acts regime on native title rights to water as 'de-facto extinguishment' of native title to water leaving only a right of compensation.[98]

The first additional future act provision affecting native title rights to water is section 24 HA of the *Native Title Act*. Section 24 HA applies to making, amendment or repeal of legislation in relation to the management or regulation of water after 1 July 1993,[99] and the grant of a lease, licence, permit or authority under legislation that relates to the management or regulation of water after 1 January 1994.[100] These acts are valid as a future act, despite their effect on native title rights to

---

[89] Ibid. s. 20.
[90] Ibid. ss. 19, 229(4).
[91] Ibid. s. 21.
[92] Ibid. ss. 22F, 232.
[93] Ibid. s. 22G.
[94] Ibid. ss. 21, 232A, 232B.
[95] Ibid. s. 233.
[96] Ibid. s. 24AA(2).
[97] *Wik Peoples* v. *Queensland* (1996) 187 CLR 1.
[98] O'Donnell, 'Briefing Paper for the Water Rights Project by the Lingiari Foundation and ATSIC', above n. 2, 104.
[99] *Native Title Act 1993*, s. 24HA(1).
[100] Ibid. s. 24HA(2).

water.[101] Crown reservations and public water works may also be valid as future acts.[102]

The non-extinguishment principle applies to the grant of inconsistent water use rights under section 24 HA. However, although native title rights which are inconsistent with such future acts are not extinguished, the native title rights and interests (and their exercise) are 'prevailed over'[103] by the inconsistent water use rights, merely leaving a right to compensation.[104] Native titleholders and claimants affected by the section 24 HA future acts are to be notified and given an opportunity to comment.[105] However, the right to comment is not a 'right to negotiate' and it is likely that the grant of water use rights to other rightholders is valid regardless of whether the right to comment is provided.[106] Together with the validation of water legislation and water use rights granted under the *Native Title Act* discussed above, section 44H of the Act confirms that the grant of water use rights to other users prevails over, and is not prevented by, native title rights and interests.

In the case of fully allocated water resources, including in Australia's largest water catchment, the Murray Darling Basin, any substantive native title rights to water would presumably be overridden by inconsistent water legislation and water use rights under it. In these circumstances, it would be impossible to recognise a native title right to use water for any purpose requiring a consumptive water use right. Again, most native title determinations explicitly spell out the priority mechanism, in words to the effect:[107]

> To the extent that the continued existence, enjoyment or exercise of the native title rights and interests referred to in paragraph 6 is inconsistent with the existence, enjoyment or exercise of the other rights and interests referred to in paragraph 8, the other rights and interests and the doing of any activity required or permitted to be done by or under the other interests, prevail over, but do not extinguish, the native title rights and interests.

---

[101] Ibid. s. 24HA(3).
[102] Ibid. ss. 24JA, 24KA.
[103] Ibid. s. 24AA(7).
[104] Ibid. s. 24HA(4); s. 238.
[105] Ibid. s. 24HA(7).
[106] The full Federal Court in *Harris* v. *Great Barrier Reef Marine Park Authority* [2000] FCA 603, 71–4 held that the opportunity to comment was precautionary only, to give information to the decision maker who may or may not take it into account.
[107] *Japalyi* v. *Northern Territory of Australia* [2014] FCA 421, cl. 9.

The cumulative effect of these continuity and priority problems on native title rights to water is that the native title process only serves to formalise traditional water interests that have continued to be practised by indigenous groups despite colonisation, and which are not inconsistent with the allocation of water use rights to other users. This outcome is 'perverse',[108] because it requires native title applicants to pretend that they have not been historically excluded from land and resource rights in order to obtain recognition. It requires them to prove that they have continued to enjoy relationships with water resources of which they have been dispossessed. As Angelis explains:[109]

> The greater the historical injustice, the harder it is to access the remedy because you can't prove what's required in terms of that continuous connection.

In essence, the native title recognition model for indigenous water rights is an incomplete response to ongoing indigenous exclusion from water law frameworks.

4.3.3 Assessing the Native Title Model

In not one of the 440-odd determinations of native title from *Mabo [No. 2]*[110] to the beginning of 2019 has an Australian court or tribunal expressly recognised a right to use water for commercial purposes.[111] Determinations typically restrict any native title rights to water to 'personal, domestic and non-commercial communal purposes'. As an example, clause 7 of the determination of the Federal Court in *Lampton on Behalf of the Juru People* v. *State of Queensland* provides:[112]

> 7. Subject to paragraphs 9, 10 and 11 below the nature and extent of the native title rights and interests in relation to the land and waters described in Part 1 of Schedule 1 are the non-exclusive rights to: . . .
>
> (e) take and use the water of the area for personal, domestic and non-commercial communal purposes;

Such an approach is not necessarily consistent with comparative native title jurisprudence. For example, the Canadian Supreme Court

---

[108] Williams, Chapter 2 above n. 44, 8–9.
[109] Interview with Sophia Angelis (Melbourne, 6 July 2018), above n. 69.
[110] *Mabo* (1992) 175 CLR 1.
[111] National Native Title Tribunal, *National Native Title Register* www.nntt.gov.au (I reviewed all native title determinations listed in the Register up until February 2019).
[112] See, e.g. *Lampton on Behalf of the Juru People* v. *State of Queensland* [2014] FCA 736, [7].

in *Tsilhqot'in Nation v. British Colombia*[113] held that, 'Aboriginal title holders of modern times can use their land in modern ways, if that is their choice'.[114] Frustration with the 'freezing' approach to proving native title rights in Australia, in comparison to other common law countries, has led some commentators to consider why exactly Australia has this particular predicament with 'continuity'.[115] Some, such as Pearson, have suggested that the problem of continuity would be addressed if native title rights were characterised as arising not out of traditional laws and customs but out of historical possession and use.[116] Pearson has made his argument with reference to the judgement of Toohey J in *Mabo [No. 2]*,[117] which raised the possibility of indigenous groups claiming a possessory title to land and resources in Australia pursuant to 'common law aboriginal title'.[118] The theory discussed by Toohey J, drawing on the work of McNiel in his book *Common Law Aboriginal Title*, provides that indigenous inhabitants in possession of their lands as at colonisation are presumed to have a fee simple estate, unless someone else can claim a better right. Because common law aboriginal title (which Pearson refers to as 'possessory title')[119] is sourced in the common law, it arises upon the acquisition of sovereignty. There is no need to prove any pre-sovereignty laws and customs, but merely to prove occupation at the time of sovereignty. Indigenous

---

[113] *Tsilhqot'in Nation v. British Colombia* [2014] SCC 44 [67].

[114] Ibid. [75].

[115] See, e.g., Simon Young, *The Trouble with Tradition: Native Title and Cultural Change* (Federation Press, 2008); Kirsten Anker, 'Law in the Present Tense: Tradition and Cultural Continuity in Members of the Yorta Yorta Aboriginal Community v Victoria' (2004) 28-File Attachments | (1) *Melbourne University Law Review* 1; Bulkan, Chapter 2 above n. 49, 836; Barcham, above n. 79, 7; Strelein, Chapter 2 above n. 49, 115; Mick Dodson, 'Mabo Lecture: Asserting Our Sovereignty' in *Dialogue about Land Justice: Papers from the National Native Title Conferences* (Aboriginal Studies Press, 2010) 13, 14; CA Zorzi, 'The "Irrecognition" of Aboriginal Customary Law' (2000) 5; Webber, Chapter 2 above n. 28, 64–6.

[116] Pearson, 'The Concept of Native Title at Common Law', above n. 45; Noel Pearson, 'Land Is Susceptible of Ownership' in Lisa Palmer, Maureen Tehan and Kathryn Shain (eds.), *Honour among Nations?: Treaties and Agreements with Indigenous People* (Melbourne University Press, 2004) 83; See also Brady Pohle, 'Possessory Title in the Context of Aboriginal Claimants' [1995] *Queensland University of Technology Law Journal*; see also Pamela O'Connor, 'Aboriginal Land Rights at Common Law: Mabo v Queensland' (1992) 18(2) *Monash University Law Review* 251.

[117] *Mabo* (1992) 175 CLR 1.

[118] Ibid. (Toohey J). See also Kent McNeil, *Common Law Aboriginal Title* (Clarendon Press, 1989).

[119] Pearson, 'The Concept of Native Title at Common Law', above n. 45; Pearson, 'Land Is Susceptible of Ownership', above n. 116, 83. Pearson argues that possessory title is not a separate ground but an alternative formulation of native title. But see Bulkan, Chapter 2 above n. 49, 823. Bulkan argues that it is a separate ground.

occupation at sovereignty is recognised by the common law, becoming a fee simple title,[120] even if the indigenous inhabitants have since lost possession.[121] Such arguments appeal to the US conceptualisation of 'Indian title' sourced in the indigenous group's exclusive use and occupation of land over a long period of time.[122] The Canadian Aboriginal title cases also emphasise occupation as the source of title, although the requirement to prove exclusive occupation is fixed at the acquisition of sovereignty, providing for a *sui generis* proprietary right (entailing rights similar to those of fee simple ownership), rather than fee simple at common law.[123] Yet despite the hopeful analyses of indigenous advocates, Australian native title jurisprudence has proved stubborn.

Some Australian commentators predict that indigenous groups may be recognised as having a right to use various resources for any (including commercial) purposes as part of evolving native title jurisprudence concerning a 'right to trade'.[124] The leading decision in the evolving jurisprudence is the High Court of Australia's 2013 decision in *Akiba on Behalf of the Torres Strait Regional Seas Claim Group* v. *Commonwealth of Australia* (*'Akiba'*).[125] The High Court in *Akiba* held that native title rights to take fish in off-shore waters for 'any purposes' were not extinguished by Queensland fisheries legislation prohibiting the taking of fish for commercial purposes without a licence, although the exercise of the right for commercial purposes might be prohibited without a licence.[126] *Akiba* was followed in two 2014 Federal Court decisions by Justice North: *BP (Deceased) on Behalf of the Birriliburu People* v. *State of Western Australia* (*'BP'*)[127] and *Willis on Behalf of the Pilki People* v. *State of Western Australia* (*'Willis'*),[128] and more recently in *Rrumburriya Borroloola Claim Group* v. *Northern Territory of Australia*.[129] Like *Akiba*, the Court in the subsequent cases emphasised

---

[120] *Mabo* (1992) 175 CLR 1 (Toohey J).
[121] Ibid. (Toohey J) on the basis that prior possession is a better right.
[122] See, e.g., *United States* v. *Santa Fe Public Railroad Company* (*1941*) 314 US 339 62 Ct 248. See generally Young, above n. 115, 45, 85, 90; *Mabo* (1992) 175 CLR 1114–5, 189 (Toohey J).
[123] See *Calder* v. *Attorney-General of British Colombia* (1973) 34 DLR (3rd) 145; (1973) 34 DLR 3rd 145185, 187–90; *Delgamuukw* v. *British Colombia* (1997) 153 DLR 4th 193241–253; *Tsilhqot'in Nation* v. *British Colombia* [2014] SCC 44 [14]; see generally Young, above n. 115.
[124] See, e.g., Michael O'Donnell, 'Indigenous Rights in Water in Northern Australia' (NAILSMA – TRaCK, 2011).
[125] *Akiba on behalf of the Torres Straight Regional Seas Claim Group* v. *Commonwealth of Australia* [2013] HCA 33.
[126] Ibid. (Hayne, Kiefel and Bell JJ).
[127] [2014] FCA 715.
[128] [2014] FCA 714.
[129] [2016] FCA 776.

that establishing a right to take resources for trade required establishing that the claim group has a right under traditional laws and customs to access and take resources for any purpose in the application area.[130]

The cases confirm that it is not necessary to prove that activity in conformity with traditional laws and customs has taken place in order to establish that a right exists, although proof of activities undertaken pursuant to laws and customs will assist in proving the existence of the right.[131] Rather, the group has to prove the existence of traditional laws and customs that would give them such a right, even if there is no evidence provided of actual trading activity.[132] However, while the *Akiba* and subsequent decisions are certainly an advance for Australian native title jurisprudence, none of the decisions directly concerned rights to water. Nor can any of the decisions be treated as a precedent for the future recognition of native title rights to water for commercial purposes. This is first because an applicant seeking commercial water rights via the native title recognition model still needs to establish a right to take and use water for any purposes under traditional (meaning pre-sovereignty) laws and customs. This is a question of fact, which turns on anthropological and historical evidence.[133] As discussed above, many groups will be unable to prove that they have continued, in a substantially uninterrupted manner since pre-sovereignty times to acknowledge traditional laws and observe traditional customs authorising the taking and use of water for any purpose. Some indigenous groups have ceased to do so because their access to water has been affected by the allocation of inconsistent land title and water use rights to others.

Akiba stands for the proposition that evidence of pre-sovereignty commercial activities is not necessary to establish a right to take and use resources for any purposes. However, the Court in *Akiba* acknowledges that evidence of such activities 'focuses attention on the right'[134] and helps to establish the existence of the right.[135] In all three cases, substantial anthropological and historical evidence was led in support of a native title right to access and take resources for any purposes.[136]

[130] See, e.g., *BP (Deceased)* [2014] FCA 715, [89]; *Willis* [2014] FCA 714.
[131] *BP (Deceased)* [2014] FCA 715 [89]; *Willis* [2014] FCA 714 [118].
[132] *BP (Deceased)* [2014] FCA 715, [89]-[90]; *Willis* [2014] FCA 714, [119].
[133] *Mabo [No. 2]* (1992) 175 CLR 1, 58 (Brennan J).
[134] *Akiba on Behalf of the Torres Straight Regional Seas Claim Group v. Commonwealth of Australia* [2013] HCA 33 [65] (Hayne, Kiefel and Bell JJ).
[135] *BP (Deceased)* [2014] FCA 715, [89]; *Willis* [2014] FCA 714, [118].
[136] Australian Law Reform Commission, above n. 65, 243 '[t]he specific native title right determined at first instance in Akiba FCA – the right to access resources and to take for any purposes resources in the determination areas- was fact specific.'

However, the extent to which that evidence concerned 'commercial' purposes was fairly limited, relating to exchange and sale of the resources themselves.[137] For example, in the *BP* case, anthropological and historical evidence was led about traditional rights to trade in ochre, shell, grindstones, ground stone axes, stone knives, wooden implements and tobacco.[138] There have been no native title cases to date in which evidence has been led about a native title right, arising from pre-sovereignty traditional laws and customs, to take and use water for commercial purposes like irrigation or industry.

Even if a native title right to take and use water for any purposes could be made out, it would be non-exercisable where inconsistent with water legislation and the grant of water use rights to other users under the *Native Title Act's* specific validation provisions in sections 24 HA and 44H. While the use of water for traditional, cultural purposes may not be inconsistent with other water use rights, the consumptive use of water for commercial activities on indigenous lands such as irrigation, agriculture, industry or tourism would almost certainly be inconsistent with water use rights held by others.[139] Given the scarce and highly contested nature of water resources in Australia, native title rights to water for commercial purposes would, in all but the most remote areas, impact on other water use rights and therefore be extinguished or non-exercisable by the native title holders.

Further, the Court in *Akiba* made it clear that while a native title right to take and use resources for any purposes may be recognised it can still be subject to regulation.[140] This means that native titleholders cannot use their rights for commercial purposes without a water use right allocated under state water law frameworks. The limited statutory rights native title holders hold to use water in enjoyment of their native title rights and interests without the need for a water use right expressly exclude the use of water for commercial purposes.

It is also important to note that most of the determinations recognising a 'right to trade', or a right to take resources for any purpose, were negotiated agreements between the parties as consent determinations rather than litigated determinations. The Federal Court in *BP (Deceased)* casts doubt on the precedent value of previous native title

---

[137] See, e.g., *Willis* [2014] FCA 714, [116], [120], [123].

[138] *BP (Deceased)* [2014] FCA 715, [59].

[139] See also O'Donnell, 'Briefing Paper for the Water Rights Project by the Lingiari Foundation and ATSIC', above n. 2, 104.

[140] *Akiba on behalf of the Torres Straight Regional Seas Claim Group* v. *Commonwealth of Australia* [2013] HCA 33.

determinations reached by consent, stating, '[t]hose determinations reflect the outcome of negotiations which doubtless involved compromises on all sides and responded to the interests rather than the rights of the parties'.[141] Other commentators have cast doubt on the precedent value of the *Akiba* decision for 'commercial' native title rights to resources. Butterly argues that the significance of the *Akiba* decision in terms of commercial fishing rights remains unclear, pointing out that native title does not pretend to allocate or reallocate commercial fishing licences, and such an important issue should be negotiated with indigenous peoples outside the 'narrow legal framework of native title'.[142] An intractable problem with the native title recognition model for indigenous water rights in Australia, is that its object is the recognition of traditional, or pre-sovereignty, interests.[143] Native title produces inevitable tensions around the continuity of indigenous rights since colonisation and the presence of other interests, because of the passage of time.

4.3.4 Native Title and Australian Water Law Frameworks

The native title characterisation of indigenous water rights has fed through to Australian water law frameworks, at the expense of the recognition and exercise of indigenous water rights for broader purposes. This is evident in the National Water Initiative, which provides:[144]

> Water planning processes will take account of the possible existence of native title rights to water in the catchment or aquifer area. The Parties note that plans may need to allocate water to native title holders following the recognition of native title rights in water under the Commonwealth Native Title Act 1993.

Under the Initiative, only 'water allocated to native titleholders for traditional-cultural purposes will be accounted for';[145] an approach

---

[141] *BP (Deceased)* [2014] FCA 715 [98].

[142] Lauren Butterly, 'Unfinished Business in the Straits: Akiba v. Commonwealth of Australia [2013] HCA' (2013) 34(8) *Indigenous Law Bulletin* 3, 5.

[143] Australian Law Reform Commission, above n. 65, 142.

[144] Commonwealth of Australia and the Governments of New South Wales, Victoria, Queensland, South Australia, the Australian Capital Territory and the Northern Territory, Introduction above n. 23, cl. 53.

[145] Ibid. cl. 54. But see O'Donnell, 'Briefing Paper for the Water Rights Project by the Lingiari Foundation and ATSIC', above Chapter 4 n. 2, 338, 185. O'Donnell argues that the states also agreed in the National Water Initiative that planning frameworks will recognise indigenous needs in relation to water access and management; 'water access entitlements' (being a perpetual or ongoing entitlement to exclusive access to a share of water from a specified

replicated in planning for the Murray Darling Basin. The *Basin Plan 2012* (Cth) provides binding limits on the quantity of water that may be taken from the Murray Darling Basin and sets out binding requirements for water resource plans. It includes an acknowledgment of the traditional owners of the Murray Darling Basin, which records the cultural, social, environmental, spiritual *and economic* connection indigenous people in the Murray Darling Basin have to lands and waters.[146] It also notes that indigenous groups have 'economic interests' in water which it defines as 'trading, hunting, gathering food and other items for use that alleviate the need to purchase similar items and the use of water to support businesses in industries such as pastoralism and horticulture'.[147]

Clause 10 of the Plan sets out the binding requirements for water resource plans in the Murray Darling Basin in providing for indigenous values and uses. Clause 10.52(1) requires water resource plans in the Basin to identify the objectives of indigenous people in relation to water resource management and the outcomes for water resource management desired by indigenous people. However, clause 10.52(2) provides that, in identifying the matters in subsection (1), regard must be had to indigenous water values and uses as 'social, spiritual and cultural'. Accordingly, despite acknowledging indigenous economic interests in the Murray Darling Basin, the Plan does not require water resource plans to provide for commercial indigenous water rights.

This assumption that indigenous water rights are inherently non-commercial and effectively non-consumptive, is supported by the characterisation of indigenous issues in the Murray Darling Basin in section 21(4) of the *Water Act 2007* (Cth) as 'social, cultural, indigenous and other public benefit issues' as opposed to the 'consumptive and other economic uses of Basin water resources'. This traditional, cultural focus might be explained by the importance of the 'cultural flows' policy pursued in the Murray Darling Basin.[148] Although the definition of cultural flows extends to economic water uses the policy has been implemented until now via in stream water protections for spiritual, cultural or environmental purposes, without the allocation of a water

---

consumptive pool as defined in the relevant water plan). Accordingly, the requirement to provide indigenous people with water access is not qualified by a requirement for the finalisation of native title rights to water or limited to recognition of traditional, cultural interests.

[146] *Basin Plan 2012* (Cth).

[147] Ibid. Sch. 1.

[148] Ibid. cl. 10.54. Clause 10.54 requires water resource plans to be prepared having regard to the views of indigenous people with respect to 'cultural flows'.

use right.[149] Limiting cultural flows in this way is likely to have been deliberate, because the agreed definition of cultural flows actually refers to both the cultural and economic objectives of indigenous water rights.[150]

Not surprisingly, under the guidance of the National Water Initiative and Basin Plan, state-based water legislation and water resource plans usually treat indigenous water interests as being covered by environmental or cultural flows, or domestic and stock rights (neither of which was designed with indigenous interests in mind). Such entitlements are not represented by a water access entitlement and cannot be used for consumptive purposes. As an example, the *Water Sharing Plan for the Coffs Harbour Area Unregulated and Alluvial Water Sources 2009* (NSW) provides that:[151]

> The Plan recognises that the environmental water provisions provide non-extractive benefits, including traditional Aboriginal spiritual, social, customary, economic, cultural and recreational benefits, and contributes to improved water quality.

Only in the case of New South Wales, considered below, does water legislation specifically provide for the allocation of water use rights to indigenous groups.

The failure of Australian water law frameworks to provide for commercial indigenous water rights was criticised on a number of occasions by the former National Water Commission before it was abolished in 2015.[152] Importantly, the shoehorning of indigenous water rights into 'environmental and public benefit' outcomes or 'domestic and stock' rights differentiates indigenous water rights from substantive water use rights that take from the 'consumptive pool' of water. This is probably a deliberate policy approach, because if indigenous water rights are managed outside of the consumptive pool, the likelihood that indigenous water use could affect consumptive third-party interests like

---

[149] See Jackson and Marcia Langton, above Chapter 2 n. 23. Jackson and Langton explain that the cultural flows policy has leveraged off the success of environmental flow allocations in the same area.

[150] See Murray Lower Darling Rivers Indigenous Nations and Northern Murray–Darling Basin Aboriginal Nations, above Chapter 2 n. 22.

[151] *Water Sharing Plan for the Coffs Harbour Area Unregulated and Alluvial Water Sources 2009* (NSW) cl. 19 on planned environmental water.

[152] National Water Commission, 'National Water Initiative : Securing Australia's Water Future : 2011 Assessment' (Commonwealth of Australia, 2011); Commonwealth of Australia, 'A Review of Indigenous Involvement in Water Planning' (National Water Commission, April 2013) http://webarchive.nla.gov.au/gov/20160615062953/http://www.nwc.gov.au/pub lications/topic/water-planning/indigenous-involvement-in-water-planning.

irrigation or industry is minimised. Yet, by failing to provide indigenous Australians with water use rights, Australian water law and policy has not responded to the distributive injustice indigenous peoples continue to experience. Other users continue to hold almost all of water use rights, and indigenous groups hold very few.

4.4 ALLOCATING INDIGENOUS WATER RIGHTS IN AUSTRALIA

It is apparent that native title has failed to recognise rights to water for all but very limited purposes and the problems with the Australian native title recognition model are intractable. However, there are discrete situations where specific laws or policies allocate indigenous rights to water despite (or perhaps in spite of) native title.

4.4.1 Statutory Land Rights and Water

Land rights legislation, which grants land title to indigenous groups to varying extents and on varying bases and establishes arrangements for decision making about land and resource use, has been implemented in Australia since the late 1970s.[153] Currently, there is land rights legislation in one form or another in all Australian states with the exception of Western Australia. Significant amounts of land in Australia have now been allocated to indigenous groups under statutory allocation models, which operate separately from native title, although the amount is obviously much less than what was held by Aboriginal people at the time of British arrival. As an example, the *Aboriginal Land Rights (Northern Territory) Act 1976* (Cth) provides for the grant of fee simple estates[154] to Land Trusts 'for the benefit of Aboriginals entitled by Aboriginal tradition to the use or occupation of the land concerned'.[155] As another example, the *Aboriginal Land Rights Act 1983* (NSW) provides for the vesting of land titles in Aboriginal Land Councils.[156]

The *Aboriginal Land Rights (Northern Territory) Act* was enacted after the failure of the courts to recognise native title in *Milirrpum and Others v. Nabalco Pty Ltd and the Commonwealth of Australia*[157] and following

---

[153] See generally Tehan, above n. 43, 532.
[154] *Northern Territory v. Arnhem Land Aboriginal Land Trust* (2008) 236 CLR 24, 63–4 ('*Blue Mud Bay*') (Gleeson CJ, Gummow, Kirby, Hayne and Crennan JJ).
[155] *Aboriginal Land Rights (Northern Territory) Act 1976* (Cth), s. 12.
[156] *Aboriginal Land Rights Act 1983* (NSW), s. 3.
[157] *Milirrpum* (1971) 17 FLR 141.

the recommendations for land rights legislation by the Aboriginal Land Rights Commission.[158] According to Tehan, land rights legislation switched the focus away from native title, which at that point had failed to provide for indigenous land rights.[159]

However, land rights legislation did not include the right for indigenous landholders to use the water on their lands.[160] Land titles granted under land rights legislation are subject to the State's power to regulate and control water.[161] The Aboriginal people entitled to use the land are only entitled to use the water on the land for domestic and stock purposes, without the need for a water use right.[162] For example, under the *Aboriginal Land Rights (Northern Territory) Act*, the fee simple land title granted does not include 'minerals', and water is included within the definition of 'minerals' excluded from the grant.[163]

The beds and banks of watercourses may, however, be included in the grant of land. The *Aboriginal Land Act 1991* (Qld), for example, provides that land that may be claimed by and granted to Aboriginal people includes a 'watercourse or lake' to the extent the watercourse or lake is within the external boundaries of land that is otherwise available State land and capable of being owned in fee simple by a person other than the State.[164] However, as mentioned earlier in this chapter, water in its natural state is not capable of being owned in fee simple. This means that grants of land under Queensland land rights legislation may include the beds and banks of rivers and lakes, but not the water in them.

---

[158] AES Woodward, 'Aboriginal Land Rights Commission: Second Report' (April 1974).
[159] Tehan, above n. 43, 529.
[160] See e.g. *Aboriginal Land Rights Act 1983* (NSW); *Aboriginal Land Trusts Act 1966* (SA); *Anangu Pitjantjatjara Yankunytjatjara Land Rights Act 1981* (SA); *Aboriginal Lands Act 1970* (Vic) 19; *Aboriginal Lands Act 1995* (Tas), which make no mention of water in relation to grants of land title.
[161] But see O'Donnell, 'Briefing Paper for the Water Rights Project by the Lingiari Foundation and ATSIC', above n. 2, 93. O'Donnell argues that this is not the case in the Northern Territory because of the interaction of Commonwealth land rights and Territory water legislation. Section 74 of the *Aboriginal Land Rights (Northern Territory) Act 1976* (Cth) provides: 'This Act does not affect the application to Aboriginal land of a law of the Northern Territory to the extent that that law is capable of operating concurrently with this Act', and it would follow, according to O'Donnell, that to the extent the Territory's right to the control and use of water under Territory water legislation is inconsistent with the exclusive possession of land under the Act, the Territory's right would be inoperative.
[162] See, e.g., *Water Act 2000* (Qld), s. 20(3); see *Water Act 1992* (NT), ss. 11, 14; *Water Act 2000* (Qld), ss. 206, 213(1)(e). In Queensland, the titleholders of land under the *Aboriginal Land Act 1991* (Qld) may also generally apply as an owner of a parcel of land for a licence to take water to use on the land for unspecified purposes, which attaches to the land
[163] *Aboriginal Land Rights (Northern Territory) Act 1976* (Cth), ss. 3, 12(2).
[164] *Aboriginal Land Act 1991* (Qld), s. 26.

In Victoria the *Traditional Owner Settlement Act 2010* (Vic) provides for the making of recognition and settlement agreements between the State and traditional owner groups as an attempt to advance reconciliation in response to the failure of native title claims.[165] The *Traditional Owner Settlement Act* provides for the allocation of 'natural resource water authorisations' to traditional owner groups, which authorise the members of a traditional owner group in relation to which a natural resource agreement has been entered into to take and use water from a waterway or bore.[166] However, in a provision that mirrors the statutory rights in section 211 of the *Native Title Act*, the *Traditional Owner Settlement Act* only allows such authorisations for 'traditional purposes', meaning 'the purposes of providing for any personal, domestic or non-commercial communal needs of the members of the traditional owner group'.[167]

O'Donnell and Tan have argued that rights to control access to water within the boundaries of indigenous land held in exclusive possession under the *Aboriginal Land Rights (Northern Territory) Act* signify potential commercial rights for indigenous people, by enabling them to control access to water.[168] They make this argument in reliance on the High Court of Australia's 2008 decision in *Northern Territory v. Arnhem Land Aboriginal Land Trust ('Blue Mud Bay')*.[169] The decision in *Blue Mud Bay* focused on section 70(1) of the *Aboriginal Land Rights (Northern Territory) Act*, which provides that '[a] person shall not enter or remain on Aboriginal land'. The High Court in *Blue Mud Bay* found that the grant in fee simple of Aboriginal land that extended to the low water mark allowed the titleholders to prevent access by holders of fishing licences to tidal waters. This was the case, the Court reasoned, because 'the expression 'Aboriginal land', when used in section 70(1), 'should be understood as extending to so much of the fluid (water or atmosphere) as may lie above the land surface within the boundaries of the grant and is

---

[165] *Traditional Owner Settlement Act 2010* (Vic), s. 1.
[166] Ibid. s. 85. Natural resource agreements may be entered into by the State and traditional owner groups in relation to land subject to a recognition and settlement agreement under s. 80 of the *Traditional Owner Settlement Act 2010* (Vic), and concern access to and use of flora, fauna, fish, forest produce and wildlife on the land.
[167] Ibid. s. 79.
[168] O'Donnell, 'Briefing Paper for the Water Rights Project by the Lingiari Foundation and ATSIC', above n. 2, 295–9; Poh-Ling Tan, 'National Indigenous Water Planning Forum – A Review of the Legal Basis for Indigenous Access to Water' (Report prepared for the National Water Commission, Griffith Law School, February 2009) 9.
[169] *Blue Mud Bay* (2008) 236 CLR 24.

ordinarily capable of use by an owner of land'.[170] Further, 'neither the license itself nor any provision of the Fisheries Act confers any permission upon the holder to enter any particular place or area for the purpose of fishing'.[171]

O'Donnell and Tan both argue that if the reasoning in *Blue Mud Bay* was held to extend to areas of Aboriginal land covered by fresh water in the form of rivers or lakes where the bed of the river or lake was included in the grant (and this is not yet the case), by analogy, a person acting in accordance with a water use right granted under Territory water legislation would be prevented from entering onto that land without the consent of the Aboriginal landholders (unless some other provision of the water legislation authorised entering the land for the purposes of exercising the right).

However, the 'right to control access'[172] to Aboriginal land under section 70(1) is not a substantive water use right, nor does it entail any sort of 'proprietary interest' in any water that may overlie the land,[173] and does not, in fact, authorise Aboriginal people to make any sort of water use within Aboriginal land, let alone the use of water for commercial purposes. It is simply a procedural right to impact upon the use of water by others within the boundaries of Aboriginal land.[174]

4.4.2 Water Law and Policy Reform – Towards an Indigenous Allocation?

There are a few situations where specific water legislation or plans have been used in Australia to allocate water use rights to indigenous groups for more than traditional, cultural purposes. Each of these operates independently from native title. Further, although each is limited in scope, the existence of allocative models brings the assumption that

---

[170] Ibid. 66 (Gleeson CJ, Gummow, Kirby, Hayne and Crennan JJ).
[171] Ibid. 59 (Gleeson CJ, Gummow, Kirby, Hayne and Crennan JJ).
[172] As explained above, where native titleholders have obtained an exclusive possession determination, they have also been accorded an accompanying right to make decisions about access to and use of land and waters, although only to the extent that this is not inconsistent with the rights and interests of others granted under legislation (such as water use rights) which take priority over native title rights. See, e.g., *Northern Territory of Australia* v. *Alyawarr, Kaytetye, Warumungu, Wakaya Native Title Claim Group* (2005) 145 FCR 442, 504–7.
[173] *Blue Mud Bay* (2008) 236 CLR 24, 64 (Gleeson CJ, Gummow, Kirby, Hayne and Crennan JJ).
[174] Interview with Lauren Butterly (Sydney, 11 July 2018), above n. 16. Butterly explains that an agreement was signed between the Northern Land Council and the Government after *Blue Mud Bay*, which allowed the government to continue to maintain the intertidal zone while negotiations and a plan were undertaken to allow for a gradual and well thought-out implementation of the decision. The agreement is renewed yearly.

indigenous water rights are (or should be) restricted to traditional, cultural purposes into question.

(a) New South Wales
The first of the statutory allocation mechanisms, under the *Water Management Act 2000* (NSW), provides for the allocation of Aboriginal 'access licences'. The objects of the *Water Management Act* include to 'recognise and foster the significant social and economic benefits to the State that result from the sustainable and efficient use of water', including 'benefits to the Aboriginal people in relation to their spiritual, social, customary and economic use of land and water'.[175]

Under New South Wales water legislation, a person may not take water other than in accordance with an 'access licence' for a purpose specified on the licence or a corresponding water use approval.[176] Access licences are held independently of land title, and are expressed as an entitlement to a share of available water in specified resources at specified times and rates.[177] The only exception to this rule is the limited statutory rights landholders enjoy to take and use water without an access licence, including domestic and stock and native title rights.[178]

Specific purpose access licences granted under the *Water Management Act* include (as a subcategory of regulated river, unregulated river and aquifer access licences), 'Aboriginal commercial', 'Aboriginal community development' and 'Aboriginal cultural' access licences.[179] A number of water resource plans made under the New South Wales legislation, which set out rules about water sharing and use for specific water resources, refer specifically to one or more of these Aboriginal access licences.[180]

Aboriginal cultural access licences only authorise water use for certain restrictive purposes. They are typically granted with the condition that 'water must only be taken by Aboriginal persons or Aboriginal

---

[175] *Water Management Act 2000* (NSW), s. 3; 'Economic' is not defined in the Act but the *Macquarie Dictionary [Electronic Resource]: Australia's National Dictionary Online* (Macquarie Library, 2003) defines 'economic' as, 'relating to the production, distribution, and use of income and wealth'.

[176] *Water Management Act 2000* (NSW), s. 60AA.

[177] Ibid. s. 56.

[178] Ibid. ss. 60A, 60C, 60D, 91A.

[179] *Water Management (General) Regulation 2011* (NSW), Sch. 3.

[180] See, e.g., *Water Sharing Plan for the Lower North Coast Unregulated and Alluvial Water Sources 2009* (NSW); *Water Sharing Plan for the Bellinger River Area Unregulated and Alluvial Water Sources 2008* (NSW).

communities for personal, domestic or communal purposes, including drinking, food preparation, washing, manufacturing traditional arte-facts, watering domestic gardens, cultural teaching, hunting, fishing, gathering and for recreational, cultural and ceremonial purposes'.[181] Aboriginal community development and Aboriginal commercial access licences, however, expressly allow water to be taken by Aboriginal persons or Aboriginal communities for (undefined) 'com-mercial purposes'.[182]

Unfortunately, the implementation of the New South Wales Aboriginal access licences has been limited. The Aboriginal cultural access licences may not be granted for more than ten megalitres per year,[183] a process which Northern Basin Aboriginal Nation Chairperson Fred Hooper describes as 'discriminatory'.[184] Aboriginal community development and commercial access licences do not typi-cally exceed 500 megalitres per year. According to a scoping study of indigenous water licence allocations by Altman and Arthur in 2009, water allocated to Aboriginal licensees in New South Wales made up no more than 0.7 per cent of total water access licence allocations.[185] The former National Water Commission reported in 2013 that only one active licence had been granted: a cultural access licence granted in 2005 in the Murrumbidgee region.[186]

(b) Queensland

The second example of a statutory allocation mechanism is in Queensland, where a number of water resource plans establish indigen-ous water reserves to support indigenous social and economic aspirations.[187] The *Water Act 2000* (Qld) predictably provides that Aboriginal or Torres Strait Islander peoples may 'take or interfere

---

[181] See, e.g., *Water Sharing Plan for the Coffs Harbour Area Unregulated and Alluvial Water Sources 2009* (NSW) cl. 75.

[182] See, e.g., *Water Sharing Plan for the Lower North Coast Unregulated and Alluvial Water Sources 2009* (NSW) cl. 79.

[183] See, e.g., *Water Sharing Plan for the Coffs Harbour Area Unregulated and Alluvial Water Sources 2009* (NSW).

[184] Interview with Fred Hooper (Queensland, 12 July 2018).

[185] JC Altman and WS Arthur, 'Commercial Water and Indigenous Australians: A Scoping Study of Licence Allocations' (CAEPR Working Paper No 57/2009, Centre for Aboriginal Economic Policy Research, September 2009) 9.

[186] Commonwealth of Australia, above n. 152.

[187] See, e.g., *Water Resource (Gulf) Plan 2007; Water Resource (Wet Tropics) Plan 2013; Water Resource (Mitchell) Plan 2007* (Qld). See generally Commonwealth of Australia, above n. 152, 12–13; Kate Cranney and Poh-Ling Tan, 'Old Knowledge in Freshwater: Why Traditional Ecological Knowledge Is Essential for Determining Environmental Flows in Water Plans' (2011) 14(2) *The Australasian Journal of Natural Resources Law and Policy* 71–114.

with' water for traditional activities or cultural purposes (excluding commercial purposes) without the need for a water access entitlement, reflecting the native title approach.[188] The *Mineral, Water and Other Legislation Amendment Bill 2017* (Qld) seeks to amend the *Water Act 2000* to include a requirement for all water plans to consider cultural outcomes for Aboriginal people and Torres Strait Islanders separately from social, economic and environmental outcomes and for more direct consideration of the values and uses of water resources in a plan area to Aboriginal peoples and Torres Strait Islanders when making state-based water plans.[189] Cultural outcomes are defined by the Bill to mean a 'beneficial consequence to an Aboriginal party or Torres Strait Islander party relating to aquifers, drainage basins, catchments, subcatchments or watercourses'.[190] Although these amendments appear to emphasise the role of indigenous peoples in water planning and will likely lead to enhanced consultation, there are no references to indigenous-specific water allocations.

However, there is specific legislation in the remote north Queensland region of Cape York that goes further. Section 27 of the *Cape York Peninsula Heritage Act 2007* (Qld) provides that water resource plans made in the Cape York Peninsula region 'must provide for a reserve of water in the area to which the plan relates for the purpose of helping indigenous communities in the area achieve their economic and social aspirations'. As an example, the *Water Resource (Wet Tropics) Plan 2013* acknowledges that 'Indigenous communities are dependent on water resources in the plan area to achieve their economic aspirations'.[191] It reserves 5,200 megalitres for indigenous social and economic purposes,[192] while 5,050 megalitres is reserved for 'indigenous economic and social aspirations' in the *Water Resource (Gulf) Plan 2007* (Qld).[193] Indigenous economic and social purposes or aspirations are not defined in either plan.

During 2018 the Queensland Government released a further draft water plan for Cape York, which is, at the time of writing, out for

---

[188] *Water Act 2000* (Qld), s. 95.
[189] *Mineral, Water and Other Legislation Amendment Bill 2017* (Qld) cl. 238, 239.
[190] Ibid. cl. 276.
[191] *Water Resource (Wet Tropics) Plan 2013* (Qld) cl. 12.
[192] Ibid. cl. 53–4.
[193] *Water Resource (Gulf) Plan 2007* (Qld) Sch. 6A. See 'Commonwealth of Australia, A Review of Indigenous Involvement in Water Planning, 2013 (Report, National Water Commission, April 2014)', above n. 152, 13, stating that it is unclear whether the reserves will be large enough to meet economic needs.

public consultation. Again the plan is intended to help Aboriginal people and Torres Strait Islanders, 'achieve their economic, social and cultural needs and aspirations'.[194] The draft plan includes economic, social, cultural and environmental water outcomes, and specifically notes the need to further economic and social outcomes for local communities.[195] The discussion of economic water outcomes in clause 17 of the draft plan acknowledges both the importance of making water available to Aboriginal users for economic development purposes, but also the economic interests other users and industry have in water in Cape York and the need to support effective and efficient water markets. Thus, the plan proposes a reserve of unallocated water for Aboriginal people.[196] The cultural water outcomes in clause 19 elaborate further on the need to allocate water for the cultural aspirations of Aboriginal people and engage Aboriginal people and knowledge in planning processes.

Although the water reserves available under the *Cape York Peninsula Heritage Act* set aside a greater volume of water than that allocated to access licences under the New South Wales regime, it is unclear how this amount is to be distributed to indigenous groups.[197] Further, there is no legislative protection of an indigenous water allocation under either of the Queensland water statutes mentioned here. This is the case despite the principle of 'sustainable management' of water resources in the *Water Act 2000* (Qld) including, 'recognising the interests of Aboriginal people and Torres Strait Islanders and their connection with the landscape in water planning'.[198]

Outside of these New South Wales and Queensland examples, it is uncommon for water resource plans to mention indigenous water use.[199] 'Strategic indigenous reserves' for commercial and cultural indigenous water use were included in some Northern Territory water resource plans, although these policy innovations do not have corresponding legislative requirements. The draft Oolloo water resource plan, for example, reserved 14,500 megalitres per year from the consumptive pool to be allocated to indigenous landowners for economic

---

[194] *Water Plan (Cape York) 2018 (Draft)* clause 2.
[195] Ibid. cl. 16.
[196] State of Queensland, 'Cape York Draft Water Plan – Statement of Intent' (June 2018) 41 www .dnrme.qld.gov.au.
[197] See O'Donnell, 'Indigenous Rights in Water in Northern Australia', above n. 124, 338, 178.
[198] *Water Act 2000* (Qld), s. 10(v).
[199] Jackson, 'Aboriginal Access to Water in Australia: Opportunities and Constraints', Chapter 2 above n. 41, 618.

development purposes to 'secure future opportunities for their livelihood'.[200]

The amount of water allocated to the Northern Territory water reserves is calculated with reference to the percentage of indigenous landholding in the Territory, under indigenous land rights legislation. The reserves represent a much larger allocation than water allocated to indigenous people under New South Wales and Queensland water legislation. However, the Northern Territory Government has declared that it will not include strategic indigenous reserves in future water resource plans.[201]

(c) Victoria

Until 2016, Victoria was lagging behind the other Australian states and territories in terms of indigenous water rights. The *Water Act 1989* (Vic) made no mention of Aboriginal interests in or relationships with water. As mentioned above, the *Traditional Owner Settlement Act 2010* (Vic) enabled the State to allocate 'natural resource water authorisations' to traditional owner groups as part of a settlement agreement, however, only for native title-like 'traditional purposes'.[202]

However, the State of Victoria now appears to be taking the lead on indigenous water rights policy development and reform in Australia. This is happening under the progressive Andrews' Labour Government's commitment to 'self-determination', and developing alongside the unfolding Treaty process. A Murray Darling Basin State, Victoria has been a key player in the national cultural flows project, seeking to influence policy development around indigenous water rights in the Basin and beyond at the State and National level. The Victorian developments have been multiple, and range from increasing involvement in water planning to greater support for economic water rights, although a specific indigenous water allocation has not been secured at this stage.

In 2016 the State released the *Water for Victoria* plan; a 'long-term direction' for the governance of Victorian water resources. The 2016–17 Victorian Budget committed $4.7 million to embedding Aboriginal values and knowledge in Victorian water management.[203]

---

[200] Northern Territory Government, 'Draft Water Allocation Plan: Oolloo Aquifer' (2012).
[201] Commonwealth of Australia, above n. 152, 26–7. Phoebe Stewart, 'Indigenous Water Reserve Policy Tap Turned Off' *ABC News (online)*, 10 October 2013.
[202] *Traditional Owner Settlement Act 2010* (Vic), s. 79.
[203] Interview with Bryony Grice (Melbourne, 9 July 2018). Grice was the Director of Environmental Policy and Community Partnerships at the Victorian State Government at the time of the interview.

Managing Aboriginal values is a key theme of the plan.[204] The plan highlights a need for greater indigenous involvement and consultation in water planning, and incorporating traditional knowledge via an 'Aboriginal Water Reference Group', with representation from traditional owners with knowledge of water management.[205]

Implementing *Water for Victoria*, the Government introduced the *Water and Catchment Legislation Amendment Bill 2017* (Vic) (still before the upper house at the end of 2018), which includes a number of amendments to the *Water Act 1989* (Vic) in relation to 'Aboriginal cultural values and uses of waterways' and 'to include specified Aboriginal parties in water resource planning and the development and review of strategies'.[206] Government official Bryony Grice explains that the Bill includes both 'top down' and 'bottom up' approaches: namely, a requirement on statutory water managers to 'partner with Aboriginal Victorians and traditional owners to incorporate values and uses into water management planning decisions' and a commitment 'to the traditional owners and Aboriginal Victorians, to build their capacity and their capability to be able to describe and articulate their values and uses'.[207] Victorian Environmental Water Holder Commissioner, Reuben Berg, explains the important role played by Aboriginal representatives on various water boards and catchment management authorities and, in particular, the Aboriginal Liaison Officers, who work in the Aboriginal community to provide for genuine engagement between water corporations and the community at grass-roots level.[208]

In 2017 the Victorian Government also passed the *Yarra River Protection (Wilip-gin Birrarung murron) Act 2017* (Vic).[209] This legislation 'recognises the intrinsic connection of the traditional owners to the Yarra River and its Country and further recognises them as the custodians of the land and waterway which they call Birrarung'.[210] Reminiscent of the *Te Awa Tupua (Whanganui River Claims Settlement) Act 2017* (NZ), the Victorian Act recognises the Yarra as

---

[204] State of Victoria, 'Water for Victoria: Securing Victoria's Future' (2016) 140 www .water.vic.gov.au/water-for-victoria.
[205] Ibid. 98.
[206] *Water and Catchment Legislation Amendment Bill 2017* (Vic) cl. 1.
[207] Interview with Bryony Grice (Melbourne, 9 July 2018), above n. 203.
[208] Interview with Reuben Berg (Melbourne, 10 July 2018).
[209] See generally Erin O'Donnell, Chapter 3 above n. 36.
[210] *Yarra River Protection (Wilip-Gin Birrarung Murron) Act 2017* preamble.

a 'living and integrated natural entity' and appoints a guardian-type entity to advise the Minister on its management.[211] This role is given to the Birrarung Council: an advisory body established by the Act to advise the Minister and to advocate for protection and preservation of the Yarra River,[212] which must have at least two traditional owner representatives on it.[213] The Act also sets out 'protection principles' for decision makers to take into account in water planning for the river,[214] and provides for the development of strategic plans for the river,[215] both of which address a range of environmental, cultural and recreational interests.

This legislation is certainly a major step forward for the recognition of cultural water interests in Australia, and for the first time an Aboriginal group, the Wurundjeri people, appeared and spoke in Parliament on the introduction of the legislation.[216] Erin O'Donnell, one of the appointed guardians for the Yarra River, emphasises the power of this model for Victoria by 'centering indigenous values as a tool for reframing, particularly white peoples' relationship to a river ... '.[217] However, with only two traditional owner members on a council of twelve, the legislation does not go nearly as far as the New Zealand model in terms of enabling indigenous self-determination. Nor does the *Yarra River Protection (Wilip-gin Birrarung murron) Act* recognise the river as a 'legal person'.

The policy developments under the *Water for Victoria* plan appear to contemplate that indigenous groups will share benefits with environmental flows and protections, and focuses on involving indigenous perspectives and knowledge in water planning and the identification of cultural values in waterways.[218] As an example, there is now an Aboriginal Commissioner on the Board of the Victorian Environmental Water Holder.[219] However, the *Water*

---

[211] Ibid. s. 1.

[212] Ibid. s. 46.

[213] Katie O'Bryan, 'New Law Finally Gives Voice to the Yarra River's Traditional Owners' [2017] *The Conversation* http://theconversation.com/new-law-finally-gives-voice-to-the-yarra-rivers-traditional-owners-83307.

[214] *Yarra River Protection (Wilip-Gin Birrarung Murron) Act 2017* Part 2.

[215] Ibid. Part 4.

[216] See generally O'Bryan, above n. 19.

[217] Interview with Erin O'Donnell (Melbourne, 9 July 2018).

[218] E Macpherson et al. 'Lessons from Australian Water Reforms: Indigenous and Environmental Values in Market-Based Water Regulation' in Cameron Holley and Darren Sinclair (eds.), *Reforming Water Law and Governance: From Stagnation to Innovation in Australia* (Springer, 2018).

[219] This is Reuben Berg. Interview with Reuben Berg (Melbourne, 10 July 2018), above n. 208.

for Victoria plan actually suggests that in the future Aboriginal Victorians may be provided with water access for economic development purposes, facilitating 'economic self-determination' via water-related Aboriginal enterprises.[220] O'Donnell explains that part of the motivation for the roadmap is an increasing awareness that indigenous people should have the ability to use their water for whatever purpose they like.[221] The plan refers to a potential to reallocate water access entitlements to traditional owners, including by acquiring water entitlements from water corporations in areas where this is allowable within sustainable limits, investing in water saving projects that create new entitlements, and buying water in water markets in areas where entitlements or use is capped at sustainable limits.[222]

Yet the aims of the plan in terms of economic water rights for Aboriginal Victorians are loose and aspirational; suggesting that any real reform may be many years away.[223] The plan advocates working with traditional owners to develop a 'roadmap for access to water for economic development', to 'consider opportunities' for Aboriginal access to water for economic development purposes and to 'notify traditional owners' when opportunities for access arise.[224] There are no concrete requirements in *Water for Victoria* to reform water legislation or create specific legal water entitlements for indigenous people in catchment-level plans.

The incremental pace of policy change under *Water for Victoria* reflects the competitiveness and political sensitivity of water distribution in Victoria, and the potential cost involved in water reallocation.[225] However, the policy developments are indicative of a tendency towards greater recognition of indigenous cultural water interests and increasing receptiveness to the idea of allocating substantive water rights to indigenous groups. If the negotiation of a State-based treaty or comprehensive agreement proceeds in Victoria, we may well see further legal and policy development around indigenous rights to water.

---

[220] State of Victoria, above n. 204, 98.
[221] Interview with Erin O'Donnell (Melbourne, 9 July 2018), above n. 217.
[222] State of Victoria, above n. 204, 106.
[223] See also Interview with Bryony Grice (Melbourne, 9 July 2018), above n. 389 who explains that this work programme is in its infancy.
[224] State of Victoria, above n. 204, 106.
[225] Jackson and Marcia Langton, Chapter 2 above n. 23.

4.5   THE FUTURE OF INDIGENOUS WATER RIGHTS IN AUSTRALIA

Indigenous Australians continue to be shut out of water law frameworks, and the purposes for which they may access and use water continue to be restricted, in a context dubbed by Aboriginal water rights scholar Virginia Marshall, '*Aqua Nullius*'.[226] As Northern Basin Aboriginal Nations chair, Fred Hooper explains:[227]

> There are some water rights for cultural purposes, but you can't use that water for any other purpose other than cultural purposes. For example, you can't trade it, its non-tradeable water. Whereas irrigators are all tradeable waters. Even water for the environment is tradeable but us poor old black fellas', we really can't do anything.

As well as being unjust, this situation is an obstacle to the economic development of indigenous communities. The need for indigenous water rights in order to support the economic development of land, contributing to indigenous and broader social outcomes, has been put by a number of indigenous groups in Australia.[228] This imperative is 'not an alternative to addressing access to cultural and customary water but an additional policy to improve the economic lives of indigenous people'.[229]

Australian governments are increasingly concerned about indigenous economic disadvantage and the need to support indigenous development, including via the productive use of indigenous lands.[230] The Council of Australian Governments' *National Indigenous Reform Agreement* (known as '*Closing the Gap*') provides that, '[a]ccess to land and native title assets, rights and interests can be leveraged to secure real and practical benefits for Indigenous people'.[231]

---

[226] Marshall, Chapter 3 above n. 59.

[227] Interview with Fred Hooper (Queensland, 12 July 2018), above n. 184.

[228] Northern Australian Land and Sea Management Alliance, 'Knowledge Series: Indigenous People's Right to the Commercial Use and Management of Water on Their Traditional Territories' (2013) www.nailsma.org.au/hub/resources/publication/indigenous-peoples-right-commercial-use-and-management-water-policy; North Australian Indigenous Land and Sea Management Alliance, above n. 17, 2.

[229] Lawlab, above n. 19.

[230] See, e.g., Commonwealth of Australia, 'Closing the Gap Prime Minister's Report 2014' (2014) 2 www.dpmc.gov.au/publications/docs/closing_the_gap_2014.doc. See, generally, Marcelle Burns, 'Closing the Gap between Policy and "Law" – Indigenous Homelands and a Working Future' (2009) 27(2) *Law in Context* 114.

[231] Council of Australian Governments, 'National Indigenous Reform Agreement (Closing the Gap)' (2012) www.federalfinancialrelations.gov.au/content/npa/health_indigenous/indigenous-reform/national-agreement_sept_12.pdf.

The need to support economic development on indigenous lands was also emphasised in the Commonwealth Government's *Paper on Developing Northern Australia ('White Paper')*.[232] The *White Paper* highlights the importance of water for the development of lands in northern Australia, including indigenous lands, particularly through irrigated agriculture.[233] It provides:[234]

> The Commonwealth Government supports northern jurisdictions taking actions that support Indigenous Australians to derive greater economic benefits from water on Indigenous land. Water can provide opportunities for Indigenous Australians in diverse areas such as aquaculture, nature based tourism and intensive horticulture. Access to water can also provide an opportunity to participate in water markets, where they exist.

The *White Paper* follows on from other Commonwealth discussion papers that underscore the need for commercial water rights for indigenous economic development, building on the *Close the Gap* agenda.[235]

The Australian Law Reform Commission recommended in 2015 that the definition of native title in section 223 of the *Native Title Act* be amended to better accommodate 'commercial rights', although it made no specific recommendation with respect to water.[236] However, the Commission did acknowledge that the native title determination process is a long, difficult, expensive and *ad hoc* approach to resolving indigenous claims to land and resources.[237] Even if the native title recognition model for indigenous water rights were to evolve to encompass rights to use water for any purposes, as some commentators predict, it could take years or even decades for any groups to benefit from recognition of such rights. The need to address indigenous disadvantage, however, is immediate. The former First Peoples' Water Engagement Council urged:[238]

---

[232] Australian Government, 'Our North, Our Future: White Paper on Developing Northern Australia' (2015) https://northernaustralia.dpmc.gov.au.

[233] Ibid. 41, 46.

[234] Ibid. 47.

[235] Australian Government, 'Position Statement: Indigenous Access to Water Resources' (National Water Commission, 2012) 1–2 www.nwc.gov.au/__data/assets/pdf_file/0009/2286 9/Indigenous-Position-Statement-June-2012.pdf.

[236] Australian Law Reform Commission, above n. 65, 249.

[237] Ibid. 15–16.

[238] Lawlab, above n. 205, 5–6.

> Delaying the elevation of indigenous rights to water for economic development on the water reform agenda will impede the effectiveness of water as a tool for improving the personal and economic wellbeing of Indigenous Australians.

Native title was never intended to be the whole story for indigenous land and resource rights in Australia, but merely part of a social justice package aimed at indigenous social and economic development. Prime Minister Keating's second reading speech for the *Native Title Bill 1993* (Cth) reveals the Commonwealth Government's intentions at that time:[239]

> The government has always recognised that despite its historic significance, the Mabo decision gives little more than a sense of justice to those Aboriginal communities whose native title has been extinguished or lost without consultation, negotiation or compensation. Their dispossession has been total, their loss has been complete. The government shares the view of ATSIC, Aboriginal organisations and the Council for Aboriginal Reconciliation, that justice, equality and fairness demand that the social and economic needs of these communities must be addressed as an essential step towards reconciliation.

When the Council of Australian Governments agreed to unbundle water use rights from landholding and introduced water trading in the early 1990s, the opportunity for a redistributive response to indigenous water injustice was not taken. Indigenous water rights receive no mention in the Council of Australian Governments' agreement.[240] The absence of debate about indigenous water rights in the early 1990s can probably be explained by a preoccupation with the fledgling native title process, which at first suggested great promise for the recognition of land and water rights alike. However, as demonstrated in this chapter, native title has proven to be an incomplete response to indigenous exclusion from water law frameworks. Moreover, to restrain indigenous water rights within the native title recognition model would disregard the vast amounts of indigenous land held outside of native title under indigenous land rights legislation. That land includes almost half of the Northern Territory held under the *Aboriginal Land Rights (Northern Territory Act) 1976* (Cth).[241]

---

[239] Commonwealth of Australia, *Parliamentary Debates*, 16 November 1993, 2877 (Keating, Prime Minister).
[240] Council of Australian Governments, above n. 19.
[241] Australian Government, 'Our North, Our Future: White Paper on Developing Northern Australia', above n. 232, 27.

There are already a few examples where indigenous groups in Australia are provided with some sort of water allocation, or policy commitments have been made towards providing water for Aboriginal economic development in the future, independent of native title. None of the recent statutory and policy developments are limited, like native title, by a threshold requirement to prove continuity with respect to their water resources since pre-sovereignty times before indigenous groups can take and use water. The undefined 'Aboriginal people' or 'Aboriginal communities' to whom the *Water Management Act 2000* (NSW) applies, for example, are not required to prove continuity of use or traditional laws and customs with respect to particular water resources since the acquisition of sovereignty before they can apply for Aboriginal access licences. Nor does section 27 of the *Cape York Peninsula Heritage Act 2007* (Qld), providing for the establishment of water reserves for economic and social purposes, requires the indigenous communities in the Cape York Peninsula area to establish continuity with respect to the water.[242] The Victorian policies appear to be open, generally, to 'Aboriginal Victorians', regardless of native title or traditional owners settlement status.[243]

The allocative models discussed above also appear to respond better to the existence of third-party interests than native title, which prioritises other users ahead of indigenous groups. The Northern Territory and Queensland indigenous water reserves set aside an 'equitable share' of the consumptive pool for indigenous economic and social aspirations.[244] In New South Wales, however, where there is intense competition for access to water, access licences are only available in low hydrological stressed, and medium to low in-stream value, water sources.[245] The licences are intended for well-watered coastal areas and will not be available in the main water catchment for the State, the Murray Darling Basin.[246] Poh Ling Tan explains that a scheme was

---

[242] Although, in the legislative scheme, indigenous communities in the Cape York peninsula area would probably be the holders of Aboriginal land titles under Queensland land rights legislation (the *Aboriginal Land Act 1991* (Qld)) who would have had to prove some sort of connection to the land.

[243] Interview with Bryony Grice (Melbourne, 9 July 2018), above n. 203.

[244] See generally William Nikolakis, 'Providing for Social Equity in Water Markets: The Case for an Indigenous Reserve in Northern Australia' in R. Quentin Grafton and Karen Hussey (eds.), *Water Resources Planning and Management* (Cambridge University Press, 2011).

[245] See Commonwealth of Australia, above n. 152, 20–3.

[246] New South Wales Government Department of Primary Industries, *How Water Sharing Plans Work* (2012) www.water.nsw.gov.au/Water-management/Water-sharing-plans/How-water-sharing-plans-work/how-water-sharing-plans-work/default.aspx.

implemented at some point to buy water rights for indigenous people in NSW, although the strategy was discontinued.[247] The New South Wales experience demonstrates the need for a statutory allocation mechanism to provide for the redistribution of water use rights where water resources are already fully allocated. Although the existing allocation models for indigenous water rights in Australia are *ad hoc* and limited,[248] they show how indigenous water rights might be provided for outside of the native title recognition model.

Various indigenous groups and commentators have suggested the establishment of an 'indigenous water fund',[249] to finance the purchase of water use rights for Aboriginal people.[250] For example, the First Peoples' Water Engagement Council, an indigenous representative advisory group established by the former National Water Commission, released an options paper recommending the establishment of an 'Indigenous Economic Water Fund' in April 2012, as follows:[251]

> The key purpose of the [Indigenous Economic Water Fund] is economic development as distinct from indigenous cultural and environmental water that should be set out in the planning process. The [Indigenous Economic Water Fund] is not an alternative to addressing access to cultural and customary water but an additional policy to improve the economic lives of indigenous people.

The First Peoples' Water Engagement Council proposed that the Indigenous Economic Water Fund would be used to acquire water use rights for 'indigenous people' to support economic development, via a range of acquisition mechanisms including 'government buyback, philanthropic buyback, self-funded buyback and gift'.[252]

The former National Water Commission also stressed the need for both a water fund and water reserve in 2012, recommending:[253]

> The Commission suggests that in water systems that are fully allocated the creation of a fund to acquire appropriate water rights should be

---

[247] Interview with Poh-Ling Tan (Brisbane, 13 July 2018).

[248] Interview with Rene Woods (New South Wales, 12 July 2018). Renee Woods is the Chairperson of the Murray Lower Darling Rivers Indigenous Nations.

[249] See, e.g., Jackson, 'Aboriginal Access to Water in Australia: Opportunities and Constraints', Chapter 2 above n. 41, 263.

[250] North Australian Indigenous Land and Sea Management Alliance, above n. 17, 2.

[251] Lawlab, above n. 19, 6.

[252] Ibid. 13.

[253] Australian Government, 'Position Statement: Indigenous Access to Water Resources', above n. 235, 1–2.

considered. In systems not fully allocated alternative approaches such as Strategic Indigenous Reserves could be set aside in water planning processes.

An Aboriginal Water Trust operated in New South Wales between 2000 and 2009 providing specific purpose grant funding for water infrastructure and offering opportunities to establish 'water based commercially viable' enterprises for Aboriginal communities.[254] The Trust did not, however, finance the acquisition of water use rights for Aboriginal groups in the market.

Funding mechanisms have been used in Australia in the past to redistribute land titles to indigenous groups, via the land market. The Aboriginal Land Fund was established to work alongside indigenous land rights legislation to 'assist Aboriginal Communities to acquire Land outside Aboriginal Reserves'[255] and operated from 1975 to 1980.[256] Then, in 1995, as part of reforms to the *Native Title Act* the Commonwealth set up the Indigenous Land Corporation. The Prime Minister originally posed the need for a land fund in his second reading speech for the Native Title Bill 1993 (Cth):[257]

> While these communities remain dispossessed of land, their economic marginalisation and their sense of injury continues. As a first step, we are establishing a land fund. It will enable indigenous people to acquire land and to manage and maintain it in a sustainable way in order to provide economic, social and cultural benefits for future generations. Addressing dispossession is essential but will not be enough to overcome the legacy of the past and achieve reconciliation.

The Indigenous Land Corporation continues to 'assist Indigenous peoples in Australia to acquire land and to manage Indigenous-specific land in a sustainable way to provide cultural, social, economic or environmental benefits for themselves and for future generations'.[258] The preamble to the *Native Title Act* notes the role this fund plays in providing for indigenous land rights where native title claims cannot be made out:

[254] New South Wales Office of Water, 'Our Water Our Country: An Information Manual for Aboriginal People and Communities about the Water Reform Process' (2012) www .water.nsw.gov.au/__data/assets/pdf_file/0004/547303/plans_aboriginal_communities_water_ sharing_our_water_our_country.pdf

[255] *Aboriginal Land Fund Act 1974* (Cth) preamble.

[256] Ibid. s. 16.

[257] Commonwealth of Australia, *Parliamentary Debates*, 16 November 1993, 2877 (Keating, Prime Minister).

[258] *Aboriginal and Torres Strait Islander Act 1995* (Cth), s. 191A.

It is also important to recognise that many Aboriginal peoples and Torres Strait Islanders, because they have been dispossessed of their traditional lands, will be unable to assert native title rights and interests and that a special fund needs to be established to assist them to acquire land.

Significantly, after a legislative amendment in 2018 the Indigenous Land Corporation now extends funding to indigenous salty and fresh-water rights, renamed the Indigenous Land and Sea Corporation, giving the Corporation acquisition and management functions with respect to water. However, it is not year clear how the Corporation will approach taking on these roles, or whether its activities will be restricted to certain water uses due to the limitations of the native title model.[259]

An allocative model, comprising both a water fund and water reserve, has in fact already been implemented in Australia with respect to environmental water rights. The Commonwealth Environmental Water Holder was established by the *Water Act 2007* (Cth) to manage environmental water holdings in the Murray Darling Basin. As well as managing in stream environmental water interests, the Commonwealth Environmental Water Holder has the capacity both to hold water use rights and purchase (and sell) water use rights in the market, in the interests of the environment.[260]

The Commonwealth Environmental Water Holder is now the single largest holder of environmental water rights in Australia.[261] Its portfolio of water use rights was acquired through a combination of government purchases and savings of water via investment in water supply infrastructure that reduced water losses and incentivised reduced water use.[262] Significantly, the portfolio of water use rights acquired by the Commonwealth Environmental Water Holder was acquired without any need for wide scale compulsory redistribution.

A way forward for Australia could be to design a mechanism to allocate and redistribute water use rights to indigenous groups: an 'Indigenous Water Holder' (or perhaps 'Indigenous Water Trust'). Further research is needed to consider the viability of an indigenous

---

[259] *Aboriginal and Torres Strait Islander Amendment (Indigenous Land Corporation) Act 2018* (Cth).

[260] *Water Act 2007* (Cth), ss. 104–15. See generally O'Donnell and Macpherson, Elizabeth, Chapter 3 above n. 84.

[261] Erin O'Donnell, 'Institutional Reform and the Victorian Environmental Water Holder' (2011) 22(2/3) *Journal of Water Law* 78.

[262] Australian Government, 'Water for the Future: Fact Sheet' (Department of Sustainability, Environment, Water, Population and Communities, 2010) www.environment.gov.au.

water holder in Australia, and the detail for its implementation, and the policy should only be pursued if desired and driven by indigenous peoples themselves.[263] Designing such a mechanism would not be straightforward, particularly when determining the representative or controlling group,[264] and devising rules about the temporary or permanent alienation of rights.[265] There is still variance of opinion from within government and between and within Aboriginal communities about the best way to structure such a model.[266] Any proposal for an indigenous water holder would also be likely to face strong opposition from other users, particularly agriculture and industry,[267] and without prioritising indigenous interests, Aboriginal Australians are often treated as 'just another stakeholder'.[268]

Yet, perhaps ironically, the basic conditions necessary for implementing an indigenous water holder in Australia have already been provided for, with the unbundling of water use rights from land title and the emergence of water markets. Because water use rights have been unbundled from land titles in Australia, and are available for purchase in water markets, they may be redistributed to indigenous landholders via voluntary purchases or savings without necessarily undermining the certainty of the rights held by others.

4.6 CONCLUSION

The recognition of ongoing native title rights to land and waters by the Australian common law in *Mabo [No. 2]* was a hugely significant step towards addressing the land and water claims of indigenous groups. Nonetheless, while the native title model plays a crucial role in recognising traditional, cultural water rights, it has failed to provide native titleholders with water rights for broader purposes.

The problems with the native title recognition model appear intractable, because the focus of native title is the historical rights of indigenous groups, pursuant to traditional laws that have been acknowledged and traditional customs that have been observed, in a substantially uninterrupted manner, since pre-sovereignty times. The idea of recognising pre-existing rights produces inevitable tensions around the

---

[263] Interview with Fred Hooper (Queensland, 12 July 2018), above n. 184.
[264] Ibid.
[265] Interview with Rene Woods (New South Wales, 12 July 2018), above n. 248.
[266] Ibid.
[267] Ibid.
[268] O'Bryan, above n. 19.

continuity of indigenous rights since pre-sovereignty times and the presence of other interests, because of the passage of time since colonisation. The impact of both of these problems is to reduce the scope for indigenous water rights.

I have argued in this chapter that if Australian law is to finally address indigenous water injustice this should best be done outside of the native title recognition model. A statutory allocation model has the potential to provide indigenous Australians with the right to use water on their lands for commercial purposes. A model of this type could take precedent from allocation models already implemented in Australia to allocate rights to land, which arose in response to the inability of the common law to recognise native title. It could capitalise on growing political attention to indigenous water injustice in places like Victoria, the emerging national treaty process, and the broadening of the Indigenous Land Corporation's functions to encompass water.

Indigenous land rights legislation seemingly overlooked the issue of water rights. However, there are already limited examples under New South Wales, Queensland and Northern Territory water legislation and policy that allocate water use rights, exercisable even for commercial purposes, to indigenous groups independent of native title. Where water use rights are held separately from landholding, and are available for purchase in water markets, they may readily be redistributed to indigenous users. This may be done without challenging the certainty of the rights of other users, because the water use rights may be purchased in the market from willing sellers. A similar mechanism already exists in Australia with respect to environmental water rights, the Commonwealth Environmental Water Holder, which has the power to hold and purchase water use rights in the market for environmental outcomes. The unbundling of water use rights from land titles and introduction of water markets may in fact provide an opportunity to address indigenous exclusion from water law frameworks, in a manner that both responds to and exploits market conditions.

CHAPTER FIVE

WATER RIGHTS FOR MĀORI IN AOTEAROA NEW ZEALAND

5.1 INTRODUCTION

The Indigenous peoples of Aotearoa New Zealand, the *iwi* (tribes) and *hapū* (subtribes) collectively known as the Māori peoples, once communally owned and managed all water in the country, pursuant to traditional laws and customs (*tikanga*) governing resource sharing and use. However, like the other indigenous peoples considered in this book, Māori have been excluded from legal frameworks that allocate rights to manage and use water resources since the British acquisition of sovereignty, with the signing of *Te Tiriti o Waitangi* (the Treaty of Waitangi), in 1840.

Although the average supply of water remains abundant in most parts of Aotearoa New Zealand, water resources are subject to increasing pressures from agriculture, industry, urbanisation and climate change.[1] The pressure on water resources affects both its available quantity for economic, social and environmental uses and its quality, challenging policy and decision makers on water allocation and management. Amidst already fierce debates about water allocation and protection, Māori are positioning claims to rights to water as both water guardians and water proprietors.

---

[1] See Sir Peter Gluckman, 'New Zealand's Fresh Waters: Values, State, Trends and Human Impacts' (Office of the Prime Minister's Chief Advisor, 12 April 2017).

Two key water developments have occurred in New Zealand in the past decade. The first is the negotiation and settlement of the claims of the Whanganui River Iwi under the Treaty of Waitangi, enabled in the *Te Awa Tupua (Whanganui River Claims Settlement) Act 2017* (NZ). Pursuant to this legislation, the Whanganui River was recognised as an 'indivisible and living whole' with 'all the rights, entitlements and obligations of a legal person', under the guardianship and management of an appointed representative.[2] The second is the Māori push for recognition of their proprietary rights to water, and the litigation and political debate around such claims, notably in the Waitangi Tribunal's National Freshwater and Geothermal Resources Inquiry. These two developments appear (at least at first sight) to have different ends, one being focused on procedural rights to recognise and provide for indigenous water relationships and management as *kaitiaki* (guardians), and the other on substantive rights to water distribution or 'ownership' backed up by the Treaty promise of *tino rangatiratanga* (highest chieftainship or authority).

However, the tendencies signify a desire for jurisdiction as a response to indigenous water exclusion, to reconstitute the relationship between the Crown and Māori in order to recognise a space for Māori as water managers, and a fair distribution of water use rights, including for development purposes. The study of indigenous rights to water in Aotearoa New Zealand in this chapter shows the variability of indigenous water demands, and a need for multifaceted responses to indigenous water injustice.

5.2 INDIGENOUS EXCLUSION FROM WATER LAW FRAMEWORKS IN AOTEAROA NEW ZEALAND

5.2.1 Māori Peoples in Aotearoa New Zealand

Prior to the British acquisition of sovereignty in 1840 a number of Māori *iwi* and *hapū* occupied and exercised sovereignty over all of Aotearoa pursuant to an intricate system of traditional laws and customs (*tikanga* Māori).[3] The Māori occupation of the territory now known as New Zealand dates back to around 1000 AD.[4] Prior to

---

[2] See *Te Awa Tupua (Whanganui River Claims Settlement) Act 2017*, ss. 12–15.

[3] Carwyn Jones, *New Treaty, New Tradition: Reconciling New Zealand and Māori Law* (New Zealand Victoria University Press, 2016).

[4] Todd Taiepa et al., 'Co-Management of New Zealand's Conservation Estate by Māori and Pakeha: A Review' (1997) 24(3) *Environmental Conservation* 236, 236.

western contact, Māori organised themselves primarily at the *hapū* (subtribe) level as the core social organisation, although land and resource management practices were similar throughout the country, depending on local geography and climate.[5]

According to Māori Lawyer and Supreme Court judge, Joseph Williams J, *tikanga* Māori determining traditional resource management and decision-making processes were (and are) underpinned by the core value of *whanaungatanga* (kinship); the idea that rights and obligations in natural resources are based on reciprocal familial relationships between people, groups and the natural world.[6] Under *tikanga* Māori, decisions are made in consensus by mandated representatives of the collective tribe for the benefit of present and future generations and the environment, in accordance with this principle of *whanaungatanga* and related concepts, such as *whakapapa* (genealogies),[7] *kaitiakitanga* (stewardship)[8] and *manaakitanga* (caring for others).[9]

Like indigenous peoples in other parts of the world, an important characteristic of rights and obligations in *tikanga* Māori is their 'intergenerational' quality; with obligations being owed to ancestors and future generations as well as current members of *iwi* and *hapū*. Further, Māori had, and continue to have, both spiritual and physical relationships with natural resources such as water.[10] Marsden explains how Māori often speak of a river's *wairua*, or spiritual existence, and its *mauri*: 'the life-force which generates, regenerates and upholds creation'.[11] Natural resources like waterways are also conceptualised as being part of territories, intrinsically interconnected with the surrounding landscape and ecosystem, rather than divisible segments capable of separation and disparate regulation, as is the case with western property law regimes for the regulation of

---

[5] Angela Ballara, *Iwi: The Dynamics of Māori Tribal Organisation from c. 1769 to c. 1945* (Victoria University Press, 1998) 282.

[6] Williams, 'Lex Aotearoa', Chapter 2 above n. 13. See also Tomas, Chapter 2 above n. 12, 228.

[7] Tomas, Chapter 2 above n. 12, 228.

[8] Māori Marsden, *The Woven Universe: Selected Writings of Rev. Māori Marsden* (Estate of Rev. Māori Marsden, 2003) 67. See Waitangi Tribunal, *Ko Aotearoa Tēnei: A Report into Claims Concerning New Zealand Law and Policy Affecting Māori Culture and Identity, Te Taumata Tuatahi* (WAI 262 Volume 1) (Legislation Direct, 2011) 23.

[9] Linda Te Aho, 'Corporate Governance: Balancing Tikanga Maori with Commercial Objectives' [2005] (2) *Yearbook of New Zealand Jurisprudence* 300; Linda Te Aho, 'Tikanga Maori, Historical Context and the Interface with Pakeha Law in Aotearoa/New Zealand' [2007] *Yearbook of New Zealand Jurisprudence* 10.

[10] Jacinta Ruru, 'Undefined and Unresolved: Exploring Indigenous Rights in Aotearoa New Zealand's Freshwater Legal Regime' (2009) 20(5–6) *Journal of Water Law* 236, 241.

[11] Marsden, above n. 8, 44.

natural resources.[12] The territorial approach to natural resources is clear in the well-known *whakataukī* (proverb): '*Ki uta ki tai*' (from the mountains to sea).[13]

The colonisation of New Zealand began with the signing of the Treaty of Waitangi in 1840. There are controversial differences between the Māori version and English translation of the Treaty. In article 1 of the document the Māori signatories ceded either 'sovereignty' or at least *kawanatanga* (government) to the Crown, and, in article 2, Māori retained 'full, exclusive and undisturbed possession' or at the most '*tino rangatiratanga*' (absolute chieftainship or independence) of lands, estates, forests, fisheries and treasures and the acquired the exclusive 'right of pre-emption' to purchase Māori land. Article 3 granted Māori the rights and protections of British subjects. However, it is now generally accepted that the Treaty relationship was made in mutual good faith,[14] and in a pragmatic approach to the textual difference in the translations, the New Zealand courts and tribunals have focused on its 'principles' and 'spirit',[15] even where the Treaty is not expressly referenced in legislation.[16] These principles include partnership, good faith, reciprocity (equality and active engagement), mutual benefit, a duty to make informed decisions (with an onus on the Crown), active protection (of rights) and redress of breaches (adequate and meaningful).[17] In developing an understanding of the commitments made under the Treaty, the courts have been guided by international indigenous rights law, particularly the *United Nations Declaration on the Rights of Indigenous Peoples*.[18] The courts have also occasionally, yet increasingly, acknowledged *tikanga* Māori as a source of law in New Zealand.[19] Of particular relevance to the management of water, is

---

[12] Lauren Butterly and Benjamin J Richardson, 'Indigenous Peoples and Saltwater/Freshwater Governance' (2016) 8(26) *Indigenous Law Bulletin* 3, 4. See also Viktoria Kahui and Amanda Richards, 'Lessons from Resource Management by Indigenous Māori in New Zealand: Governing the Ecosystems as a Commons' (2014) 102 *Ecological Economics* 1, 5. Jim Williams, 'Resource Management and Māori Attitudes to Water in Southern New Zealand' (2006) 62(1) *New Zealand Geographer* 73, 25.

[13] 'Our Fresh Water 2017' (Ministry for the Environment & Statistics New Zealand, 2017) 21.

[14] Alan Ward, *An Unsettled History: Treaty Claims in New Zealand Today* (Bridget Williams Books, 1999) 37.

[15] Ibid. 662 (Cooke P).

[16] *Huakina Development Trust v. Waikato Valley Authority* [1989] 3 NZLR 257.

[17] See Te Puni Kokiri, *He Tirohanga o Kawa Ki Te Tiriti o Waitangi = A Guide to the Principles of the Treaty of Waitangi as Expressed by the Courts and the Waitangi Tribunal* (Te Puni Kokiri, 2002).

[18] *New Zealand Māori Council v. Attorney-General* [2013] NZSC [97] (Elias CJ, McGrath, William Young, Chambers and Glazebrook JJ). See generally Valmaine Toki, 'Rights to Water an Indigenous Right?' (2012) 20 *Waikato Law Review: Taumauri* 107, 107.

[19] See, e.g., *Takamore v. Clarke* [2012] 2 NZLR 733 SC.

the Māori right to govern their own affairs, typically represented by the concept of *tino rangatiratanga* (highest authority or chieftainship), recognised in article 2 of the Treaty.

The New Zealand courts have invariably recognised the continuance of native or aboriginal title in New Zealand despite colonisation, referred to in New Zealand as Māori customary title.[20] This title derives from and exists as a matter of Māori custom, can only be extinguished by clear and plain legislative intent and is only alienable to the Crown (under the principle of pre-emption recorded in article 2 of the Treaty).[21] However, there remains very little Māori customary land in New Zealand. Most of this land was purchased by the Crown, confiscated by legislation[22] or converted into ordinary fee simple by the Native Land Court (now the Māori Land Court)[23] and subsequently alienated as part of the colonisation and settlement project.[24] Those Māori who retained their communal lands, and those who acquired communal land titles via the Native Land Court process, began over time to use corporations and trusts to organise as multiple owners and enable collective decision making.[25]

The commitments by the British Crown under the Treaty of Waitangi were, for much of the time since colonisation, not observed by New Zealand governments. The status of the Treaty was originally accepted by the courts in 1847,[26] but by 1877 relegated to the status of a 'simple nullity'.[27] However, from the 1960s onwards, with the emergence of the international indigenous rights movement, New Zealand has actively engaged in a process of reparative or restorative justice, involving attempts to settle historical and contemporary Māori grievances. In 1975, the *Treaty of Waitangi Act* established the Waitangi

---

[20] See Williams, 'Lex Aotearoa', Chapter 2 above n. 13; Andrew Erueti, 'Translating Maori Customary Title into a Common Law Title' [2003] *New Zealand Law Journal* 421(3); Jones, above n. 3; Paul G McHugh, *Aboriginal Title [Electronic Resource]: The Modern Jurisprudence of Tribal Land Rights* (Oxford University Press, 2011); Paul Havemann (ed.), *Indigenous Peoples' Rights: In Australia, Canada & New Zealand* (Oxford University Press, 1999).

[21] *Attorney-General v. Ngati Apa* (2003) 3 NZLR 643 [13] (Elias J).

[22] *New Zealand Settlements Act 1863* (NZ).

[23] *Native Lands Act 1862* (NZ). *Native Lands Act 1865* (NZ).

[24] See Richard Boast, *The Native Land Court 1862–1887: A Historical Study, Cases, and Commentary* (Thomson Reuters, 2013); Richard Boast, *Buying the Land, Selling the Land: Governments and Maori Land in the North Island 1865–1921* (Victoria University Press, 2008).

[25] These Māori lands are typically held in Māori land incorporations and trusts and subject to the jurisdiction of the Māori Land Court under the *Te Ture Whenua Māori Act 1993* (NZ). See Elizabeth Macpherson, 'Iwi Companies' in Susan Watson et al., *Corporate Law in New Zealand* (Thomson Reuters, 2018).

[26] *R v. Symonds* [1847] NZPCC 387.

[27] *Wi Parata v. The Bishop of Wellington* [1877] 3 NZLR 72.

Tribunal, a permanent Commission of Inquiry charged with addressing Māori grievances and educating the public. In 1985, an amendment to the Act extended the Tribunal's historical jurisdiction back to 1840, for acts or proposed acts omitted or committed by the Crown.[28] The Tribunal hears claims from *iwi* and *hapū* of Crown breach of the principles of the Treaty, enquires into and reports on those claims and recommends certain compensation by the Crown including the transfer of assets (money and property).[29] Māori grievances under the Treaty have traditionally related to the loss of land and resources but may address a range of historical and contemporary issues including education and healthcare, culture and language, and economic disadvantage.

Māori groups can pursue 'comprehensive settlement' of their Treaty of Waitangi grievances, either after presenting their claims before the Waitangi Tribunal or as part of 'direct negotiations' with the Crown through the Office of Treaty Settlements. The Crown treats with 'large natural groupings' of claimants, or multiple claimant collectives,[30] who elect 'mandated' representatives to negotiate a settlement. Settlement redress is usually *iwi*-specific, although there have been regional or thematic settlements,[31] and is implemented by legislation after the group establishes a 'post-settlement governance entity' to receive and manage settlement assets (property and cash).[32]

New Zealand's Treaty settlement process is underpinned by a desire to repair historical grievances and re-establish the relationship between the Crown and Māori.[33] However, the process is also intended to be redistributive and reallocate assets to Māori in order to support their

---

[28] *Treaty of Waitangi Act 1975* (NZ), s. 6(1)(d).

[29] Ibid. (NZ), s. 6(4A)(a). Crown redress cannot include the return of privately held property. See ibid. (NZ) ss. 8A–8E state that the Tribunal only has the power to make recommendations on whether land can be returned. However, the Tribunal does have binding powers in the *State-Owned Enterprises Act 1986* (NZ).and the *Crown Forest Assets Act 1989* (NZ) (ss. 35–40) to return land.

[30] See, e.g., *Tauranga Moana Iwi Collective Redress and Ngā Hapū o Ngāti Ranginui Claims Settlement Bill* (NZ).

[31] See, e.g., *Treaty of Waitangi (Fisheries Claims) Settlement Act 1992* (NZ); *Central North Island Forests Land Collective Settlement Act 2008* (NZ).

[32] New Zealand and Office of Treaty Settlements, *Ka Tika ā Muri, Ka Tika ā Mua: He Tohutohu Whakamārama i Ngā Whakataunga Kerēme e Pā Ana Ki Te Tiriti o Waitangi Me Ngā Whakaritenga Ki Te Karauna = Healing the Past, Building a Future: A Guide to Treaty of Waitangi Claims and Negotiations with the Crown* (2015).

[33] Joe Williams, Australian Institute of Aboriginal and Torres Strait Islander Studies and Native Title Research Unit, *Confessions of a Native Judge: Reflections on the Role of Transitional Justice in the Transformation of Indigeneity* (Native Title Research Unit, Australian Institute of Aboriginal and Torres Strait Islander Studies, 2008) 3.

economic development amid ongoing social and economic disadvantage and inequity. Māori continue to emphasise their Treaty right not only to their lands and resources but, in line with international law, the right to their 'development'.[34] By 2017 the New Zealand Government had allocated over $1.5 billion worth of assets to Māori under Treaty settlements and the Māori economy in New Zealand was estimated at $50 billion.[35] However, the Treaty settlement framework has been widely criticised as being *ad hoc,* lengthy[36] and inconsistently applied,[37] especially with respect to the Crown's 'large natural groupings' policy, which is said to disenfranchise overlapping claimants and internally divide *iwi* and *hapū*.[38] The process internalises power imbalances between the Crown and *iwi* and *hapū,* and provides a forum for tradeoffs and compromises to be made about both historical grievances and contemporary redress. In this context, former Treaty of Waitangi Settlements Minister, Christopher Finlayson, explains how even the most factual aspects of Treaty settlements are carefully negotiated and mediated between the parties:[39]

> Apologies are not pro forma; they are carefully drafted by Crown and Māori historians. They can be the source of intense debate between the parties.

5.2.2 Water Law Frameworks and Māori Rights

Access to water has not, historically, been a major concern in New Zealand, as a country with high annual rainfall and a 'clean and green' image. Increasingly, though, New Zealand rivers suffer from irregular flow, poor water quality and high nutrient loads as a consequence of more than 150 years of intensive agriculture and expanding industry.[40] The variability of water conditions in different parts of the country present challenges for regulators.[41] Like the indigenous peoples considered in other parts of this book, Māori have been excluded from legal

---

[34] See, e.g., Waitangi Tribunal, *The Stage 1 Report on the National Freshwater and Geothermal Resources Claim: WAI 2358* (Legislation Direct, 2012).

[35] Chapman Tripp, *Te Ao Māori: Trends and Insights* (Piripi 2017) www.chapmantripp.com.

[36] Interview with Chris Finlayson (Wellington, 4 October 2018).

[37] Iorns Magallanes, Chapter 3 above n. 4, 306.

[38] Margaret Mutu, 'Māori Issues' (2017) 29(1) *The Contemporary Pacific* 144, 147.

[39] Interview with Chris Finlayson (Wellington, 4 October 2018), above n. 36.

[40] OECD, *OECD Environmental Performance Reviews: New Zealand 2017* (OECD Publishing, 2017).

[41] New Zealand and Ministry for the Environment, *Next Steps for Fresh Water: Consultation Document* (2016) 6–7.

frameworks providing for the management and use of water for much of the time since sovereignty.

The British Crown brought with it the common law doctrine of riparian rights and soon began to exercise what it saw as its sovereign right to govern water. As early as 1903 the New Zealand Government passed legislation confirming that all water resources were vested in the Crown.[42] The Government intensified Crown control over water regulation with the *Water and Soil Conservation Act 1967*, reflecting national interest in legislating governmental authority over drainage, flood control and water supply.[43] Section 21 of the Act provided that water resources were vested in the Crown and that the Crown has the sole right to regulate and allocate water, without any acknowledgement of the ongoing water use by, and water relationships of, Māori. The Crown continues to resist Māori proprietary claims for water rights on the basis that 'no one can own water', because that would be inconsistent with the vesting of all waters in the Crown on behalf of the New Zealand public.[44]

Due to the Crown's inflexible position about the ownership of water, Māori legal action over rivers and lakes has, until recently, focused on the land under the water: their beds. Section 14 of the *Coal Mines Amendment Act 1903*[45] declared that the beds of all navigable rivers 'shall remain and shall be deemed to have always been vested in the Crown'.[46] Despite the enduring uncertainly around non-navigable rivers, the Crown then simply assumed that it owned the land under the water (but not the water itself, which is vested in the Crown on behalf of the public),[47] without issuing any certificate of title.[48] In 1983 the Property Law and Equity Reform Committee recommended the statutory vesting of rivers in the Crown, however,

---

[42] *Coal Mines Amendment Act 1903* (NZ), s. 14; *Water Power Act 1903* (NZ), ss. 2, 5; the vesting carries through to *Water and Soil Conservation Act 1967* (NZ), s. 21 and; *Resource Management Act 1991* (NZ), s. 354.

[43] Anne Salmond, 'Tears of Rangi: Water, Power, and People in New Zealand' (2014) 4(3) *HAU: Journal of Ethnographic Theory* 285, 298.

[44] *Resource Management Act 1991* (NZ), s. 354.

[45] Later, Section 261 of the *Coal Mines Act 1979* (NZ). Now Section 354(1)(c) of the *Resource Management Act 1991* (NZ).

[46] Recently, the Supreme Court found that a whole of river approach to navigability was inappropriate. See *Paki v. Attorney General* (No. 1) [2012] NZSC 50.

[47] James Morris and Jacinta Ruru, 'Giving Voice to Rivers: Legal Personality as a Vehicle for Recognising Indigenous Peoples' Relationships to Water' (2010) 14(2) *Australian Indigenous Law Review* 49, 53.

[48] Mick Strack, 'Land and Rivers Can Own Themselves' (2017) 9(1) *International Journal of Law in the Built Environment* 4, 8.

this recommendation was not taken up, leaving the question open to the courts.[49] With the common law, New Zealand also imported the *ad medium filum aquae* rule that ownership of a riverbank extends in the bed to the middle of the river's flow (the mid-point presumption).[50] In *Paki (No. 2)*, it was held that the mid-point presumption applied to non-navigable rivers only if there was no 'uninvestigated' contiguous Māori customary land, and the presumption was rebutted if it was shown to be contrary to the custom and use of the original Māori grantees (of lands converted from customary title to fee simple through the Native Land Court process between 1887 and 1899).[51] The case signalled an openness to claims for recognition of customary title to the bed of rivers. Yet, ironically, if it is the custom or usage (or *tikanga*) of a particular *iwi* or *hapū* that ownership of a river is inseparable from ownership of land, as is often asserted, then the group would necessarily have intended to alienate the riverbed at the same time as they sold the land. As will be discussed below, the question of native title to the water on top of the bed remains unanswered in New Zealand.

New Zealand now has a fairly comprehensive, integrated natural resource management regulatory regime in the *Resource Management Act 1991*, underpinned by an overarching principle of 'sustainable management'.[52] Unlike Australia and Chile, the New Zealand approach is still a 'planned' rather than integrated market water allocation model, with low incidence of water trading. Under the *Resource Management Act* consent authorities, usually local municipal councils, make decisions to grant a 'resource consent' to take and use water on a 'first come, first served' basis,[53] with the addition, if considered necessary, of specified conditions. Water allocations for take and use are temporal, applying up to thirty-five years, although they may be renewed.[54] The consents to use resources granted under the *Resource Management Act* are not 'property rights',[55] although they may have many of the incidents of property as an excludable right to exploit

---

[49] Ibid. 6.
[50] For earlier cases, see, e.g., *Mueller v. Taupiri Coal Mines* [1990] 20 NZLR 89 CA; *Tamihana Korokai v. Solicitor General* [1912] 32 NZLR 321 CA; and *Re the Bed of the Wanganui River* [1962] NZLR 600 CA.
[51] *Paki v. Attorney General (No. 2)* [2015] 1 NZLR 67 SC [142], [25] (Elias CJ).
[52] *Resource Management Act 1991* (NZ), s. 5.
[53] Ibid. (NZ), s. 14; *Fleetwing Farms v. Marlborough* [1997] 3 NZLR 257 CA.
[54] *Resource Management Act 1991* (NZ), ss. 123, 124, 124B(2).
[55] *Aoraki Water Trust v. Meridian Energy Ltd* (2005) 2 NZLR 268 [34]–[35].

resources. In practice, water resource consents are immensely valuable, allowing for commercial use and benefits from activities including agriculture, tourism and hydropower development.

There is no specific recognition in the Resource Management Act of Māori rights to water, or provision for a Māori water licence or allocation. Instead, Māori water relationships have been accounted for as part of planning processes under the *Resource Management Act*, reflected in acknowledgements of Māori cultural relationships or values, or involvement in water governance via consultation or co-management arrangements, in accordance with the Crown's bottom-line that 'no one can own water'.

There are a number of protections of environmental and cultural values built into the *Resource Management Act*, which have been used at times to recognise and provide for Māori rights to water. These include the requirement for decision makers considering applications for consents to use water to: recognise and provide for the 'relationship of Māori and their culture and traditions with their ancestral lands, water, sites, *waahi tapu* (scared sites), and other *taonga* (treasures)';[56] 'have particular regard' to *kaitiakitanga*;[57] and 'take into account' the principles of the Treaty.[58] These requirements have been characterised as 'strong directions, to be borne in mind at every stage of the planning process'.[59] However, these 'considerations' are merely part of the other, non-indigenous specific interests in this 'Part 2 trilogy', and without any priority, the indigenous interests are defeasible.[60] The courts have made clear that they do not amount to a right of veto and should an unfavourable decision be issued, it is enough that the decision maker has considered the Treaty principles,[61] to which the Waitangi Tribunal has reacted with concern.[62]

The *Resource Management Act* also provides for a number of collaborative governance mechanisms including joint management agreements between local authorities with Māori *iwi* and *hapū* concerning natural resources,[63] and the devolution of decision making on resource

---

[56] *Resource Management Act 1991* (NZ), s. 6(e).
[57] *Resource Management Act 1991* (NZ), s. 7(a).
[58] Ibid. s. 8.
[59] *McGuire v. Hastings District Council* [2000] UKPC 43; *McGuire v. Hastings District Court* [2002] NZLR 577 PC594 (Lord Cooke of Thorndon).
[60] Waitangi Tribunal, *The Ngawha Geothermal Resource Report 1993*, WAI 304 (Legislation Direct, 2006) 145.
[61] *Watercare Services Ltd v. Minhinnick* (1998) 1 NZLR 63, 79 (HC).
[62] Waitangi Tribunal, *Ngawha Geothermal Resource Report 1993*, above n. 60.
[63] *Resource Management Act 1991* (NZ), s. 36B.

management to Māori groups (although the latter mechanism has not been used for Māori).[64] As a result of a 2017 amendment to the Act, *iwi* and *hapū* may enter into voluntary mana whakahono ā rohe agreements, which are intended to increase Māori participation in collaborative governance.[65] An agreement is currently being negotiated between Ngā Puna Wai o Te Tokotoru and the Bay of Plenty Regional Council and is due for completion in October 2019. The hope is that these novel agreements will succeed where the previous collaborative provisions have 'almost completely failed to deliver partnership outcomes'.[66]

Part 9 of the Act also allows any person to apply to the Minister for the Environment for a 'water conservation order' to protect environmental or cultural water values, including protecting water bodies of 'outstanding significance in accordance with *tikanga* Māori'.[67] In 2011, Te Runanga o Ngāi Tahu successfully amended a water conservation order for Te Waihora (Lake Ellesmere), submitting that the lake was of utmost important to Ngāi Tahu, both culturally and spiritually, and to ignore this significance would erode the customary rights of Ngāi Tahu.[68]

Despite these protections, the general consensus is that Māori water concerns are inadequately managed under the *Resource Management Act*.[69] The legislation has been described by the Waitangi Tribunal as having 'disappointingly unrealised potential',[70] and being ultimately non-compliant with the Treaty.[71] The inadequacies of the *Resource Management Act* are exacerbated by the resource consent granting and appeals process, whereby resource consents may be issued with inadequate or no consultation with Māori.[72] The Environment Court, responsible for hearing cases concerning the application of the Act, has been

---

[64] Ibid. (NZ), s. 33.

[65] Ibid. (NZ), ss. 58L, 58U.

[66] Waitangi Tribunal, *Ko Aotearoa Tēnei* (WAI 262, Volume 1), above n. 8, 115.

[67] *Resource Management Act 1991* (NZ), s. 199(2)(e).

[68] Environment Canterbury, 'Report of the Hearing Committee on an Application to Vary a National Water Conservation Order for Lake Ellesmere/Te Waihora in Canterbury' (2011) [43].

[69] Interview with Riki Ellison (Wellington) (5 February 2019).

[70] Waitangi Tribunal, *Ko Aotearoa Tēnei* (WAI 262, Volume 1), above n. 8, 116.

[71] Waitangi Tribunal, *Ko Aotearoa Tēnei: A Report into Claims Concerning New Zealand Law and Policy Affecting Māori Culture and Identity, Te Taumata Tuarua* (WAI 262 Volume 2) (Legislation Direct, 2011) 705.

[72] Section 36A provides that that a local authority does not have to consult any person about an application. However, a consent application must include an assessment of environmental effects. As part of that assessment, any effect on resources having cultural and spiritual value must be considered (Sch. 4, (7)(1)(d) and Sch. 4 (6)(1)(h)).

described as 'ill-suited to the constructive inclusion of stakeholders in decision-making'.[73] Māori applicants before the Environment Court are often perceived as anti-development and treated as 'just another stakeholder'.[74] Although the Environment Court has accepted that the Waitangi Tribunal is a 'highly persuasive' authority,[75] and increasingly admits stories and lore as evidence,[76] it has been wary about the weight, content and application of *tikanga* Māori.[77]

The overarching national policy framework for the management of water in New Zealand under the *Resource Management Act*, the *National Policy Statement for Freshwater Management 2014*, includes some limited guidance about the incorporation of Māori interests in water planning. The relevant provisions, called *Te Mana o Te Wai* (translating as 'the power of water'), were introduced in September 2017 as part of the *Next Steps for Freshwater* review and reflected feedback gathered around a need to incorporate Māori perspectives. The National Policy Statement provides for ecosystem and human health and makes them compulsory national values with minimum standards. Within these, *Te Mana o Te Wai* provides for 'the integrated and holistic well-being of a freshwater body' and 'acknowledges and protects the mauri of the water' to further the health of its environment and people.[78] Local authorities are required to 'consider and recognise' *Te Mana o Te Wai*,[79] take Māori values into account and take 'reasonable steps' to involve *iwi* and *hapū* in resource management so that their values and interests are reflected in decisions.[80] Whether these requirements will fundamentally change decision making at the local level, including through conditions on resource consents to protect ecosystem and cultural values, remains to be seen.[81]

What is clear, is that Māori are deeply concerned about the state of water resources in Aotearoa New Zealand, especially declining water

---

[73] James Lennox, Wendy Proctor and Shona Russell, 'Structuring Stakeholder Participation in New Zealand's Water Resource Governance' (2011) 70(7) *Ecological Economics* 1381, 1391.

[74] M Durette et al., *Māori Perspectives on Water Allocation* (Wellington: Ministry for the Environment, 2009) 16.

[75] *Bleakley v. Environmental Risk Management Authority* [2003] 3 NZLR 213 HC 233.

[76] See, e.g., *Outstanding Landscape Protection Society v. Hastings District Council* [2008] NZRMA 8 EC 29–30.

[77] *Beadle v. Minister of Corrections* [2002] BCL 701 BC200269088 EC [436], [497].

[78] Ministry for the Environment, 'National Policy Statement for Freshwater Management (Amended 2017)' 7. www.mfe.govt.nz/sites/default/files/media/Fresh%20water/nps-freshwater -ameneded-2017_0.pdf.

[79] Ibid. 11.

[80] Ibid. 24.

[81] Interview with Jacinta Ruru (Christchurch, 7 December 2018).

quality.[82] Many argue for the incorporation of *mātauranga Māori* (Māori knowledge) in approaches to the management of water, in order to support the health and *mauri* (life force) of water resources and the people connected to them.[83] In the absence of adequate legislative and policy protections, Māori have continued internal processes of resource management, including *iwi*-specific cultural health index reports,[84] and practical monitoring systems.[85] They have also in some cases, secured *ad hoc* political settlements concerning water, such as the Waikato River Settlement and the Te Awa Tupua Settlement for the Whanganui River discussed below. These settlements enable co-management of important water resources between local authorities and Māori groups, in partial recognition of indigenous water relationships and jurisdictions.

5.3 LEGAL RIGHTS FOR RIVERS AND MĀORI WATER RIGHTS

Outside of the *Resource Management Act*, a number of Treaty settlements have provided for cultural redress in relation to water, such as statutory acknowledgements (recognition of association with an area and right to be heard for decisions taken regarding it), place name changes, limited rights to participate in collaborative governance, or 'co-management', of rivers or lakes, and the vesting of their beds.[86] These settlements are partial attempts to resolve historical grievances about the management and ownership of water, and reflect Māori water conceptualisations such as the territorial approach to natural resources perceiving a river as an 'integrated and living whole'.[87]

---

[82] Interview with Tania Gerrard (Wellington, 28 August 2018). Gerrard was the Director of Water at the Ministry for the Environment at the time of the interview.

[83] Gary Brierley et al., 'A Geomorphic Perspective on the Rights of the River in Aotearoa New Zealand: Geomorphology and the Rights of the River' [2018] (Special Issue Paper) *River Research and Applications* 3.

[84] See, e.g., NIWA, '2016 Pilot Waikato River Report Card: Methods and Technical Summary – Prepared for Waikato River Authority' (NIWA, March 2016).

[85] See, e.g., the Ngā Waihotanga Iho estuarine monitoring toolkit developed by NIWA/Taihoro Nukurangi (National Institute of Water and Atmospheric Research).

[86] See, e.g., *Te Arawa Lakes Settlement Act 2006* (NZ), s. 23; *Ngāti Awa Claims Settlement Act 2005* (NZ), s. 164; *Ngaa Rauru Kiitahi Claims Settlement Act 2005* (NZ), s. 29. Where the beds of lakes were vested, it was commonly subject to the Crown's ownership of the 'stratum', the space of water and air above the bed, held on behalf of the New Zealand people. Controversially, it has been suggested that this idea lacks precedent, and has simply been assumed. See the Parliamentary debate to the *Te Arawa Claims Settlement Bill* (12 September 2006) 634 NZPD 5370 (NZ).

[87] See generally, O'Donnell and Macpherson, Introduction above n. 24.

One of the earlier examples involved the vesting of the bed of Te Waihora (Lake Ellesmere) in Te Rūnanga o Ngāi Tahu, as part of their 1998 Treaty settlement,[88] although the vesting did 'not of itself confer any rights or impose any obligations on Te Rūnanga o Ngāi Tahu of ownership, management, or control' of the lake's water, aquatic life or structures.[89] Nor did the vesting affect public access,[90] private ownership or lawful commercial rights.[91] More accurately, the vesting of Te Waihora provides Ngāi Tahu with participation rights in the collaborative governance of the lake.[92]

The first major river settlement concerned the Waikato River in 2009.[93] The Waikato is a major navigable river, in fact New Zealand's longest river, and supplies drinking water for New Zealand's largest city, Auckland. The river also flows through the traditional territory of the Waikato-Tainui and Raukawa Iwi, and a number of *iwi* and *hapū* hold a particular cultural and strategic relationship with the river. The *iwi* claim a relationship with the river as an 'ancestor' and an 'indivisible' whole,[94] similar to the conceptualisation of Te Awa Tupua by the Whanganui Iwi, considered later in this chapter, albeit without the grant of legal personality for the river. The settlement legislation established the Waikato River Authority to oversee the vision for the river, promote its holistic management, and fund the newly established Waikato River Clean-Up Trust.[95] Despite the intention that the governance arrangements established by the Waikato River Settlement would 'ensure that the highest level of recognition is given to the restoration and protection of the Waikato River',[96] the values of the Waikato River and its people are not necessarily a priority consideration for decisions taken under the *Resource Management Act*.[97] According to Muru-Lanning, the Settlement has been the subject of internal conflict within Waikato-Tainui and contests over power between Māori and government.[98]

---

[88] Ngai Tahu was also granted the power to appoint 'guardians' to advise hydropower and general development in lakes Manapouri, Monoawai and Te Aanu. See the *Conservation Law Reform Act 1990* (NZ), s. 5.

[89] *Ngāi Tahu Claims Settlement Act 1998* (NZ), s. 171.

[90] Ibid. s. 173.

[91] Ibid. s. 174.

[92] Memon and Kirk, Chapter 3 above n. 63, 954.

[93] *Waikato-Tainui Raupatu Claims (Waikato River) Settlement Act 2010* (NZ).

[94] Ibid. preamble (1), (11), (17)(g).

[95] Ibid. s. 2.

[96] Ibid. Sch. 2, s. 2(a).

[97] Ibid. s. 17.

[98] Marama Muru-Lanning, *Tupuna Awa: People and Politics of the Waikato River* (Auckland University Press, 2016) 176.

More recently, the Ngāti Rangi Treaty Settlement, currently before Parliament as the *Ngāti Rangi Claims Settlement Bill 2018* (NZ), includes redress designed to enable Ngāti Rangi to fulfil their *kaitiaki* responsibilities over rivers, including the Te Waiū-o-Te-Ika framework. The framework establishes a joint river entity comprised of all *iwi* connected to the Whangaehu River, who will work for the benefit of the river. The river is recognised as an indivisible and living whole from its source (Crater Lake Mount Ruapehu) to the sea, although, like the Waikato, the river is not declared to be a legal person.[99] As will be discussed below, the New Zealand Government broke new ground in 2017 by declaring that the Whanganui River was a legal person as part of a full and final settlement of the claims of local Māori concerning the river.

5.3.1 Te Awa Tupua – the Whanganui River

The Whanganui River is iconic in Aotearoa New Zealand. It flows from the foothills of Mt Tongariro, through the remote and wild King Country and Whanganui region, out to the ocean at the city of Whanganui. It has been epitomised in New Zealand colonial and twentieth-century history and has been the subject of film, art and literature, with prominent New Zealand poet James K Baxter having established a community along its banks at Jerusalem. The river is a resource of utmost importance to the Whanganui *iwi* and *hapū* who depend on it for their physical and spiritual sustenance and see it as an indivisible and living whole and revered ancestor.

The river has also been a site of protest, contestation and conflict, in a long-running dispute between the Whanganui Iwi and the Crown about the right to own and control the river territory. Whanganui Māori made their first petitions about the river to the Crown in the 1870s, and an extended series of court cases concerning the river followed from 1938 to 1962.[100] Initially, the concerns of Whanganui Iwi about the river were for access to *mahinga kai* (food gathering), *wāhi tapu* (sacred sites) and rights of navigation. Later, their concerns related to the impacts of settler tourism, resource extraction, and the establishment of the Tongariro Power Scheme in 1958 and consequential environmental degradation.[101]

---

[99] *Ngāti Rangi Claims Settlement Bill 2018* (NZ), s. 107(2).
[100] See, e.g., *In re the Bed of the Wanganui River* [1962] NZLR 600 CA.
[101] See generally, Waitangi Tribunal, *The Whanganui River Report* (WAI 167) (GP Publications, 1999).

The Whanganui Iwi took their claims about the river to the Waitangi Tribunal in the 1990s, and the Tribunal released its Whanganui River Report in 1999. In the Report, the Tribunal spent some time describing the physical and spiritual relationship of the Whanganui Iwi with the river as follows:[102]

> The river was central to Atihaunui lives, their source of food, their single highway, their spiritual mentor. It was the aortic artery of the Atihaunui heart. Shrouded in history and tradition, the river remains symbolic of Atihaunui identity. It is the focal point for the Atihaunui people, whether living there or away. Numerous marae still line its shores.

As to the legal nature of the *Iwi's* interests in the river, the Tribunal framed the interests as 'ownership' of the entire river territory:[103]

> In Māori terms, the Whanganui River is a water resource, a single and indivisible entity comprised of water, banks and bed. There is nothing unexpected in that. It is obvious that a river exists as a water regime and not as a dry bed. The conceptual understanding of the river as a tupuna or ancestor emphasises the Māori thought that the river exists as a single and undivided entity or essence. Rendering the native title in its own terms, then, what Atihaunui owned was a river, not a bed, and a river entire, not dissected into parts.

However, the Tribunal's findings produced a 'deadlock'[104] for settlement negotiations between the *Iwi* and the Crown, because of the Crown's position that 'no one can own water' that is vested in the Crown on behalf of the New Zealand public.[105] The dispute over the river finally settled in 2017 when the Crown recognised that the river is a legal person, and designed a complicated collaborative governance framework for its management in the *Te Awa Tupua (Whanganui River Claims Settlement) Act 2017 (Te Awa Tupua Act)*. In this model, neither the *Iwi* nor the Crown own the river – it owns itself.

As explained above, in the New Zealand context, political settlements about rivers are not new.[106] However, the settlement for the Whanganui River was the first to use the concept of legal personality,

---

102 Ibid. xiii.
103 Ibid. 337.
104 Sanders, Chapter 3 above n. 64, 5.
105 *Water and Soil Conservation Act 1967* (NZ), s. 21; *Coal Mines Amendment Act 1903* (NZ), s. 14; *Water Power Act 1903* (NZ), ss. 2, 5; *Resource Management Act 1991*, s. 354.
106 See Jacinta Ruru, 'Indigenous Restitution in Settling Water Claims: The Developing Cultural and Commercial Redress Opportunities in Aotearoa, New Zealand' (2013) 22(2) *Pacific Rim Law & Policy Journal* 311 at 340.

following the recognition of Te Urewera Forest as a legal person in another political settlement between Māori and the Crown in 2014.[107] The *Te Awa Tupua Act* is, therefore, a new approach to resolving indigenous water claims in Aotearoa New Zealand, as a compromise allowing greater indigenous involvement in governance, while putting off the more difficult question of ownership.

5.3.2   Te Awa Tupua Model

The *Te Awa Tupua Act* introduces a new regime for the governance of the Whanganui River, which (somewhat clumsily) sits alongside and interacts with *Resource Management Act* processes for decision making about rivers, and enables input to planning and decision making about the use of the river from Māori as well as a range of other users. The Act purports to introduce a territorial approach to the management of the river; recognising the river and its tributaries as 'an indivisible and living whole, comprising the Whanganui River from the mountains to the sea, incorporating all its physical and meta-physical elements'.[108] As well as reflecting the territorial approach to natural resources in indigenous rights theory and international law, discussed in other parts of this book, this approach is informed by the *tikanga* of the Whanganui Iwi. The territorial approach contradicts western conceptions of property in natural resources; whereby resource components are segmented in their regulation and use. This is different to the usual application of state-based New Zealand law, where the bed, water, banks, minerals and wildlife of a river are often owned and regulated separately.

The Act then 'declares' Te Awa Tupua to be a 'legal person', with all the 'rights, powers, duties and liabilities of a legal person'.[109] As discussed in Chapter 3, the rights of legal persons are understood to be the rights to sue, be sued, enter into legal contracts, and the right to hold property. The use of 'declares' in section 14 is interesting, where section 12 in contrast 'recognises' the river as 'Te Awa Tupua'. Accordingly, the river is recognised as having always been Te Awa Tupua, but the legal person status is something new. However, by using 'declares', instead of, perhaps, 'grants', the Act avoids being seen to confer legal personality on the river, inviting a possible argument that legal

---

[107] *Te Urewera Act 2014* (NZ), s. 11. This is part of a tendency towards the recognition of natural resources as legal persons under Treaty settlements, with Mount Taranaki and Te Mata Peak subsequently being earmarked for the approach.
[108] *Te Awa Tupua (Whanganui River Claims Settlement) Act 2017* (NZ), s. 12.
[109] Ibid. (NZ), s. 14.

personality (clearly a western law construct) may be an approximation for the nature of river interests in *tikanga* Māori, regardless of whether the government recognises it as such. The fact that Te Awa Tupua is a legal person means that it may, if necessary, go to court in its own name to protect itself from damage caused to the river by others. Because the legislation recognises and seeks to protect both the physical and metaphysical qualities of the river, it seems possible that the river could take action to protect its *mauri* from damage by river and riparian users, as well as strict physical damage.[110]

Because the river requires a human representative to enforce its rights, the Act establishes the office of Te Pou Tupua, to be its 'human face'.[111] In the spirit of Treaty partnership, two representatives make up the *Pou*, one nominated by the Crown and the other by the *Iwi*. The initial representatives, former Member of Parliament Dame Tariana Turia and educator Turama Hawira, were appointed for a three-year term in September 2017.[112] Both have links to Whanganui and do not currently work for the Crown, suggesting that their authority, expertise as indigenous knowledge holders,[113] and contextual familiarity are important markers of perceived legitimacy. Christopher Finlayson, the Minister responsible for pushing through the Whanganui River Settlement describes Turia as 'an obvious and necessary appointment', given that 'no New Zealander has a greater understanding of the aspirations of the Whanganui Iwi'.[114] Although the Act does not use the word 'guardian', the model is commonly described as an application of Stone's 'guardianship'[115] proposal for the rights of nature.[116] In the New Zealand context, however, Te Pou Tupua may be better understood as reflecting the idea of *kaitiakitanga*, as the concept has evolved in Treaty jurisprudence and the *Resource Management Act*, discussed above.[117]

---

[110] Interview with Baden Vertongen (Wellington, 28 August 2018).

[111] *Te Awa Tupua (Whanganui River Claims Settlement) Act 2017*, ss. 18–19.

[112] Zaryd Wilson, 'Whanganui River Representatives Appointed' *New Zealand Herald*, 5 September 2017 www.nzherald.co.nz/nz/news/article.cfm?c_id=1&objectid=11916893.

[113] See generally Maria Hepi et al., 'Enabling Mātauranga-Informed Management of the Kaipara Harbour, Aotearoa New Zealand' (2018) 52(4) *New Zealand Journal of Marine and Freshwater Research* 497, 506.

[114] Interview with Chris Finlayson (Wellington, 4 October 2018), above n. 36.

[115] Stone, Chapter 3 above n. 19.

[116] Abigail Hutchison, 'The Whanganui River as a Legal Person' (2014) 39(3) *Alternative Law Journal (Gaunt)* 179; Mari Margil, 'The Standing of Trees: Why Nature Needs Legal Rights' (2017) 34(2) *World Policy Journal* 8, 8.

[117] Iorns Magallanes, Chapter 3 above n. 4, 281.

Te Pou Tupua is required to act in the interests of the Whanganui River, consistent with statutory management values.[118] These values, called Tupua te Kawa, are the 'intrinsic values that represent the essence of Te Awa Tupua',[119] and help us to understand the content of the river's rights. For example, a value requiring responsibility towards the river's health recognises the interdependency between the river's health and that of the Whanganui Iwi.[120] Curiously, as a legal person Te Awa Tupua may be entitled to human rights protections under the *New Zealand Bill of Rights Act 1990*, which under section 29 bestows certain human rights protections on legal as well as natural persons.[121]

The Act also sets up a number of other advisory, consultative and decision-making entities as part of a complex collaborative governance approach accounting for a number of river interests and not just those of Māori. These include Te Karewao, the advisory body to Te Pou Tupua, made up of local indigenous representatives and relevant local authorities.[122] The recommendations of Te Karewao help Te Pou Tupua determine how to spend the NZD $30 million fund (Te Korotete) in furtherance of the river's 'health and wellbeing'.[123] A separate strategy group (Te Kōpuka) comprises representatives with interests in the river that may include *iwi*, local authorities, government departments, commercial and recreational users, organisations and environmental groups.[124] With similar stakeholder involvement, Te Heke Ngahuru is the strategy group specific to the advancement of the river's wellbeing.[125]

How, exactly, the various governance arrangements will operate in practice remains unclear, as the surrounding policy framework has not been fully developed and the institutions are in their infancy. Yet, the 'institutional depth' of the Te Awa Tupua model suggests it may be more effective than comparable models from other jurisdictions arising out of judicial decisions,[126] like that of the Atrato River in Colombia

---

[118] *Te Awa Tupua (Whanganui River Claims Settlement) Act 2017* (NZ), s. 19.
[119] Ibid. s. 13.
[120] Ibid. (NZ), s. 13(c).
[121] *New Zealand Bill of Rights Act 1990* (NZ), s. 29.
[122] See *Te Awa Tupua (Whanganui River Claims Settlement) Act 2017* (NZ), ss. 27–8.
[123] Ibid. s. 57. Note that the Crown has contributed the initial funding but Te Pou Tupua can receive other funding, see s. 58(3).
[124] Ibid. ss. 29, 32(1).
[125] Ibid. s. 35.
[126] Erin O'Donnell and J Talbot-Jones, 'Creating Legal Rights for Rivers: Lessons from Australia, New Zealand, and India' (2018) 23(1) *Ecology and Society* 13.

considered in Chapter 6. Finlayson is 'confident' that the model will be successful, as both the river and the Whanganui Iwi are well funded to enable them to do their job properly, and great pains were taken to align the model with existing *Resource Management Act* processes.[127] There are thirty-four cross-references to the *Resource Management Act* in the *Te Awa Tupua Act*.[128] Decision makers under the *Resource Management Act* 'must recognise and provide for' the river's status and have 'particular regard' to the Tupua te Kawa values,[129] in language reminiscent of the *Resource Management Act's* Part II trilogy. However, Te Pou Tupua does not have a right to veto resource consents authorising use of the river. Te Awa Tupua is a 'public authority' under the *Resource Management Act*,[130] and the Whanganui Iwi Trustees have the right to be heard or make any submission on any matter relating to the river.[131] Their interests are recognised as 'greater than, and separate from, any interest in common with the public generally'.[132]

The *Te Awa Tupua Act* may also increase the likelihood that the river will be accorded status as an 'affected person' under the *Resource Management Act*, meaning that Te Pou Tupua will need to be notified about resource consent applications related to the river. This is not the case for the general public, provided the activity's adverse effects are minor or less than minor.[133] Interestingly, if Te Pou Tupua is deemed an affected person and consents to a development proposal, the consent authority cannot disregard any effect of that activity on the river,[134] unlike the case of affected persons who give consent in other cases.[135] This seeks to protect the river's interests over and above the guardianship of Te Pou Tupua. A major constraint, however, is that Te Awa Tupua's status and the Tupua te Kawa values are not necessarily determining factors in any decision about the river or its use.[136] Ultimately, the *Te Awa Tupua Act* is ordinary legislation and the basic framework of the *Resource Management Act* continues to

---

[127] Interview with Chris Finlayson (Wellington, 4 October 2018), above n. 36.
[128] This also includes the legislation specified under Schedule 2 of the *Te Awa Tupua (Whanganui River Claims Settlement) Act 2017* (NZ).
[129] *Te Awa Tupua (Whanganui River Claims Settlement) Act 2017* (NZ), s. 15.
[130] Ibid. s. 17(c).
[131] Ibid. ss. 72(d), 73. Note that Whanganui Iwi Trustees are representatives under the post-settlement governance entity, distinct from Te Pou Tupua.
[132] Ibid. s. 72(d).
[133] *Resource Management Act 1991* (NZ), s. 95E.
[134] *Te Awa Tupua (Whanganui River Claims Settlement) Act 2017* (NZ), s. 63.
[135] *Resource Management Act 1991* (NZ), ss. 95D(c), 104(3)(a)(ii).
[136] *Te Awa Tupua (Whanganui River Claims Settlement) Act 2017* (NZ), s. 15(5).

determine the process for decision making, which, as explained above, has often been applied to the detriment of Māori.

In essence, the *Te Awa Tupua Act* recognises *tikanga* Māori as principles for decision making about the river, and enhances Māori input into planning and decision making for the river, thereby providing a place for the *iwi* and *hapū* with interests in the river to exercise their indigenous jurisdiction. Sanders has argued that the Te Awa Tupua model is essentially a 'constitution', which restructures the relationship between the Crown and Māori,[137] although remaining outside of New Zealand's core constitutional documents.[138] For Whanganui Iwi, the Act protects the river's interests, to which they belong (as opposed to rights they have over the river), exemplified in their *whakataukī*, 'I am the river, and the river is me'.[139]

However, a major limitation of the *Te Awa Tupua* framework is its complete failure to engage with *iwi* demands for substantive rights to own the river territory, which remains a significant grievance for Whanganui iwi'.[140] Whanganui Iwi interests in the river, and as interpreted by the Waitangi Tribunal, have always been framed in the nature of 'ownership' of river territory. Te Awa Tupua is at once an 'indivisible and living whole', yet it is still to be carved up into various rights and regulatory regimes for its bed, water and wildlife.[141] The *Te Awa Tupua Act* vests the Crown-owned parts of the bed of the river in Te Awa Tupua;[142] however, this vesting does not transfer any rights in the water to the river. The vesting does not affect public rights of use, fishing or navigation, and nor does it interfere with private resource consents and permits for water use.[143] The hydroelectric power generator's rights are unaffected by the Act under a protection of the rights of State-Owned Enterprises and Mixed Ownership (partially privatised) Model companies.[144] The failure to address Māori substantive water claims in the Whanganui River model raises serious questions about distributive justice, and the ability for Māori to influence decision making, given that Genesis Energy Ltd, the operator of the

---

[137] Sanders, Chapter 3 above n. 64, 25.
[138] Unlike most countries, New Zealand does not have one formal written constitution, and constitutional norms can be found across a range of legislation and common law principles.
[139] *Te Awa Tupua (Whanganui River Claims Settlement) Act 2017* (NZ), s. 13(c).
[140] Ibid. s. 69(15), (16).
[141] Macpherson and O'Donnell, Chapter 4 above n. 21.
[142] *Te Awa Tupua (Whanganui River Claims Settlement) Act 2017* (NZ), ss. 41, 46.
[143] Ibid. s. 46.
[144] Ibid. s. 46(2)(c).

Tongariro Hydroelectric Power Scheme based on the river, and a member of Te Kōpuka, continues to divert 82 per cent of the river's headwaters for its operations.[145] Exactly how much power Te Pou Tupua has to agitate for the river among competing users is debateable, where the river itself is not a water rights holder and Te Pou Tupua is not a decision maker with respect to water. There is yet to be any litigation over the legislation, however, there is clearly potential for conflict over water sharing among competing users and, as a legal person, Te Awa Tupua enjoys liabilities as well as rights. A 'silver lining' may remain in the fact that the Act similarly has no impact on any 'existing private property rights, including customary rights and title',[146] leaving open the possibility of a native title claim to the water in the river in the future.

5.4 MĀORI CLAIMS TO WATER 'OWNERSHIP'

While the Crown has focused on settling Māori water grievances through cultural redress models involving 'co-management', vesting of beds, and now the construct of the 'legal person',[147] Māori continue to emphasise their ongoing rights to a 'fair share' of substantive water rights, which may amount to rights of 'ownership'.[148] As Counsel for the New Zealand Māori Council explains, the Crown's existing approach to increasing *iwi* involvement in water planning via collaborative governance models is a 'band aid' approach, and while the Crown sees developments like co-management of rivers and the *new mana whakahono ā rohe agreements to* be 'significant improvements',[149] they push the issue of 'proprietorship' to the back of the queue.[150] Māori demands for recognition of their proprietary rights in water have, until recently, been met with strict 'bottom lines' from New Zealand governments, including the legal position that 'no one can own water'.

---

[145] O'Donnell and Talbot-Jones, above n. 126, 10.

[146] *Te Awa Tupua (Whanganui River Claims Settlement) Act 2017*, s. 46(2)(b).

[147] Jacinta Ruru, 'Māori Rights in Water – the Waitangi Tribunal's Interim Report' [2012] *Māori Law Review* http://maorilawreview.co.nz/2012/09/maori-rights-in-water-the-waitangi-tribunals-interim-report/.

[148] See Interview with Riki Ellison (Wellington), above n. 69. Ellison has worked with Māori water rights since 1997, been a representative on the Land and Water Forum and most recently the Kahui Wai Māori.

[149] Crown Law, 'Closing Submissions on Behalf of the Crown in the National Freshwater and Geothermal Resources Inquiry' (20 November 2018) https://forms.justice.govt.nz at [19.4].

[150] Interview with Richard Fowler (Wellington, 17 August 2018). See also Taihakurei Durie, 'Ngā Wai o Te Māori: Ngā Tikanga Me Ngā Ture Roia. The Waters of the Māori: Māori Law and State Law' (Paper Prepared for the New Zealand Māori Council)' (2017), paras. 199, 200.

However, such rhetoric clouds the legal reality that water rights are of immense practical and economic value,[151] have been vested without consultation or compensation in the Crown and are unfairly distributed throughout the country.[152]

The idea of a Māori right of 'ownership' in water has a patchy history in New Zealand law. In 1929, Acheson J discussed the ownership of *Lake Omapere* in customary law, based on traditional occupation and use:[153]

> the usual signs of ownership would be the unrestricted exercise of fishing rights over it, the setting up of eel weirs at its outlets, the gathering of raupo or flax along its borders, and the occupation of villages or fighting pas close to its shore.

The Waitangi Tribunal traditionally observed that a Māori right of ownership (reflected in the promise of *tino rangatiratanga*) may mean 'control' over water resources.[154] Outside of the water context, the Tribunal in its *Ko Aotearoa Tenei* (Wai 262) Report made explicit that a right to control or manage is potentially severable from ownership:[155]

> [Article 2 of the Treaty] guarantees rights in the nature of ownership, the Māori text uses the language of control – tino rangatiratanga – not ownership. Equally, kaitiakitanga – the obligation side of rangatiratanga – does not require ownership. In reality, the kaitiakitanga debate is not about who owns the taonga but who exercises control over it.

The Tribunal has acknowledged that individual ownership may not have been possible in *tikanga* Māori for many *taonga* (treasures), however, individual ownership might be appropriate in certain cases.[156] As will be discussed below, the Tribunal has most recently framed Māori rights to water in *tikanga* as 'more than ownership'.

---

151 Interview with Riki Ellison (Wellington), above n. 69.
152 Durie, above n. 152 para. 198.
153 *Lake Omapere* [1929] 11 MB BI 253 MLC 262–3.
154 For Tribunal reports relating to water bodies, see, e.g., The Waitangi Tribunal, *Report on the Kaituna River Claim (WAI 4)* (Department of Justice, 1984); The Waitangi Tribunal, *Report on the Manukau Claim (WAI 8)* (Department of Justice, 1985); The Waitangi Tribunal, *The Mohaka River Report (WAI 119)* (Department of Justice, 1992); The Waitangi Tribunal, *Te Ika Whenua Rivers Report (WAI 212)* (GP Publications, 1998); Waitangi Tribunal, *Whanganui River Report*, above n. 101.
155 Waitangi Tribunal, *Ko Aotearoa Tēnei (WAI 262, Volume 1)*, above n. 8, 112.
156 Waitangi Tribunal, *Stage 1 Report on National Freshwater and Geothermal Resources*, above n. 34, 69.

As pointed out by Ruru, it is entirely arguable that Māori *iwi* and *hapū* could make out a native title-based claim to water, although such a question is yet to come, directly, before the courts.[157] So much was left open by the Court of Appeal in the *Ngati Apa* case on the foreshore and seabed in 2003.[158] Although the country contexts differ, it is clear in Australia, as discussed in the previous chapter, that native title can apply to water as well as land, depending on the traditional laws and customs of any particular group concerning any particular resource. As Ruru explains, water and land are typically seen by Māori as one holistic entity, so Maori Māori would not have been aware of any distinction between rights to land and water at the time when the Crown assumed sovereignty. Neither the statutory vesting of water in the Crown, nor the legal treatment of water as a public or common good by the common law, removes the potential for native title to be made out.[159]

However, because of the Crown's unwavering position that 'no one can own water', the prospect of pursuing a native title claim to water faces considerable uncertainty.[160] The Australian study in Chapter 4 has demonstrated, in any event, the limited utility of native title in responding to indigenous water injustice. Any native title claim to water in New Zealand would still need to satisfy a requirement for continuity of connection between people and resource, something often difficult to make out after centuries of colonisation and settlement, and adjust to the fact that many water resources are fully allocated to others. All of these issues, point squarely to the need for the New Zealand Government to consider some form of comprehensive water settlement with Māori.

Māori claims to water ownership were brought to a head in 2011, when the re-elected National Government proposed to partially privatise certain state assets, including three hydro-electric power generators. The asset sale triggered an action for judicial review by the New Zealand Māori Council,[161] and an urgent application to the Waitangi Tribunal. The claimants argued that the alienation of shares to private

---

[157] Jacinta Ruru, 'Māori Legal Rights to Water: Ownership, Management, or Just Consultation?' [2011] *Resource Management Theory & Practice*; Ruru, 'Undefined and Unresolved: Exploring Indigenous Rights in Aotearoa New Zealand's Freshwater Legal Regime', above n. 10; interview with Jacinta Ruru (Christchurch, 7 December 2018), above n. 81.

[158] *Ngati Apa* v. *Attorney-General* (2003) 3 NZLR 643 [89].

[159] Section 354 of the *Resource Management Act 1991* (NZ) states that the new rules of the Act shall not affect any rights that existed prior to the Act's passage in 1991.

[160] O'Bryan, Introduction above n. 19, 170.

[161] *New Zealand Māori Council* v. *Attorney-General* [2013] NZSC.

owners who were not bound by the Treaty would prejudice them, by removing the Crown's ability to provide relief to water claims in the future, in the way of shares, governance rights or royalties.

The Waitangi Tribunal, in its State One Report on the National Freshwater and Geothermal Resources Claim recommended that the Crown should delay the sale until an agreement had been reached with Māori, and urged the Crown to call an urgent national *hui* (gathering) to discuss the issue with Māori. The Tribunal considered that the sale of shares was detrimental to Māori interests in water and the Crown's ability to provide a remedy in the future, constituting a Treaty breach.[162] Importantly, the Tribunal during State One provided in-depth analysis of the nature and extent of Māori water interests. The Tribunal found Māori water rights to be equivalent to the common law concept of 'ownership' as follows:[163]

> Our generic finding is that Maori had rights and interests in their water bodies for which the closest English equivalent in 1840 was ownership rights, and that such rights were confirmed, guaranteed, and protected by the Treaty of Waitangi, save to the extent that there was an expectation in the Treaty that the waters would be shared with the incoming settlers.

The Tribunal found that water bodies were *taonga* (treasures) over which *hapū* or *iwi* exercised '*tino rangatiratanga*' (highest authority or chieftainship) and customary rights in 1840, and with which they had a physical and metaphysical relationship under *tikanga* Māori.[164] In fact, the Tribunal reasoned, Māori interests in water, as *rangatiratanga*, were 'more than ownership', although ownership was the nearest western equivalent.[165] It characterised the nature and extent of the Māori proprietary right to water as an 'exclusive right to control access to and use of the water while it was in their *rohe*' (emphasis added – meaning area of influence) and set out a list of 'indicia of ownership' to evidence such rights.[166]

Significantly, the Tribunal also found at Stage One that the Māori right to water included a right to develop the resources and benefit from them commercially, in line with international law around the indigenous 'right to development'. The Tribunal reasoned:[167]

---

[162] Waitangi Tribunal, *Stage 1 Report on National Freshwater and Geothermal Resources*, above n. 34, 143.
[163] Ibid. 81.
[164] Ibid. 75.
[165] Ibid. 76.
[166] Ibid. 81.
[167] Ibid. 35.

The interested parties emphasised the Treaty right of development and the choice of Māori to work in two worlds: to resist assimilation and protect their mātauranga Māori and tikanga (knowledge and law) but also to benefit commercially from development, as guaranteed by the Treaty and affirmed by the United Nations Declaration on the Rights of Indigenous Peoples.

The Crown maintained its 'bottom lines' before the Tribunal: that '(1) no-one owns fresh water, including the Crown; (2) there will be no generic share of freshwater resources provided for *iwi*; (3) there will be no national settlement of *iwi/hapū* claims to freshwater resources; (4) freshwater resources need to be managed locally on a catchment-by-catchment basis within the national freshwater management framework; and (5) the next stage of freshwater reform will include national-level tools to provide for *iwi/hapū* rights and interests'.[168] Critically, the Crown was opposed to a national water settlement, or 'Waterlords Deal' reminiscent of national settlements for fisheries ('Sealords Deal') or Crown forests ('Treelords Deal'). The Crown's 'preferred option' to resolve Māori claims was to enhance the role of Māori in water governance and management as cultural redress,[169] through settlements like that for the Whanganui River. The Tribunal, on the other hand, considered the Crown's 'preferred option' to be insufficient, because it does not recognise a residual property right from which Māori could obtain a commercial benefit, pursuant to their (Treaty) right to development.[170]

The Crown went ahead with the hydro asset sales, and the New Zealand Māori Council applied to the Court for judicial review relying on the Treaty provision in section 45Q of the *Public Finance Act 1989*.[171] The Supreme Court ultimately found that the partial privatisation would not undermine the possibilities for the Crown to remedy water claims in the future.[172] However, the Court emphasised the need for the Government to resolve the issue of Māori water rights as follows:[173]

---

[168] Ministry for the Environment, 'Briefing to the Incoming Minister for the Environment' (December 2017) www.beehive.govt.nz/sites/default/files/2017–12/Water.pdf [61].

[169] Waitangi Tribunal, *Stage 1 Report on National Freshwater and Geothermal Resources*, above n. 34, 136.

[170] Ibid.

[171] This was an application of an earlier case about the privatisation of Crown assets in *New Zealand Māori Council v Attorney-General* [1987] 1 NZLR 641.

[172] *New Zealand Māori Council v. Attorney-General* [2013] NZSC [149]–[150].

[173] Ibid. [148].

This is not to say that the water claims should be parked. The Waitangi Tribunal has emphasised the need for urgency in addressing proprietary claims. It appears from the policy initiatives and from the assurances given in the litigation that the message that there is need for action on these claims has been accepted.

5.5 THE FUTURE OF MĀORI WATER RIGHTS IN AOTEAROA NEW ZEALAND

Despite New Zealand's growing water worries, neither the National Government nor the subsequent 2017 Labour-led coalition have been able to secure any comprehensive freshwater reform. The former National Government foreshadowed changes to the water allocation system towards increasing use of water trading. This position was based on the advice of a 'Land and Water Forum', a large stakeholder group including businesses, environmental NGOs, *iwi* representatives, and scientists, which endorsed a move to a market-based allocation system. At present, water transfers occur only within limited frameworks such as hydro-transfer forums, or informal local resource consent-holder networks.[174] The Land and Water Forum maintains that the current policy does not realise the economic value of water at its highest potential, and has recommended that resource consents should be flexible so that water is used for its most efficient purposes.[175] The Forum also recommended that the government introduce a market-based transfer system for discharge contaminants on the same principles.

Any wholesale water reform would, of course, require the government to resolve outstanding Māori water claims. The Land and Water Forum proposed a number of mechanisms for Māori interests and rights in water involving input from a Freshwater Iwi Leaders Group.[176] These included: priority access to unallocated water (or discharge); commercial arrangements with consent holders; a set quantum or share of the consumptive pool of water (potentially a quota management system like the Sealord's Deal);[177] and freeing

---

[174] The Land and Water Forum, 'The Fourth Report of the Land and Water Forum' (2015) 62 www.landandwater.org.nz/Site/About_Us/default.aspx.
[175] Ibid. 5.
[176] See *Iwi Chairs Forum – Fresh Water* https://iwichairs.maori.nz/our-kaupapa/fresh-water/.
[177] *The Māori Fisheries Act 2004* (NZ) implemented the 1992 Fisheries Claims Settlements or 'Sealord deal' providing for Māori quota rights to fisheries in a market-based model for commercial fishing rights.

up water from current allocations with better practices.[178] Research done for the Iwi Leaders Group, citing the success of the Murray Darling Basin in Australia, called for an 'equitable, permanent share of water allocated for commercial use'.[179]

Stage Two of the Waitangi Tribunal's National Freshwater and Geothermal Resources Inquiry, heard in 2018, considered how Māori rights of water ownership can be best accommodated in legal and policy frameworks, including the *Resource Management Act*. Speculation remains about the possibility of a national comprehensive 'Waterlords Deal' comprising commercial and cultural redress. In closing submissions before the Tribunal, in November 2018, Counsel for the Freshwater Iwi Leaders Group noted the difficulty of advancing freshwater reform because of 'politics getting in the way'.[180] They continue to pursue reform via direct dialogue with the Crown at the highest level, rather than litigation.[181] Counsel for the New Zealand Māori Council argued for protection of both Māori cultural water rights through co-management approaches and proprietary water rights via, potentially: a redistributive model (referred to as 'reverse grandparenting'); some sort of quota management system (perhaps like the Sealord's Deal); royalties or taxes for commercial water use; or cash compensation.[182] The Māori Council's barrister also argued for the establishment of an independent Water Commission, made up of at least half Māori membership, to manage and regulate water in New Zealand.[183]

Settling Māori water grievances will be no easy task.[184] Some catchments are already over allocated; third parties who hold commercially valuable water rights may be opposed to a Māori water settlement; there will be difficult questions to answer around the nature, content and duration of a Māori allocation,[185] who might

---

[178] The Land and Water Forum, above n. 174, 42.
[179] Marcus Sin, Kieran Murray and Sally Wyatt, *The Costs and Benefits of an Allocation of Freshwater to Iwi* (Sapere Research Group, 2014) 7 www.iwichairs.maori.nz/Kaupapa/Fresh-Water/.
[180] Kahui Legal, 'Closing Submissions on Behalf of the Freshwater [Te Pou Taiao] Iwi Leaders Group in the National Freshwater and Geothermal Resources Inquiry' (14 November 2018) https://forms.justice.govt.nz at [14].
[181] Ibid. at [17], [18].
[182] See Woodward Law, Capital Chambers & Thorndon Chambers, 'Closing Submissions on Behalf of the Claimant (New Zealand Māori Council) in the National Freshwater and Geothermal Resources Inquiry' (26 October 2018) 22 https://forms.justice.govt.nz.
[183] See ibid. 22, 24.
[184] Interview with Tania Gerrard (Wellington, 28 August 2018), above n. 82.
[185] Interview with Riki Ellison (Wellington), above n. 69.

hold and manage a Māori water allocation at a national, *iwi* or *hapū* level, and whether water allocations should be linked in some way to landholding, or be tradeable;[186] and some relationships between Māori and regional authorities are more contentious than others.[187] Former Minister of Treaty Settlements, Christopher Finlayson, suggested that in order to protect existing third-party rights, one option may be to 'shave' a Māori allocation for economic development purposes off water resource consents as they are renewed at the end of their thirty-five year term.[188] Others ask why should Māori wait so long for distributive justice, when other users hold immensely valuable resource consents, allocated free of charge, with no regard for the Treaty.[189]

In the lead-up to the 2017 general election, the Labour party proposed a water royalty on commercial water users (like irrigators and water bottling operators) to help resource the restoration of degraded waterways,[190] raising further questions as to the effect that this would have on Māori claims to ownership.[191] The elected Labour-led coalition Government has continued with the *Next Steps for Freshwater* review of water law and policy, although, in the complicated political context, the government has struggled to reach an agreed policy position on water reform. There is increasing frustration around the inability or unwillingness of the government to resolve issues of Māori water 'ownership' or devise a comprehensive water settlement.[192]

In 2018, the Labour Government released 'Shared Interests in Freshwater: A New Approach to the Crown/Māori Relationship for Freshwater',[193] in which it appears to have softened its resistance to Māori water claims and finally agreed to abandon the five

---

[186] Ibid.

[187] Interview with Baden Vertongen (Wellington, 28 August 2018), above n. 110.

[188] Interview with Chris Finlayson (Wellington, 4 October 2018), above n. 36.

[189] Interview with Riki Ellison (Wellington), above n. 69.

[190] See The New Zealand Labour Party, *Clean Water for Future Generations* (2018) www .labour.org.nz/water.

[191] The New Zealand Labour Party, 'Manifesto – Water Policy' (2017) 9. In 2017, Labour proposed a manifesto with a twelve-point plan for Improving Freshwater Quality. Within that manifesto, Labour state they 'will seek to resolve the issue of Māori rights and interests in freshwater'.

[192] Interview with Richard Fowler (Wellington, 17 August 2018), above n. 150. Fowler is legal counsel for the New Zealand Māori Council in the National Freshwater and Geothermal Resources Inquiry.

[193] Ministry for the Environment and Māori Crown Relations Unit, 'Shared Interests in Freshwater: A New Approach to the Crown/Māori Relationship for Freshwater' (2018).

'bottom lines' in favour of a more constructive approach. Under the new approach, the government continues to consult with the Iwi Leaders Forum but has appointed another consultative group to discuss Māori water issues, called 'Kahui Wai Māori', with broader representatives including the New Zealand Māori Council and other Māori land corporations and *hapū* groups.[194] While the Crown continues to maintain that 'no one owns water', the new approach provides:[195]

- 'The Crown and Māori have a key shared interest in improving the quality of New Zealand's freshwater, including the ecosystem health of our waterways.
- The Crown and Māori have a shared interest in ensuring sustainable, efficient, and equitable access to and management of freshwater resources.
- No one owns freshwater – it belongs to everyone, and we all have a guardianship role to look after it.
- The Crown acknowledges that Māori have rights and interests in freshwater, including accessing freshwater resources to achieve their fair development aspirations for underdeveloped land.
- The Crown acknowledges that existing users also have interests that must be considered.'

Cabinet has agreed to take a 'phased approach' to Māori freshwater concerns, starting with a focus on water-quality issues rather than water allocation, and then engaging on broad policy parameters regarding Māori desires for access to freshwater resources to allow development of under-developed land.[196] The Crown's expressed preference in response to the 'issue of allocation' is to 'find a mechanism to more equitably share the resources over time through a "regulatory" route: in scarce catchments this proposal could require the generation of "headroom" between the total allocated quantum of "use rights" and the sustainable limit in order to give Māori (and other new users) the opportunity to obtain a share of those use rights'.[197] The new approach appears to

---

[194] See Ministry for the Environment, *Kahui Wai Māori Group to Work on Freshwater Press Release* (3 August 2018) www.beehive.govt.nz/release/kahui-wai-m%C4%81ori-group-work-freshwater.

[195] Ministry for the Environment and Māori Crown Relations Unit, above n. 193, 7.

[196] Ministry for the Environment and Māori Crown Relations Unit, above n. 193. See paras. 8–9 of the Cabinet paper attached to the policy.

[197] Ibid. See paras. 10–11 of the Cabinet paper attached to the policy.

recognise the need to maintain and provide certainty for existing water users and the public interest, but that Māori have a shared interest in water governance, guardianship and distribution, including for economic development. Yet as Riki Ellison points out, the Crown and Māori alike face a huge challenge, in terms of how to transition from an existing, broken, water allocation framework to one that will adequately address Māori concerns.[198]

5.6 CONCLUSION

Cutting a path to water reform will be immensely difficult for the New Zealand Government, requiring trade-offs to be made between cultural, environmental and commercial interests. What is clear is that Māori are seeking both: (a) recognition of their distinctive water relationships and influence and control over water governance as indigenous water jurisdiction; and (b) a substantive distribution of the consumptive pool of water for any purpose including economic development. These two approaches appear (at least at first sight) to have different ends, one being focused on procedural rights to recognise and provide for indigenous water relationships and management as *kaitiaki* (guardians), and the other on substantive rights to water distribution, *tino rangatiratanga* or 'ownership'. However, both tendencies signify a desire to reconstitute the relationship between the Crown and Māori in order to recognise a space for Māori as both water managers and water owners. The *Te Awa Tupua Act* is likely to remain the gold standard for water settlements that attempt to provide such jurisdiction as cultural redress, although changes to the *Resource Management Act*, including provision for mana whakahono ā rohe agreements may also provide opportunities for greater Māori authority and control.[199] However, legal person models are *ad hoc* and situation-specific. Why is the Whanganui River alone, and not other sacred and threatened rivers in Aotearoa New Zealand, accorded this special status? If the Crown is committed to redressing the historical injustice Māori have experienced by being excluded from water law frameworks, it

---

[198] Interview with Riki Ellison (Wellington), above n. 69.
[199] Incidentally, mana whakahono ā rohe agreements between local authorities and *iwi/hapū* under the *Resource Management Act* could themselves envisage the recognition or treatment of resources as 'legal persons'. In the United States, rights for nature have been provided for most successfully at municipal level.

must engage with Māori substantive and commercial water rights claims. The study of indigenous rights to water in Aotearoa New Zealand in this chapter shows the variability of indigenous water demands, and a need for multifaceted responses to indigenous water exclusion.

CHAPTER SIX

RIVERS AS SUBJECTS AND INDIGENOUS WATER RIGHTS IN COLOMBIA

6.1 INTRODUCTION

It might seem unlikely for Colombia to be included in a book about novel responses to indigenous water exclusion, after experiencing decades of civil war, drug trade, and economic and political instability. Yet, in a landmark decision of November 2016, following the example of the Whanganui River (*Te Awa Tupua*) in Aotearoa New Zealand, Colombia's highest court declared that the Río Atrato (Atrato River) is an *'entidad sujeto de derechos'* ('legal subject'). This 'watershed' case was an action for protection of constitutional rights brought in the Colombian Constitutional Court on behalf of a number of indigenous and afrodescendent communities,[1] in response to serious environmental and humanitarian damage caused by illegal mining in the region of Chocó.

Although legal rights for rivers have developed in Colombia via the courts rather than the parliament, the decision has been welcomed by the government, and is currently in a detailed process of implementation. The Río Atrato case must be considered in its context, which is characterised by extreme social, cultural and environmental degradation and suffering. However, the Colombian experience reveals

---

[1] I have referred to the Afrocolombian communities as 'tribal', because of their classification as such under international law, discussed further below.

invaluable lessons about the historical production of indigenous water exclusion and the objectives of legal person models.

What is clear from studying the Colombian experience is that legal rights for rivers are one way to respond to indigenous water injustice. Against a history of ignoring the concerns of indigenous and afrodescendent 'river communities', the Colombian Court attempted to recognise and provide for their water relationships by enabling their participation in bespoke collaborative water governance arrangements. The Constitutional Court acknowledged that the perspectives and needs of indigenous and tribal peoples have thus far been overlooked in the distribution of decision-making powers with respect to the river, with catastrophic consequences. Faced with this predicament, the Court found a new way to recognise and provide for indigenous and tribal water rights in partnership with the state, thereby beginning to repair the state's relationship with the river communities. With a significant amount of judicial activism, the Court looked abroad to Aotearoa New Zealand, where legal person models have been used in an attempt to recognise distinct Māori relationships with natural resources, pursuant to reparative settlements with the Crown.

The Colombian case is 'culturally located', in the sense that river rights are a consequence of the recognition of the rights of indigenous and tribal peoples as river communities.[2] While legal rights for rivers are presented in the dominant literature as an 'earth-centred' model, the Colombian case shows how human rights norms are co-opted and adapted to confer legal personality on the river, as 'biocultural rights'. Legal personality is deemed necessary, not simply as recognition the intrinsic value of nature, but as a necessary condition for the exercise of 'the third generation of human rights'.[3]

6.2 INDIGENOUS EXCLUSION FROM WATER LAW FRAMEWORKS IN COLOMBIA

6.2.1 Indigenous Peoples in Colombia

Like all the countries considered in this book, the indigenous peoples of Colombia have suffered widespread dispossession and disenfranchisement since the time of colonisation, including the loss of traditional

---

[2] Macpherson and Clavijo Ospina, Felipe, Introduction above n. 19, 284.
[3] *Tierra Digna* [2016] Corte Constitucional [Constitutional Court], Sala Sexta de Revision [Sixth Chamber] (Colombia) No T-622 of 2016 (10 November 2016).

territories and disrupted access to water resources. The evidence of indigenous occupation of the land now known as Colombia is estimated to go back more than 20,000 years.[4] Prior to the arrival of the Spanish, a number of diverse indigenous peoples lived in Colombia with their own distinct cultures, languages and laws.[5] During this time the indigenous peoples exercised full sovereignty over their territories, lands and natural resources, which they held pursuant to common property regimes,[6] under customary law.

The imperial relationship with Colombia began in 1499, when it was 'discovered' by Alonso de Ojeda. Active colonisation by the Spanish Empire occurred from the mid-sixteenth century. Since its early colonisation Colombia has been regarded as a nation rich in natural resources, including plants, minerals and rivers.[7] Colombia was attractive to the colonisers because of its mineral wealth, in particular the presence of gold, with deposits in major rivers, like the Río Atrato in the region of Chocó, earning it much renown and a mythical status as a land of hidden treasure or 'El Dorado'.[8]

One of the first Spanish settlements in Latin America was in the Chocó. In 1510, Santa María la Antigua del Darién was founded in the Gulf of Urabá. The colonisers founded other settlements along the Caribbean coast at Santa Marta (1526) and Cartagena (1533) before expanding into the interior to Santa Fe de Bogotá (1538) via successive 'entradas' or 'incursions'.[9] Santa Fe would become the capital of the Nuevo Reino de Granada, formed in 1539.

After the arrival of the Spanish, the conquistadors began to encroach on indigenous territories. Colonisation brought with it the typical scourge of disease, violence and dislocation.[10] Indigenous territories were enclosed within the 'encomienda' system, pursuant to which indigenous peoples were confined and required to pay 'tribute' (gold) to colonial settlers ('encomenderos') who were themselves given 'mercedes

---

[4] *Colombia: A Country Study* Library of Congress, Washington, DC 20540 USA 4 www.loc.gov /item/2010009203/.
[5] Ibid. 5, including notably two Chibcha groups, the Taironas (based in the Sierra Nevada of northern Colombia) and the Muiscas (in the area now encompassing the capital city of Bogotá).
[6] See Edella Schlager and Elinor Ostrom, 'Property-Rights Regimes and Natural Resources: A Conceptual Analysis' (1992) 68(3) *Land Economics* 249 for a discussion of common property regimes.
[7] Oscar Darío Amaya Navas, *La Constitución Ecológica de Colombia: Análisis Comparativo con el Sistema Constitucional Latinoamericano* (Universidad Externado de Colombia, 2002) 37–9.
[8] Caroline Williams, *Between Resistance and Adaptation: Indigenous Peoples and the Colonisation of the Chocó, 1510–1753* (Liverpool University Press, 2004) 10.
[9] Ibid. 10–11.
[10] Ibid. 10–14.

de tierra' (land concessions) by the Spanish Crown.[11] Tribute was typically paid via a system of enforced indigenous labour and, in return, the indigenous peoples received the colonist's protection and the 'civilising' impact of Christianity.[12] The *encomiendas* were later discontinued and tribute paid as tax to the Spanish Crown rather than individual landholders, but the idea of indigenous occupation and foreign protection was replaced by the idea of the '*resguardo*', or reservation. *Resguardos* allowed indigenous peoples to remain on a portion of their ancestral lands in collective ownership, again for their own protection and control, although some indigenous *resguardos* were overseen by indigenous councils called '*cabildos*'.[13]

The territory now known as Colombia first obtained independence from the Spanish Crown in 1810, under the leadership of the liberator Simon Bolivar. The *Colombian Act of Confederation 1811* classified vast area of lands considered isolated and uninhabited as *terra nullius*, in much a similar way as occurred in Australia.[14] However, the Act clarified that the status as *terra nullius* did not mean that 'wandering tribes' or 'barbarous indians' in those territories could be removed, but should be 'respected as legitimate and ancient owners' whilst noting the benefit they would enjoy from civilisation, religion and commerce, provided they did not, by hostility, provoke other treatment.[15] In fact the 'civilisation' of the indigenous and the freeing up of perceived wastelands continued to be a key concern for Colombian governments throughout the 1800s.[16]

In 1820, Bolivar issued a decree protecting the collectively owned *resguardos*, although a subsequent law abolishing the requirement for tribute and entitling indigenous peoples to equal protection before the

---

[11] Comisión Económica para América Latina y el Caribe, *Los pueblos indígenas en América Latina. Avances en el último decenio y retos pendientes para la garantía de sus derechos. Síntesis* [Indigenous Peoples in Latin America. Advances in the last decade and current objectives for the guarantee of their rights] (27 October 2014) 24 www.cepal.org.

[12] Maria Clara van der Hammen, 'The Indigenous Resguardos of Colombia: Their Contribution to Conservation and Sustainable Forest Use' (Netherlands Committee for IUCN The World Conservation Union, NC-IUCN / GSI Series 1) 12 cmsdata.iucn.org/downloads/the_indigen ous_resguardos_of_colombia.pdf; *Colombia: A Country Study*, above n. 4, 9; Caribe, above n. 11, 23–4.

[13] van der Hammen, above n. 12, 13; *Colombia: A Country Study*, above n. 4, 11.

[14] *Acta de Confederación de la Provincias Unidas de la Nueva Granada* [Act of Confederation of the United Provinces of New Granada] *1811* (Colombia) ('*Colombian Act of Confederation*'), art. 23.

[15] Ibid. art. 24.

[16] See *Constitución de Rionegro 1863* [Constitution of Rionegro 1863] (Colombia) ('*Constitution of Rionegro (Colombia)*'), art. 78, which provided that sparsely populated indigenous areas would be regulated by the government with the object of supporting colonisation and improving conditions.

law also led to further dispossession.[17] In this period many *resguardos* were distributed among private indigenous owners, as well as settlers and local authorities. Their indigenous occupants, now often landless peasants, abandoned traditional territories and were assimilated into creole Colombia.[18] However, some *resguardos*, particularly in remote areas, remained in communal ownership and were amplified under policies responding to the pan-American indigenous rights movement of the 1960s and 70s.

Today, Colombia has the largest indigenous territories in the world, with as many as 710 *resguardos* covering approximately 24 per cent of the country's territory.[19] The *resguardos* are protected in article 329 of the *Constitución Política de Colombia* (Political Constitution of Colombia or 'Colombian Constitution'), which provides that '[t]he *resguardos* are collective property and inalienable'.[20] Article 330 of the *Constitution* explains that the indigenous territories are governed by indigenous '*consejos*' (or boards), acting in accordance with community uses and customs, which have the following regulatory and planning functions:

1. To ensure the application of land use and settlement laws in their territories.
2. To design policies, plans and programmes for economic development within their territory, in harmony with the National Development Plan.
3. To promote public investments in their territories and ensure their proper execution.
4. To oversee and distribute resources.
5. To ensure the preservation of natural resources.
6. To coordinate programmes and projects promoted by different communities in the territory.
7. To support the maintenance of public order within the territory in accordance with the instructions and rules of the National Government.
8. To represent the territories before the National Government and other governmental entities, and
9. Such other purposes prescribed by the Constitution and the law.

---

[17] van der Hammen, above n. 12, 13–14.
[18] Ibid. 14.
[19] *Colombia: A Country Study*, above n. 4, 84 and 423.
[20] See also *Constitución Política de Colombia 1991*, art. 63.

With respect to the exploitation of natural resources in the indigenous territories, the Constitution requires that to be done 'without prejudice to the cultural, social and economic development of indigenous communities' and 'in decisions that are adopted with respect to said exploitation, the Government will promote the participation of the representatives of the respective communities'.

In fact, the *resguardos*, recognise indigenous jurisdiction to produce law and custom for self-government. This is reinforced in article 21 of *Decree 2164 1995*, which defines the *resguardo* as:[21]

> ... a legal and sociopolitical institution of special character, comprised of one or more indigenous communities, who with a collective title enjoys the guarantee of private property, possesses and manages its territory and internal life by an autonomous organisation protected by an indigenous jurisdiction with its own normative system.

In terms of people, according to the 2005 Census, more than 14 per cent of the Colombian population belongs to an 'ethnic group'.[22] Within this ethnic grouping 3.4 per cent identify as indigenous (1,392,623 people), and there are around ninety different indigenous 'peoples' in Colombia, distributed across its territory.[23] 10.5 per cent of the population (4,273,722) identifies as 'Afrocolombian', the other minorities being Raizal, Roma (gypsy) and Palenquero.[24]

The afrodescendent communities were brought to Colombia from Africa by the Spanish colonisers from the sixteenth century to boost the slave economy. The African slaves were first put to work extracting gold from the Santa María de la Antigua del Darién mine in northern Chocó in 1510.[25] Today, the Afrocolombian communities exist much like the indigenous, living off the land and taking only limited part in mainstream Colombian society. They also occupy reservations similar to the *resguardos,* called *'consejos mayores'.* *Consejos mayors* are also representative boards, adopting the concept of 'councils', or local authority areas. Afrocolombians, and afrodescendent communities in other parts of Latin America, are

---

[21] See also *Decree 2164 1995* (Colombia), art. 22.
[22] A new census was taken in 2018, although results about the indigenous population are yet to be released.
[23] Gobierno de Colombia, 'Una Nación Multicultural – Su Diversidad Étnica' (DANE – Departamento Administrativo Nacional de Estadísticas, May 2007) www.dane.gov.co/files/censo2005/etnia/sys/colombia_nacion.pdf.
[24] Ibid.
[25] Robert C West, *La Minería de Aluvión en Colombia Durante el Período Colonial* [Alluvial Mining in Colombia during the Colonial Period] (Imprenta Nacional de Colombia, 1972).

contemplated by the concept of 'tribal' peoples, under the International Labour Organisation's *Convention 169 Concerning Indigenous and Tribal Peoples in Independent Countries*.[26] At a domestic level, the Colombian Constitutional Court has recognised the special character of the afrodescendent communities (which they describe as 'multiethnic and multicultural') in equal status to any other ethnic community and as part of the social and ethnic diversity of the Colombian nation.[27]

6.2.2 Colombian Water Law Frameworks and Indigenous Rights

Despite the limited historical recognition of indigenous rights to terri-tory and self-government discussed above, Colombian law has, until very recently, ignored the issue of indigenous rights to water. When the Spanish Crown acquired territory in Latin America, including the lands now comprising Colombia, they brought with them the Spanish system of *regalias*, pursuant to which kings or lord granted concessions to use water, referred to as '*mercedes*', upon payment of a fee.[28] Since colonisation, a series of historical laws have been passed to regulate the use of water. The *Acta de Confederación de la Provincias Unidas de la Nueva Granada 1811 [Act of Confederation of the United Provinces of New Granada 1811]* (Colombia) was the first statute in Colombia to speci-fically address rights to water. Article 34 provided that a number of 'common goods' including rivers belonging to the provinces come within the domain of Congress. After a failed Bolivarian experiment for a *Gran Colombia* ('Great Colombia') encompassing New Granada, Venezuela and Ecuador in the early 1800s, a series of constitutions of New Granada and Colombia were enacted between 1832 and 1888.[29] Many of these constitutions vested responsibility for the management of inter-province canals and navigable rivers in the central government,[30] with a concern to facilitate navigation across the

---

[26] International Labour Organization, *Convention Concerning Indigenous and Tribal Peoples in Independent Countries* (opened for signature 27 June 1989, 1650 UNTS 383 (entered into force 5 September 1991) ('*ILO Convention 169*').

[27] *Consejo Comunitario Mayor Cuenca Río Cacarica* v. *The Ministry of Environment and Others*, No T-955 of 2003, Corte Constitucional [Constitutional Court], Sala Octava de Revision [Eighth Chamber] (Colombia) (17 October 2003) 81.

[28] See Alejandro Vergara Blanco, *Derecho de Aguas* [Water Law] (Editorial Jurídica de Chile, 1998), Vol. II, 309. See generally the discussion of *regalías* in the Chilean context in Chapter 7.

[29] Oscar Darío Amaya Navas, above n. 7, 48–62.

[30] *Constitución Política de la Nueva Granada 1853* [Political Constitution of New Granada 1853] (Colombia) ('*Colombian Constitution 1853*'), art. 10; *Constitución Política de la Confederación Granadína 1858* [Political Constitution of the Granadine Federation 1858] (Colombia) ('*Constitution of the Grenadine Confederation 1858*'), art. 15; *Constitution of Rionegro*

country to the sea by river.[31] None of these early laws acknowledged any pre-existing indigenous use of water, nor any particular need to recognise or provide for indigenous water access or management or compensate for their loss.

Today Colombia is divided into five diverse hydrographic areas, and water is unequally distributed and utilised throughout the country.[32] Colombia is traditionally understood, like Aotearoa New Zealand, as being hydro-rich, although water resources are subject to increasing stress from agriculture, industry and urbanisation and supply pressure from contamination and climate change. The contamination of water resources from poor wastewater management and sanitation is a particularly big problem in Colombia, and the Río Bogotá running through the capital is known to be one of the most polluted rivers in the world.[33] The diversity of water conditions throughout the country presents a particular challenge for regulators.

The main demand for water in Colombia, as in all of the countries studied in this book, is for agriculture and industry.[34] However, despite being of immense economic and strategic value, water also has social and cultural importance in Colombia.[35] The well-watered and demand-intensive regions adjoining the Magdalena River have been the subject of evocative idolatry both in folklore and contemporary literature, including the writings of Nobel Laureate Gabriel García Marquez.[36] Waters are certainly of particular cultural and economic importance to indigenous communities. There are, therefore, difficult decisions to be made about sharing water between productive and social, environmental and cultural uses and managing water quality.

Unlike other Latin American countries such as Chile, Colombia does not have a discrete water code, being managed instead as a component of constitutional environmental protections and associated laws.[37] The basis for contemporary water rights in Colombia is

---

(Colombia), arts. 17, 49; *Constitución Política de Colombia 1886* [Political Constitution of Colombia 1886] (Colombia), art. 185.

[31] See generally Oscar Darío Amaya Navas, above n. 7, 50–7.

[32] García Pachón, Introduction above n. 24, 23.

[33] See García Pachón, Introduction above n. 24; Luis Felipe Guzmán Jiménez, *Las Aguas Residuales en la Jurisprudencia del Consejo de Estado: Periodo 2003–2014* [Wastewater in the Jurisprudence of the Administrative Court of Colombia: 2003–2014] (Universidad Externado de Colombia, 2015), 18–20.

[34] García Pachón, Introduction above n. 24, 23.

[35] Gloria Amparo Rodríguez, Carlos Lozano Acosta and Andrés Gómez Rey, *Protección Jurídica del Agua en Colombia* (Grupo Editorial Ibáñez, 2011), 34–5.

[36] Gabriel García Márquez, *Love in the Time of Cholera* (Camberwell, Victoria Penguin, 2008).

[37] García Pachón, Introduction above n. 24, 31.

found in the *Código Civil Colombiano* ('Colombian Civil Code)';[38] a Code strikingly similar to Chile's Civil Code, having been based on the Chilean law and drafted by the same person (Andres Bello) on a Roman law foundation. With respect to water, the Colombian Civil Code adopts the Roman law principle that water is a '*bien de uso publico*' (public good), belonging to all the nation, although the State has the right of '*dominio*', to use, enjoy and dispose of water as it sees fit:[39]

> Rivers and other waters that flow in natural courses are property of the Union, for public use in their respective territories
> That is the case except for waterways that are born and die within the same property: their ownership, use and enjoyment belongs to riparian owners, and passes from the owners to their successors.

The Colombian system of water regulation can be characterised as a centralised system of government control,[40] where the government has powers of regulation and distribution (namely the *Ministerio del Medio Ambiente y Desarollo Sostentible* or 'Ministry for the Environment and Sustainable Development', and regional and local authorities) and water trading is not envisaged.[41] Principally, the *Natural Resources and Environmental Protection Code 1974* (Colombia) regulates the use of water. It confirms the public vesting of water, and recognises that all people have a right to take and use water to satisfy their basic needs, and those of their families and animals, provided they do not cause prejudice to other users.[42] This provision is similar to 'domestic and stock' rights in the Australasian context, discussed in Chapter 4. However, private use rights can be acquired pursuant to law,[43] by way of an administrative concession,[44] subject to water availability and intended

---

[38] *Código Civil* [Civil Code] (Colombia) 1887 ('*Colombian Civil Code*').

[39] Defensoría del Pueblo de Colombia, *Avance del Derecho Humano al Agua en la Constitución, la Jurisprudencia y los Instrumentos Internacionales 2005–2011* [The Advance of the Human Right to Water in the Constitution, Jurisprudence and International Instruments 2005–2011] (2012) 16. See Colombian Civil Code 1887, art. 674 for an explanation of 'bienes de la unión'.

[40] See María Cecilia Roa-García, Patricia Urteaga-Crovetto and Rocío Bustamante-Zenteno, 'Water Laws in the Andes: A Promising Precedent for Challenging Neoliberalism' (2015) 64 *Geoforum* 270 who discuss the difficulty of characterising the Colombian system as either centralised or neoliberal.

[41] Lawrence J MacDonnell and Neil S Grigg, 'Establishing a Water Law Framework: The Colombia Example' (2007) 32(4) *Water International* 662.

[42] *Código Nacional de Recursos Naturales Renovables y de Protección al Medio Ambiente (Decreto 2811 del 18 de Diciembre de 1974)* [National Code of Renewable Natural Resources and Protection of the Environment (Decree 2811 of 18 December 1974] (Colombia) ('*Natural Resources and Environmental Protection Code 1974* (Colombia)'] art. 86.

[43] Ibid. arts. 80, 85.

[44] Ibid. arts. 88, 51.

use.[45] In contrast to Chile, those who use water for profit are required to pay taxes to the government towards infrastructure and maintenance.[46] However, like the Chilean model, private water users within a catchment can organise as water user associations, offering an opportunity for local, private governance and influence.[47]

Some provision has been made in the Colombian system of water regulation to protect the water access of the vulnerable. Aside from the right to take water for domestic or stock needs under the *Natural Resources and Environmental Protection Code*, responsible authorities are required to prioritise water allocations for human and domestic use in *Decree 1541* of 1978.[48] However, Roa-Garcia and Urteaga-Crovetto estimate that approximately 70 per cent of small water users do not hold a water concession, relying instead on 'informal' or customary use, and despite intentions to protect water access by the poor and marginalised, the system has been appropriated by large elites reinforcing the exclusion of local and indigenous users and encroaching on their territories.[49]

Nowhere in the Colombian water laws is any specific provision made for indigenous peoples to use water on their territories. In fact, *Decree 2164 1995* (Colombia) explicitly provides that the creation, amplification or restructuring of an indigenous *resguardo* does not modify in any way the legal regime for the use of water.[50] Faced with limited regulatory tools within existing water law frameworks to protect indigenous water rights, indigenous and environmental activists have turned to environmental and human rights protections under Colombia's Constitution to project indigenous territorial rights, including the right to water.

The Colombian Constitution was passed in 1991, at the height of the Colombian drug wars. It is a progressive document, and is exceptional within the Latin American context, in spite of the obvious domestic challenges at the time.[51] The Colombian Constitution places an emphasis on social justice and human rights, and its founding principle is the idea of the *Estado Social de Derecho*, meaning a social welfare state based on the rule law, encompassing the guarantees of human dignity

---

[45] Ibid. art. 89.
[46] Ibid. art. 151.
[47] Ibid. art. 161.
[48] *Decree 1541 1978* (Colombia), art. 41.
[49] Roa-García, Urteaga-Crovetto and Bustamante-Zenteno, above n. 40, 274.
[50] *Decree 2164 1995* (Colombia), art. 24.
[51] See Roa-García, Urteaga-Crovetto and Bustamante-Zenteno, above n. 40, 274.

('*vida digna*') and common welfare ('*bienestar general*').[52] Although, in line with the neoliberal influence of the 'Washington Consensus' at the time, the Constitution also permits the privatisation of certain public services.[53]

The Constitution includes a discrete section known as '*la Constitución Ecológica*' (the 'Ecological Constitution'); a series of more than thirty provisions intended to protect environmental interests including both rights and obligations.[54] In particular, articles 79 and 80 recognise the collective right of all people to a healthy environment and the responsibility of the State to: protect the diversity and integrity of the environment; conserve areas of special ecological importance; plan the management and use of natural resources to guarantee their sustainable development, conservation, restoration or substitution; and prevent and control environmental deterioration.[55] The Ecological Constitution has been relied on to protect other water interests,[56] including the Río Bogotá, in which a superior Court, *el Consejo del Estado* (the Council of State), also made a series of very prescriptive orders in response to serious environmental contamination of the river, although without recognising that the river was a legal subject.[57] The Ecological Constitution has provided fertile ground for the reconceptualisation of natural resources like rivers as legitimate rights-bearing entities in Colombia, and the recognition of the particular relationships indigenous communities hold with the natural world.

The Constitutional Court has played a key role in upholding constitutional rights in Colombia, and is particularly activist in its application, seeing itself as both a creator and an enforcer of laws,[58] in stark contrast to other Latin American countries of the civil law tradition. Despite there being no specific recognition of a right to water in the Constitution, the Constitutional Court has developed a line of

---

[52] *Constitución Política de Colombia 1991*, arts. 1–2, 366.
[53] See generally Rene Urueña, 'The Rise of the Constitutional Regulatory State in Colombia: The Case of Water Governance' (2012) 6(3) *Regulation & Governance* 282; Roa-García, Urteaga-Crovetto and Bustamante-Zenteno, above n. 40.
[54] Oscar Darío Amaya Navas, above n. 7.
[55] *Constitución Política de Colombia 1991* arts. 1, 2, 8, 49, 79, 86, 88, 95, 333 and 366.
[56] See generally Oscar Darío Amaya Navas, above n. 7.
[57] *Gustavo Moya Ángel y otros v. Empresa de Energia de Bogota y Otros* [*Gustavo Moya Angel and Others v. the Bogotá Energy Company and Others*] [2014] Consejo de Estado, Sala de Contencioso Administrativo, Sección Primera AP-25000–23-27–000-2001–90479-01 (28 March 2014). See Luis Felipe Guzmán Jiménez, above n. 33, 18.
[58] *José Manuel Rodríguez Rangel v. Enrique Chartuny González* [1992] Corte Constitucional de Colombia [Constitutional Court of Colombia] T-406/92 (5 June 1992).

jurisprudence attempting to protect the human right to water,[59] including for indigenous communities.[60] These are the building blocks for the Court's response to indigenous exclusion from water law frameworks in the case of the Río Atrato.[61]

6.3 LEGAL RIGHTS FOR RIVERS AND INDIGENOUS WATER RIGHTS

6.3.1 El Río Atrato

The Atrato River is located in the poorest and most disadvantaged region of Colombia; El Chocó, located on the Pacific coast.[62] Chocó is abundant in natural resources and biodiversity, being known as one of the most biodiverse places on the planet. It has also been a place of conflict and contestation involving extractive industries, organised criminals, drug traders and (unfortunately) indigenous and tribal peoples. El Chocó and its inhabitants have been caught up in Colombia's long and bitter civil war, and *guerrilla* groups, including the FARC-EP, ELN, paramilitary organisations and drug dealers of the 'Clan del Golfo', have all had a presence in the region at one time or another.[63] Today almost half the population of Chocó lives in abject poverty.

Colombian governments have historically ignored Chocó, and had little regulatory or physical presence there. Bonet argues that the region's poverty and the government's neglect are the product of a colonial legacy of weak or absent public institutions; geographic

---

[59] See generally Roa-García, Urteaga-Crovetto and Bustamante-Zenteno, above n. 40; Gloria Amparo Rodríguez and Andrés Gómez Rey, 'La Participación Como Mecanismo de Consenso para la Asignación de Nuevos Derechos [Participation as a Consensus Mechanism for Assignment of New Rights]' (2013) 37 *Pensamiento Jurídico* 71.

[60] *Marcos Arrepiche contra el Alcalde del Municipio de Puerto López y el Gobernador del Meta* [*Marcos Arrepiche v. the Mayor of Puerto López and the Governor of Meta*] [2010] Corte Constitucional de Colombia [Colombian Constitutional Court] T-143/10 (26 February 2010).

[61] Interview with Eugenia Ponce (Bogotá, 1 September 2017); interview with Andres Gomez Rey (Bogotá, 30 August 2017).

[62] According to Colombia political and administrative organisation, the Chocó region (Departamento de Chocó) has an area of nearly 46,530 square kilometres. Chocó also has thirty 'municipios' (counties) in five subregions: 'Atrato, San Juan, Pacífico Norte, Baudó (Pacífico Sur) and Darién'. Departamento Nacional de Estadística, 'Censo General 2005: "Proyecciones Nacionales y Departamentales de Población 2005–2020"' (2010).

[63] César Rodríguez Garavito, *Etnicidad.Gov – Los Recursos Naturales, Los Pueblos Indígenas y el Derecho a la Consulta Previa en los Campos Sociales Minados* [Natural Resources, Indigenous Peoples and the Right of Prior Consultation in Mining Territories] (Centro de Estudios de Derecho, Justicia y Sociedad, Dejusticia, 2012) 6–9; Alfredo Molano Bravo, *De Río en Río: Vistazo a Los Territorios Negros* (Editorial Aguilar, 2017).

conditions unsuited to communication and access; an economic reliance on extractive industries (particularly gold mining); and the isolation of the region from the rest of the country.[64]

Chocó is also ethnically distinct from the rest of Colombia, with 97 per cent of its approximately 500,000 inhabitants belonging to an indigenous or Afrocolombian community.[65] The indigenous communities (belonging to the Embera-Dóbida, Embera-Katío, Embera-Chamí, Wounan and Tule peoples) live on their ancestral territories, within *resguardos* and the Afrocolombian communities live on '*consejos mayores*'. Both live in much the same way as they have for centuries, in isolation from the dominant Colombian culture, relying on artisanal mining of gold and silver, traditional agriculture and hunting and fishing.[66]

The Río Atrato is the longest and third most navigable river in Colombia. It winds its way 750 kilometres through Chocó from the Andes Mountains ('*Cerro Plateado*') to the gulf of Urabá in the Caribbean Sea. The river is part of a massive basin covering 40,000 square kilometres (60 per cent of the territory of Chocó) and is fed by more than 15 rivers and 300 streams.[67] The people of Chocó who live alongside the Atrato River, much like the Whanganui Iwi in Aotearoa New Zealand, depend on the river for their physical and spiritual sustenance.[68] They claim to have distinct relationships with the river not just as their ancestral territory, but as a 'space to reproduce life and recreate culture'.[69] For the Atrato communities the river is a social, economic, logistical, spiritual and territorial space, forming the core of their distinct cultural identity.[70]

---

[64] Jaime Bonet, '¿Por Qué es Pobre El Chocó? Documentos de Trabajo Sobre Economía Regional [Why Is Chocó Poor? Working Paper on Regional Economics]' (90, Banco de la República, Bogotá, Centro de Estudios Económicos Regionales, 2007).

[65] Departamento Nacional de Estadística, above n. 62.

[66] Bonet, above n. 64; *Tierra Digna* [2016] Corte Constitucional [Constitutional Court], Sala Sexta de Revision [Sixth Chamber] (Colombia) No T-622 of 2016 (10 November 2016) 2.

[67] *Tierra Digna* [2016] Corte Constitucional [Constitutional Court], Sala Sexta de Revision [Sixth Chamber] (Colombia) No T-622 of 2016 (10 November 2016) 2–3; Juan Carlos Bello, *Atlas de la Biodiversidad de Colombia* (Instituto Alexander von Humboldt, 2000).

[68] Camilo Antonio Hernandez, Chapter 2 above n. 19, 12.

[69] *Tierra Digna* [2016] Corte Constitucional [Constitutional Court], Sala Sexta de Revision [Sixth Chamber] (Colombia) No T-622 of 2016 (10 November 2016) 165.

[70] Acta Final de Inspección Judicial, Appendix 1, Part B and [6.2] in *Tierra Digna* [2016] Corte Constitucional [Constitutional Court], Sala Sexta de Revision [Sixth Chamber] (Colombia) No T-622 of 2016 (10 November 2016); interview with Pilar Garcia (Bogotá, 30 August 2017), Chapter 2 above n. 11; interview with Viviana González Moreno (Bogotá, 2 September 2017).

Chocó is referred to as being 'megabiodiverse', with 90 per cent of the region being protected forest area.[71] However, extractive industries are increasingly threatening natural resources in the region. Artisanal mining has been practised in El Chocó for centuries (including by some afrodescendent and indigenous communities). Yet, since the 1990s mechanised illegal mining and logging (without permit or concession from the state or the communities) has intensified along the river. The illegal mines use dredges and excavators in their activities (called 'dragons'), which are often operated by armed organised criminals (including both *guerrilla* groups and narco-traffickers).[72] The illegal miners use mercury and cyanide in the mining process.

Mining activities have led to the extreme degradation of the Atrato River.[73] They have destroyed the natural course of the river and flooded the rainforest in many parts, and the river has become contaminated with dangerous chemicals killing fish and vegetation. Aside from the impact on the environmental condition of the river, the effect of illegal mining on the river communities has been catastrophic.[74] The people have suffered as the chemicals make their way into the food chain, traditional subsistence practices are disrupted, and villages are displaced. All of this has occurred under complacent central government and regional and local authorities who have been unwilling, or unable,[75] to respond adequately to the environmental and humanitarian consequences, despite the plight of the river communities. The Court refers to the Atrato situation as 'a humanitarian, social and environmental crisis without precedent'.[76]

6.3.2 The Atrato Model

In early 2015, the human rights NGO *Centro de Estudios para la Justicia Social 'Tierra Digna'* initiated proceedings in the Constitutional Court of Colombia against the President of the Republic, Ministry for the Environment and Sustainable Development and others concerning the

---

[71] Bello, above n. 67. Ley 2 de 1959 'Por la cual se dictan normas sobre economía forestal de la Nación y conservación de recursos naturales renovables'.

[72] César Rodriguez Garavito, above n. 63, 6–9; Molano Bravo, above n. 63.

[73] Interview with Andres Gomez Rey (Bogotá, 30 August 2017), above n. 61.

[74] Interview with Viviana González Moreno (Bogotá, 2 September 2017), above n. 70.

[75] Interview with Cristian Carabaly (Bogotá, 31 August 2017).

[76] *Tierra Digna* [2016] Corte Constitucional [Constitutional Court], Sala Sexta de Revision [Sixth Chamber] (Colombia) No T-622 of 2016 (10 November 2016) 109; see also Defensoría del Pueblo, 'Crisis Humanitaria en el Chocó: Diagnóstico, Valoración y Acciones de la Defensoría del Pueblo 'The Humanitarian Crisis in Chocó: Diagnosis, Analysis and Action by the Public Defender's Office' (2014) www.defensoria.gov.co/public/pdf/crisisHumanitariaChoco.pdf.

Río Atrato on behalf of a number of Indigenous, afrodescendent and peasant communities in Chocó.[77] The case was an 'acción de tutela' under article 86 of the Colombian Constitution; a writ for protection of the constitutional rights of the river communities. Article 86 allows all Colombians to apply to any judge or the Constitutional Court for an order for protection of their fundamental rights when they are made vulnerable or threatened by an act or omission of a public or private authority. The communities alleged that the activities of illegal miners in Chocó violated their fundamental human rights to life, health, water, food security, healthy environment, culture and territory under the Constitution.

In its decision, the Court found that the Colombian Government had violated all of the fundamental constitutional rights alleged to have been breached by the river communities through its omission to control and eradicate illegal mining in Chocó. Then, significantly, the Court recognised that the Río Atrato (together with its basin and tributaries) is an 'entidad sujeto de derechos'. This result came as something of a surprise to the parties to the case because, according to one of the lawyers from the NGO Tierra Digna, legal rights for the river had not been argued for as part of the applicants' submissions, although the Court's position was welcomed by the NGO and the communities:[78]

> What we wanted to establish was the relevance that the river had amongst the Afro-Colombian and Indigenous communities of Chocó and to focus on that interdependence between them to win the action for protection of the rights of the communities. All for the purpose of protecting the river. We didn't think that it would happen via the category of *sujetos de derecho*, as was adopted by the Court. But I think that it was the best decision possible in order to respond to the situation and to respond to the cosmovision of the communities themselves.

The term 'entidad sujeto de derechos' translates directly as 'an entity the subject of rights'; however, the concept is amorphous in law. In Latin American civil law, a *sujeto de derechos* is an entity with rights and obligations.[79] Typically, the only *sujetos de derechos* are humans or

---

[77] The Centro de Estudios para la Justicia Social 'Tierra Digna' represented the following river communities: Consejo Comunitario Mayor de la Organización Popular Campesina del Alto Atrato (Cocomopoca), el Consejo Comunitario Mayor de la Asociación Campesina Integral del Atrato (Cocomacia), la Asociación de Consejos Comunitarios del Bajo Atrato (Asocoba), el Foro Inter-étnico Solidaridad Chocó (FISCH).

[78] Interview with Viviana González Moreno (Bogotá, 2 September 2017), above n. 70.

[79] Varsi Rospigliosi, Enrique, *Tratado de Derecho de las Personas* (Coedición Universidad de Lima – Gaceta Jurídica, 2014).

natural persons. However, the application of the concept has spread in recent years via Latin American jurisprudence to animals. This includes the case of chimpanzee 'Cecilia' recognised as an *entidad sujeto de derechos* and therefore capable of pleading the writ of *habeas corpus* in order to secure her release from the Mendoza Zoo in Argentina to a Brazilian wildlife sanctuary.[80] The concept has also been used to protect the rights of persons of reduced capacity, such as children, the unborn foetus, or the mentally unwell. The idea of recognising something as a *sujeto de derechos* is to treat it as a rightholder deserving of respect, and not simply the property of others to be managed.[81]

The concept *entidad sujeto de derechos* is distinct, in actual fact, from the concept of legal personality, or *personalidad jurídica* in Spanish, and a *sujeto de derechos* may or may not have legal personality.[82] Latin American civil law makes a distinction between physical and legal persons, the latter of which encompasses 'virtual' persons like companies. In New Zealand, the same concept of the legal person used to recognise the rights and obligations of companies has been adopted for natural resources like the Whanganui River. In Colombia, despite widespread misunderstanding, the concept of the *sujeto de derechos*, is not equivalent to a legal person, although both physical and legal persons are *entidades sujetos de derechos*, or legal subjects.

If the river is a legal subject, the next logical question is what rights or obligations does it receive? In *Tierra Digna*, the Court determined that the Atrato River's rights (as distinct from the communities' human rights) are for its protection, conservation, maintenance and restoration by the State and ethnic communities. However, the Court did not go much further to detail what the rights, or for that matter obligations, of the river might mean in practice. As one of the lawyers for the communities explains:

> What about obligations, how has that been understood? Because that was one of the criticisms [of the decision]. Or not criticisms, but comments that I have heard made by the lawyers. What will be required of the river? I mean, does this 'personality' give rights but not obligations?

---

[80] *Acción de hábeas corpus presentada por la Asociación de Funcionarios y Abogados por los Derechos de los Animales (AFADA)* [2016] Tercer Juzgado de Garantías, Mendoza (Argentina) P-72.254/15 (3 November 2016).

[81] Amparo Rodríguez and Gómez Rey, above n. 59, 78.

[82] Gabriel Andrés Suárez Gómez and Giovanni José Herra Carrascal, 'El Agua Como Sujeto de Derechos [Water as the Subject of Rights]' in *Tratado de Derecho de Aguas* [Water Law Treatise] (Universidad Externado de Colombia, 2018), Vol. I, 99–102.

The 163-page decision of the Constitutional Court of Colombia in *Tierra Digna* is a watershed moment for indigenous and environmental rights in Latin America.[83] It includes detailed analysis of domestic and international human rights norms and environmental laws and even engages with theoretical literature on environmental law, indigenous rights and the rights of nature, reflecting the involvement of human rights NGOs and academics in the case.[84]

In its judgement the Court traverses the rich Colombian jurisprudence on indigenous, afrodescendent and environmental rights. These include cases where the Constitutional Court has protected the collective rights of ethnic communities to survive, to have a territory, to manage natural resources, for legal autonomy, to be consulted in accordance with ILO *Convention 169*, to participate in development projects, and to enjoy a healthy environment. The Court's analysis is underpinned by the 'Ecological Constitution', and the guarantees of human dignity, common welfare and social justice at the core of the Colombian constitutional model: el *Estado social de derecho* (a state of 'social development').[85]

The Court also relied on conceptions of indigenous resource rights in international law and decisions of the Inter-American Court of Human Rights. In declaring that the river is an *entidad sujeto de derechos*, the Court points to indigenous and environmental rights protections in *ILO Convention 169*, the United Nations Declaration on the Rights of Indigenous Peoples, the Convention on Biodiversity, the American Declaration on the Rights of Indigenous Peoples, the UNESCO Convention on Intangible Property and the human right to water. The Court also discusses the key Inter-American Court decisions on indigenous territorial rights: *Sawhoyamaxa* v. *Paraguay*,[86] *Awas Tingni*[87] and *Saramaka* v. *Suriname*.[88] This outward-looking approach to constitutional 'lawmaking' also references domestic comparative law from within and beyond the Latin American civil law perspective, including the then *Te Awa Tupua (Whanganui River Claims*

---

[83] Macpherson and Clavijo Ospina, Felipe, Introduction above n. 19.

[84] Interview with Viviana González Moreno (Bogotá, 2 September 2017), above n. 70.

[85] *Tierra Digna* [2016] Corte Constitucional [Constitutional Court], Sala Sexta de Revision [Sixth Chamber] (Colombia) No T-622 of 2016 (10 November 2016) 25.

[86] *Sawhoyamaxa* v. *Paraguay*, IACHR Series C No 146 (29 March 2006) ('*Sawhoyamaxa* v. *Paraguay*').

[87] *The Mayagna (Sumo) Awas Tingni Community* v. *Nicaragua*, IACHR Series C No 79 (31 August 2001) ('*Mayagna* v. *Nicaragua*').

[88] *Saramaka People* v. *Suriname* (Preliminary Objections, Merits, Reparations, and Costs) IACHR Series C No 172 (28 November 2007).

Settlement) Bill in New Zealand, along with *Te Urewera Act 2014* (NZ) and laws from Ecuador (2008)[89] and Bolivia (2009).[90] This 'internationalisation' of rights for nature in the Atrato decision shows how comparative 'precedents' help courts develop and make law. The specific adoption of a similar approach to that for the *Te Awa Tupua Act* can, in fact, be explained by the simple curiosity that the judge's clerk in the case had done postgraduate research on indigenous rights in New Zealand and had visited the country becoming fascinated by the Whanganui River Settlement.[91]

The Court made several prescriptive orders to implement its decision, including that the rights of the river will be represented by a guardian, with one representative from government and one from the claimant communities.[92] Again, in doing so the Court took inspiration from the *Te Awa Tupua* model from New Zealand, discussed in Chapter 5, which utilises a guardian with dual membership from the Crown and Whanganui River Iwi (*Te Pou Tupua*). The Court gave the President and the communities one month to each choose their representative.

The Court then ordered the guardians to design and form a *Comisión de Guardianes del Río Atrato* ('Commission of Guardians of the Atrato River') within three months, involving the designated guardians plus an *'equipo asesor'* (advisory group). That group included the Humboldt Institute and WWF Colombia, both of whom developed a project for the protection of the River Bita in another part of the country (Vichada, Orinoquía region). This advisory group is similar in many respects to representative collaborative governance institutions for the Whanganui River discussed in Chapter 5: *Te Karewao* and *Te Kōpuka*.

Significantly, the Court designed a special interdisciplinary body to verify and evaluate the proper execution of the orders called the *'panel de expertos'* (panel of experts) headed by the main public sector regulatory body in the country, the *Procuraduría General de la Nación*.[93] This panel is comprised of several experts from the river communities and public, private, academic and non-government organisations and is charged with ensuring that the Court's orders are carried out promptly and correctly.

---

[89] *Constitución de la República del Ecuador 2008* (Ecuador).
[90] *Ley de Derechos de la Madre Tierra (Ley 071) 2010* (Bolivia).
[91] Interview with Pilar Garcia (Bogotá, 30 August 2017), see Introduction above n. 11.
[92] *Tierra Digna* [2016] Corte Constitucional [Constitutional Court], Sala Sexta de Revision [Sixth Chamber] (Colombia) No T-622 of 2016 (10 November 2016) 162.
[93] Ibid. 157.

The Court also ordered a number of government departments and entitles and universities to get together to design and implement a plan to decontaminate the river.[94] This order is similar to the collaboration of persons with interests in the Whanganui River in order to address and advance the health and well-being of *Te Awa Tupua* contemplated by the strategy *Te Heke Ngahuru*, discussed in Chapter 5. Additional orders required other government bodies, including the police, military and Ministry of Defence to develop a plan to neutralise and permanently eradicate illegal mining in Chocó. Finally, the judgement included orders intended to support the recuperation of traditional forms of subsistence and food gathering and further research into the effects of the contamination. Importantly, the Court's mandate requires the participation of the river communities at every stage of execution of the orders.

Such an activist decision of a court of law, particularly within a civil law jurisdiction where legislated law is paramount, might be expected to generate resistance from the executive government. Indeed there is a mechanism under the Colombian Constitution by which the government could have applied for the decision to be 'nullified'.[95] However, the Colombian government appears to have embraced the decision,[96] with the President's named representative being the Ministry for the Environment and Sustainable Development.[97] The Ministry has started its role in implementing the orders, although the process has not been easy, as a Ministry representative explains:[98]

> All of this now has been costed, but honesty the country was not prepared for the judgment. We cannot say that we celebrate the sentence. It creates complexity for us. Implementing a court decision always generates complexity. But what we do celebrate obviously is that it gives us another way to legally view the environment. And that form is, to reiterate, to give it certain rights.

[94] Ibid. 159 order 5.
[95] *Constitución Política de Colombia 1991* art. 237; interview with Eugenia Ponce (Bogotá, 1 September 2017), above n. 61.
[96] The President has not made a public statement about the decision but the various ministries have released a number of plans and resolutions including Ministerio de Ambiente y Desarollo Sostenible [Ministry for the Environment and Sustainable Development, 'Decreto No. 1148 Por el Cual se Designa al Representante de los Derechos del Río Atrato en Cumplimiento de la Sentencia de T-622 de 2016 de La Corte Constitucional [Decree No. 1148 Designating the Representative of the Rights of the Atrato River Giving Effect to Sentence T-622 of 2016 of the Constitutional Court'.
[97] Ibid. art. 1.
[98] Interview with Cristian Carabaly (Bogotá, 31 August 2017), above n. 75.

In January 2018 the Ministry for the Environment and Sustainable Development passed a Resolution about the internal departmental arrangements for implementing the decision.[99] Article 2 of that resolution confirms in detail the distribution of functions within the Ministry to implement the Court's orders. First, the Ministry as a legal representative of the river's rights must, together with the river communities, form the commission of guardians. Although taking significantly longer than the one month specified in the judgement, the fifteen river guardians were eventually all appointed by the last week of August 2017. The seven river communities each appointed one male guardian and one female guardian to ensure gender equity.[100]

The Resolution specifies that the '*Despacho del Viceministro*' ('Deputy Minister's Office') is responsible for ensuring that various institutes connected to the Ministry join in carrying out the orders. The *Dirección de Gestión Integral de Recurso Hídrico* ('Directorate for Integrated Water Management'), referred to as the *Gerente del Río Atrato* ('Atrato River Manager'), has a number of functions. These include to develop and manage the workplan for the project, manage internal coordination, act as intermediary between the Ministry and the river communities and environmental authorities, and manage information and documentation.

The Resolution then sets up four technical committees involving various departmental offices to concentrate on specific orders and requires them to meet at least twice per month. They are: a technical committee on illegal mineral extraction; a technical committee on deforestation; a technical committee on decontamination of waterways; and a technical committee on community and environmental health. Finally the Resolution details the various functions of the different directorates in implementing the decision, namely the Directorate of Integrated Water Management, the Directorate of Forests, Biodiversity and Ecosystem Services, the General Directorate of the Territorial Environmental Authority and Coordination of the

---

[99] *Ministerial de Ambiente y Desarrollo Sostenible Resolución n. 0115 de 26 Enero 2018 Por Medio de la Cual Se Asignan Funciones al Interior del Ministerio de Ambiente y Desarollo Sostenible a Efectos de Dar Cumplimiento a lo Dispuesto en la Sentencia T'622 de 2016* [Ministry of Environment and Sustainable Development Resolución No 0115 of 26 January 2018 Which Assigns Functions Internal to the Ministry of Environment and Sustainable Development in Compliance with That Provided for in Sentence T-622 of 2016] (Colombia).

[100] *¡Todas y Todos Somos Guardianes del Atrato!* [We Are All the Guardians of the Atrato!] (4 September 2017) https://co.boell.org/es/2017/09/04/todas-y-todos-somos-guardianes-del-atrato.

National Environmental System (SINA), the Directorate of Environmental and Urban Affairs and the Office of Green and Sustainable Commerce.

The Ministry for the Environment has started to implement the orders, including developing a plan for decontamination of the river, including amplifying environmental monitoring in the catchment, constructing wetlands for elimination of mercury and encouraging economic activity to shift from mining.[101] In May 2018, the Ministry ordered the temporary suspension of all mining in the Río Quito, a major tributary of the Atrato, due to extreme contamination.[102] However, the government has already been accused of excessive delay in implementing the orders, failing to actively involve the guardians in decision making about the river and its advocacy, and inadequately protecting the guardians from constant threats from armed groups involved in illegal mining.[103] After the change of government in 2018 to the conservative Duque administration, the future of the Atrato River as a legal subject, and the institutions necessary to implement the Court's wide-ranging orders, is uncertain.

6.4 THE PRESENT AND FUTURE OF INDIGENOUS WATER RIGHTS IN COLOMBIA

6.4.1 River Communities, Legal Subjects and Biocultural Rights

The recognition by the Constitutional Court of Colombia that the Río Atrato is a legal subject is a response to the continual ignorance by consecutive governments, since the arrival of the Spanish in Latin America, of the rights and experiences of indigenous and tribal peoples living alongside the river.[104] The region of Chocó, remote and mysterious, has been forgotten by Colombian governments. The communities, although permitted to live collectively on indigenous territories, have not enjoyed recognition of any legal rights to use

---

[101] *El Gobierno Colombiano Recibe Propuestas para el Plan de Descontaminación del Río Atrato* www.iagua.es/noticias/minambiente/gobierno-colombiano-recibe-propuestas-plan-descontaminacion-rio-atrato.

[102] *El Ministerio de Ambiente Ordena Suspender la Minería en el Río Quito, Chocó* [The Minister for the Environment Orders the Suspension of Mining in the Quito River, Chocó] *El Espectador* www.elespectador.com/noticias/medio-ambiente/ordenan-suspender-toda-la-mineria-en-uno-de-los-rios-mas-importantes-de-choco-articulo-754898.

[103] *Guardianes del río Atrato: amenazados e ignorados* [Guardians of the Atrato River: threatened and ignored] Colombia 2020 https://colombia2020.elespectador.com/territorio/guardianes-del-rio-atrato-amenazados-e-ignorados.

[104] Interview with Felipe Clavijo (Bogotá, 12 August 2018).

water. Their specific uses of and interests in water are not represented by any sort of water licence or concession. Nor have they been involved in water resource management and planning, or even consulted about the impact of water development on them. The indigenous and tribal peoples of Chocó have, like the indigenous peoples of the other countries considered in this book, been excluded from water law frameworks in Colombia. The decision of the Constitutional Court is an attempt to repair centuries of historical injustice against the indigenous and afrodescendent communities by recognising their river relationships and jurisdiction.

Can the Court's decision properly be characterised as an 'ecocentric' effort to recognise the river as a rights-bearing entity? Certainly the decision does so, in bestowing the river with the status of an *entidad sujeto de derechos* or legal subject. The Court explicitly recognises that it is taking an ecocentric approach in its analysis, referring to the interconnectedness of humans with nature, the superior interest of the environment, and human obligations to protect nature's rights. In this sense the Court explains:[105]

> Nature and environment is a cross-cutting theme in the Colombian constitutional order. Its importance is obviously due to the human beings who inhabit it and the need to have a healthy environment to lead a dignified life in conditions of well-being, but is also for the other living organisms with which the planet is shared, *understood as entities worthy of protection in of themselves*. This means being conscious of the interdependence that connects all living things on the Earth; that is, to recognise that we are integral parts of the global ecosystem and biosphere, before applying normative categories of domination, simple exploitation or utility.

The decision also extends the burgeoning line of jurisprudence about the rights of nature in Latin America.[106] Colombia is a pioneer in rights of nature law or 'earth jurisprudence'. Aside from the Río Atrato decision, the Colombian courts have also recognised animals as *sujetos de derechos* in a number of cases, including the well-known case of the *Oso 'Chucho'* (a bear called 'Chucho') which, like the chimpanzee in Argentina, was granted the constitutional writ of *Habeas Corpus* in order to be freed from captivity in the Barranquilla

---

[105] *Tierra Digna* [2016] Corte Constitucional [Constitutional Court], Sala Sexta de Revision [Sixth Chamber] (Colombia) No T-622 of 2016 (10 November 2016) 42.

[106] Interview with Eugenia Ponce (Bogotá, 1 September 2017), above n. 61; interview with Pilar Garcia (Bogotá, 30 August 2017), Chapter 2 above n. 11.

Zoo.[107] The Amazon rainforest was recently recognised by the Colombian Supreme Court as a legal subject in response to the threat of deforestation and climate change.[108] Other cases are anticipated concerning rivers, ecosystems and wetlands.[109]

However, the rights of the river communities, as indigenous and tribal peoples who have been excluded from river management, are at the heart of the decision.[110] The Court is able to take its 'earth-centred' approach by co-opting and adapting constitutional environmental and indigenous human rights protections (the right to life, to water, to culture, to live in a clean environment, etc.).[111] The Court extends the understanding in previous Colombian and Latin American constitutional jurisprudence around the right to water beyond drinking water and sanitation to a broader right to water for its natural presence in rivers, creeks or lakes.[112] It even refers to the rights of nature as the 'third generation' of human rights, after civil and political and then economic social and cultural rights.[113]

The Atrato judgement is, ultimately, about recognising the communities' ancestral, territorial, communal and 'biocultural' rights. According to Bavikatte and Bennett, the term 'biocultural rights' denotes a community's long established right, in accordance with its customary laws, to steward its lands, waters and resources, being increasingly recognised in international environmental law.[114] As they explain, biocultural rights are not simply claims to property, in

---

[107] *Fundación Botánica y Zoológica de Barranquilla (Fundazoo) and Others v. Ministeria de Ambiente y Desarollo Sostenible and Others* (Unreported, Corte Suprema de Justicia [Supreme Court of Justice], 16 August 2017). Animals are already recognised as 'sentient beings' under the *Ley de Protección Animal Número 1774 de 2016* [Animal Protection Act 2016] (Colombia).

[108] *Andrea Lozano Barragán, Victoria Alexandra Arenas Sánchez, Jose Daniel y Felix Jeffry Rodríguez peña y otros v. El President de la República y otros*, No. STC4360-2018, Corte Suprema de Justicia [Supreme Court of Justice], Sala de Casación Civil [Appeals Chamber] (Colombia) (4 April 2018).

[109] Interview with Viviana González Moreno (Bogotá, 2 September 2017), above n. 70; interview with Felipe Clavijo (Bogotá, 12 August 2018), above n. 104.

[110] Interview with Viviana González Moreno (Bogotá, 2 September 2017), above n. 70.

[111] The Court also relies on the 'precautionary principle' of environmental law in reaching its result. *Report of the United Nations Conference on Environment and Development* (UN Doc A/ CONF.151/26 (Vol. I) (12 August 1992) annex I ('Rio Declaration on Environment and Development') principle 15. See *Tierra Digna* [2016] Corte Constitucional [Constitutional Court], Sala Sexta de Revision [Sixth Chamber] (Colombia) No T-622 of 2016 (10 November 2016) 4.

[112] Interview with Felipe Clavijo (Bogotá, 12 August 2018), above n. 104.

[113] *Tierra Digna* [2016] Corte Constitucional [Constitutional Court], Sala Sexta de Revision [Sixth Chamber] (Colombia) No T-622 of 2016 (10 November 2016) 137.

[114] See K Bavikatte and T Bennett, 'Community Stewardship: The Foundation of Biocultural Rights' (2015) 6(1).

the typical market sense of property being a universally commensurable, commodifiable and alienable resource, but collective rights of communities to carry out traditional stewardship roles vis-à-vis nature, as conceived of by indigenous ontologies. The Court explains its approach as follows:[115]

> This approach has a special relevance in Colombian constitutionalism, keeping in mind the principle of cultural and ethnic pluralism that supports it, together with the ancestral knowledge, use and customs of indigenous and tribal peoples. Accordingly in the following paragraph we explore an alternative vision of the collective rights of the ethnic communities in relationship to their cultural and natural surroundings, which are called, 'biocultural rights'.

The Constitutional Court recognises the inescapable connection between the rights of nature and the rights of humans as biocultural rights, the basis of which is the 'profound unity between nature and the human species'.[116] In fact, the concept of 'biocultural rights' is the key concept in the case.[117] As former clerk of the Constitutional Court involved in drafting (and subsequently, via the *Procuraduría General de la Nación*, implementing), the decision explains:[118]

> In the Atrato ruling ecocentric and biocultural perspectives have been developed to allow new interpretations on the relationship between humans (ethnic communities) and nature (a river, a mountain), giving each of them equal character as *one* entity subject of the *same* rights. This characterisation is the birth of the declaration of the Atrato River as a subject of rights; a whole new interpretation in our constitutional law.

Biocultural rights are quintessentially cultural,[119] and the adoption of the legal subject model is an attempt by the Court to translate, accommodate and approximate river interests and relationships

---

[115] *Tierra Digna* [2016] Corte Constitucional [Constitutional Court], Sala Sexta de Revision [Sixth Chamber] (Colombia) No T-622 of 2016 (10 November 2016) 42.

[116] Ibid. 47; see also Bavikatte and Bennett, above n. 114.

[117] Interview with Eugenia Ponce (Bogotá, 1 September 2017), above n. 61.

[118] Interview with Felipe Clavijo (Bogotá, 12 August 2018), above n. 104.

[119] The development of the rights of nature in Ecuador and Bolivia has also occurred within a context of legal and cultural pluralism, via the concept of *Buen Vivir* – to live a good life. See generally Tom Perreault, 'Tendencies in Tension: Resource Governance and Social Contradictions in Contemporary Bolivia' in *Governing Resource Extraction* (2017); Craig M Kauffman and Pamela L Martin, 'Can Rights of Nature Make Development More Sustainable? Why Some Ecuadorian Lawsuits Succeed and Others Fail' (2017) 92 *World Development* 130.

existing in indigenous and tribal customs and laws.[120] These biocultural rights, the Court acknowledges, 'are not new rights for the ethnic communities', but rather a category that unifies their interconnected rights in natural resources and to culture.[121] The Court describes the river communities' rights as the right 'to administer and exercise trusteeship in an autonomous manner over their territories – in accordance with their own laws and customs'.[122] This is a right for the communities to river governance and livelihoods. Clavijo explains:[123]

> ... what it does, precisely, is legitimate all those communities as indigenous and afrodescendent ethnic entities (or even farmers/peasants) so that they may depend not just on a territory, but on the biodiversity of a territory; the whole environment, beginning with these nature relationships which are every day clearer.

Gomez Rey agrees with the focus of the decision on the rights of the people, although at the same time the Court is giving a voice to the river:[124]

> What happened with the case of the Atrato River? It's not that the river itself is given rights as such. They are recognising that there is an indispensable relationship between people and nature. And that relationship must be, by one way or another, known and respected. And this case, what it has in common with the cases of New Zealand is that they are resources that are very threatened, so sometimes there are moments in which the river or the area needs its own voice because people are not protecting its interests.

The process of determining and accommodating biocultural rights in Colombia resembles other recognition models like native or aboriginal title, where state law recognises and gives effect to pre-existing indigenous laws and customs.[125] However, we know from the legal pluralism literature that when states recognise indigenous rights and interests, there is an inevitable process of translation, accommodation

---

[120] See Chapter 2 for a discussion of the accommodation and translation of indigenous rights via recognition models.

[121] *Tierra Digna* [2016] Corte Constitucional [Constitutional Court], Sala Sexta de Revision [Sixth Chamber] (Colombia) No T-622 of 2016 (10 November 2016) 44.

[122] Ibid. 43.

[123] Interview with Felipe Clavijo (Bogotá, 12 August 2018), above n. 104.

[124] Interview with Andres Gomez Rey (Bogotá, 30 August 2017), above n. 61. See Chapter 3 for a discussion of the role of legal rights for nature as giving 'voice' to natural resources.

[125] See Chapters 4 and 5 for a discussion of native title.

and mediation.[126] Legal personality is a mechanism used to recognise indigenous and tribal relationships and jurisdictions to manage the natural world. However, the indigenous rights are not recognised in their complete form, and are actually limited via the process of recognition. As an example of this, while the Atrato communities' biocultural rights are positioned as being territorial in nature,[127] and although the indigenous and afrodescendent communities successfully claimed a failure to protect their right to 'territory', the Court does not recognise a right of property for the communities in the river, nor for the river to own itself. In this regard the Atrato case, like that of the Whanganui River is about sharing water management rather than sharing water.[128]

The Court in *Tierra Digna* dedicated page upon page of the judgement to discussing the human rights of the river communities, and their biocultural basis. However, the Court gave very little space to discussing the content of the river's own rights, referred to briefly as rights of protection, conservation, maintenance and restoration. Much conceptual ambiguity remains as to how far the river's rights will reach, or in what situations they may be protected.[129] The main outcome of the Atrato decision is that it resets the relationship between the state and river communities.[130] It does this in recognition of the culturally specific guardianship relationship between people and nature, as biocultural rights, and as an attempt to compensate for centuries of neglect and abuse of the people and their territory. The decision, therefore, can best be characterised as 'constitutional' in the sense that it reconstitutes the deteriorated relationship between the state and river communities.[131] The significance of this act of recognition is that it enables an institutional, as well as cultural, shift. As part of the new relationship between the Government and the people, the Court gives the river communities a new jurisdiction as guardians responsible for defending the river's rights.

---

[126] See Chapter 2 for a discussion of legal pluralism and the limiting impacts of recognition policies.

[127] *Tierra Digna* [2016] Corte Constitucional [Constitutional Court], Sala Sexta de Revision [Sixth Chamber] (Colombia) No T-622 of 2016 (10 November 2016) 45, 48–56.

[128] Interview with Andres Gomez Rey (Bogotá, 30 August 2017), above n. 61.

[129] Ibid.

[130] See Katherine Sanders, '"Beyond Human Ownership"? Property, Power, and Legal Personality for Nature in Aotearoa New Zealand' (2018) 30, Issue 2(2) *Journal of Environmental Law* 207.

[131] See generally Sanders, above n. 130 for a characterisation of legal rights for the Whanganui River as a 'constitutional' act.

6.4.2   Guardianship and 'River Jurisdiction'

A common criticism of the Río Atrato decision to recognise the river as an *entidad sujeto de derechos*, much like *Te Awa Tupua* in Aotearoa New Zealand, is that the model is overly ambitious, idealistic and impractical.[132] Rights for nature laws have been poorly implemented in comparative contexts. In Ecuador, for example, protections of the rights of nature (and the concept of *buen vivir*) have been erratically implemented by courts,[133] and are often described as 'weak', although they are gradually having greater policy impact in decision-making processes.[134] The broad protection of the rights of Mother Earth in Bolivia, too, has had little practical impact in preventing environmentally damaging development.[135]

I have argued elsewhere with Erin O'Donnell that in order for grants of legal personality to provide opportunities for improved outcomes they must be accompanied by strong and independent institutions, clearly expressed river values, and legal rights to use the river.[136] If other users have legal or proprietary interests in rivers and the river does not, the scope for enhancing environmental values in river regulation is seriously undermined. However, the Court in the Atrato decision appears conscious of the risk of ambivalence around implementation, and is probably wary given the government's record of inaction in the region.[137] Accordingly, the detail and strength of the orders are intended to obligate the executive branch of government to take action to respond to the serious environmental and humanitarian crisis in the river territory. As Clavijo explains:[138]

> The implementation procedure is probably one of the best parts of the ruling. This mechanism allows communities, experts, universities, institutions and State authorities to share knowledge, mistakes and best practice to protect the river. It is also a very powerful tool to ensure the fulfilment of the orders.

---

[132] Interview with Pilar Garcia (Bogotá, 30 August 2017), Chapter 2 above n. 11.
[133] Mary Elizabeth Whittemore, 'Problem of Enforcing Nature's Rights under Ecuador's Constitution: Why the 2008 Environmental Amendments Have No Bite' [2011] (3) *Pacific Rim Law & Policy Journal* 659.
[134] Kauffman and Martin, above n. 119, 138.
[135] Perreault, 'Tendencies in Tension: Resource Governance and Social Contradictions in Contemporary Bolivia', above n. 119, 18.
[136] Elizabeth Macpherson and Erin O'Donnell, Chapter 3 above n. 62.
[137] Interview with Andres Gomez Rey (Bogotá, 30 August 2017), above n. 61.
[138] Interview with Felipe Clavijo (Bogotá, 12 August 2018), above n. 104.

The adoption of the legal subject as a model for recognising rights of natural resources is itself a response to the problem of enforceability. The Ecuadorian and Bolivian laws have been poorly implemented because it is not clear who has standing to take action to uphold nature's rights.[139] The Colombian model tackles this ambiguity by enabling the river to take legal action in its own name if necessary, via the guardians, to respond to environmental threat or promote environmental outcomes. If the Colombian government fails to properly implement the orders, the Guardians of the Río Atrato could return to the Constitutional Court on the river's behalf to voice its concerns.

Yet, much more significant than the creation of the legal fiction of rights for rivers, is the institutional framework accompanying it to protect and manage river rights by partnership, consensus and collaboration. The Atrato case engages directly with the difficulty of enforcing the rights of nature, by making prescriptive orders about how the river's rights must be protected and implemented and giving the Court itself an ongoing role, via regular reports on implementation from the government. It is remarkable just how far the Court has gone in designing institutions to represent the river; inspired by the Te Awa Tupua model, principle of guardianship, and collaborative governance approach in New Zealand law. The Ministry for the Environment is aware, clearly, of the importance of strong institutional collaboration across government, and the difficulty of implementing a decision of such importance.[140]

Thus, the Court in *Tierra Digna* enables river communities to participate in decision making about the river via a number new river institutions. Foremost, it creates the commission of guardians, responsible for furthering the river's interests, although a range of other entities are created to support the work of the guardians. The decision gives powers to the guardians, thereby carving out a space for the indigenous and afrodescendent communities where they can practice decision making and autonomy over the river, from which they have been excluded since the arrival of the Spanish.

The flipside of the need for strong institutions for legal rights for rivers to be effective leads to the obvious question: if there are strong institutions managing the use of the river, are legal rights for rivers in fact necessary? Many have asked the same question, including the

---

[139] See generally Kauffman and Martin, above n. 119.
[140] Interview with Cristian Carabaly (Bogotá, 31 August 2017), above n. 75.

Director of the Centre for Environmental Law at the Universidad Externado who, although acknowledging the profound cultural importance of the decision, put it this way:[141]

> I believe that [in the Atrato case] there is a transformation of third generation rights towards a mix with legal personhood. For example, there is a right to a healthy environment. But who holds that right? Of course: people! But how do you make sure the right can be enjoyed. So we will give legal personality to nature or one of its elements. What is the point of that? What are the legal consequences of that? That is what I ask, and what can be debated from a legal standpoint. Can't you make the same decisions about protecting the river without the necessity to recognise legal personality in a thing? I think that yes, you can. Furthermore, if the administration would work; if the executive via its environmental or catchment authorities that already exist, could develop the provisions, obligations and conditions in the legal framework, the work of the Judge in the Atrato case would be absolutely unnecessary. And because of that it would be unnecessary to recognise the river as the holder of rights.

6.5 CONCLUSION

In Colombia, a country with complex and interrelated human rights and environmental challenges, the *Tierra Digna* decision on the Atrato River presents new hope for how governments might reconstitute relationships with indigenous groups around access to and management of water. The Atrato decision is much more about the management of water than it is about water use, given that the decision in no way impacts on water allocation frameworks, which continue to ignore indigenous peoples. Yet in a country where most indigenous uses of water are customary, and continue in spite of state-based water allocation frameworks, the Atrato decision goes some way towards recognising indigenous water relationships and jurisdictions. The chapter has shown that the recognition of the river as a legal subject is not determinative. In fact, the biocultural rights of the river communities could have been recognised and the institutions for managing the river could have been created, without recognising that the river is an *entidad sujeto de derechos*. Still, the normative power of the legal subject model cannot

---

[141] Interview with Pilar Garcia (Bogotá, 30 August 2017), Chapter 2 above n. 41.

be ignored, and should not be underestimated. Commentators should avoid taking a Eurocentric view as to its relevance.

Legal rights for rivers are clearly not a complete answer to indigenous water exclusion. Indigenous river communities also need substantive water allocations, in order to have a sufficiently strong voice in lobbying for water access and influencing decision making about river management. Yet, because the river is a subject it has representatives from the community, or guardians, and they have a voice on behalf of the river, where previously they had none. The Colombian study is highly significant, in that it underscores the strength of legal person models in creating new jurisdictions for indigenous peoples in which to participate in river sharing, governance and use.

CHAPTER SEVEN

RECOGNISING AND ALLOCATING INDIGENOUS WATER RIGHTS IN CHILE

7.1 INTRODUCTION

In the fourth and final country study in this book, indigenous Chileans have been excluded from water law frameworks, both in terms of substantive rights to use water and procedural rights to be consulted about or involved in state-based water governance, since the Spanish acquisition of sovereignty. Indigenous Chileans have enjoyed only partial recognition of their territorial rights to land and water by successive governments, often simply via permission for them to remain on their ancestral lands as occupiers or possessors. However, until the late twentieth century, there was no specific recognition of indigenous rights to use water.

Since the passage of the *Indigenous Law 1993* (Chile) indigenous Chileans have been able to 'regularise' their '*ancestral*'[1] use of water as a formal right, or '*derecho de aprovechamiento*'. Interestingly, and in line with the productivist logic of Chilean water law frameworks, indigenous water rights have not been restricted to traditional, cultural uses, as has been assumed in Aotearoa New Zealand and Australia. However, because water use rights were unbundled from landholding with the introduction of water markets in the early

---

[1] I discuss the meaning of '*ancestral*' in the Chilean context later in this chapter. For present purposes, 'ancestral' water rights are water use rights that have been possessed historically, arising from pre-sovereignty times.

1980s, other users had already accumulated the majority of available water use rights by the time the *Indigenous Law* was introduced. Accordingly, the *Indigenous Law* included another mechanism: an Indigenous Land and Water Fund, intended to support the redistribution of *derechos de aprovechamiento* to indigenous communities, where necessary, via the market. In this chapter, I examine both the process for recognition of *ancestral* water rights and the redistribution of *derechos de aprovechamiento* via the Fund. The recognition of *ancestral* water rights in the *Indigenous Law* offers only limited potential to respond to indigenous water exclusion in Chile. In these circumstances, a well-funded redistributive mechanism is crucial.

In the first part of this chapter, I show how indigenous Chileans have been excluded from legal frameworks determining access, use and management of water. That exclusion intensified after the unbundling of water use rights from landholding, and the introduction of water markets, in the early 1980s, enabling other users to accumulate water use rights at the expense of indigenous groups.

I then examine the recognition of the *ancestral* water rights of northern indigenous groups under article 64 of the *Indigenous Law*, finding it to be only a partial response to indigenous exclusion from water law frameworks. Indigenous landholders have only benefited from the recognition of *ancestral* water rights pursuant to article 64 where they have been able to prove to the courts that they have continued to exercise these *ancestral* water rights since pre-Columbian times, and where other users do not hold *derechos de aprovechamiento* that would be prejudiced as a result of the recognition.

Towards the end of this chapter, I discuss the Indigenous Land and Water Fund; a mechanism for allocation (and reallocation) of *derechos de aprovechamiento* to indigenous landholders. The Fund is able to reallocate water rights for any purposes, in addition to the more limited recognition of *ancestral* water rights. This analysis reveals important lessons around legal and policy mechanisms for water rights distribution for all countries considered in this book (and beyond), in particular, the need for such a redistributive mechanism to be adequately resourced. I finish this chapter with a prognosis of the uncertain future of indigenous water rights in Chile, discussing the *Water Code Amendment Bill* (Chile), still languishing before the Chilean Senate at the beginning of 2019.

7.2 INDIGENOUS EXCLUSION FROM WATER LAW FRAMEWORKS IN CHILE

7.2.1 Indigenous Peoples in Chile

Like the indigenous peoples considered in other parts of this book, the indigenous peoples living in the territory now within the Republic of Chile have suffered a long history of physical, social and economic disadvantage including the loss of, or failure by the state to recognise their rights to own and govern their traditional territories, including their water resources. Chile was colonised as a settled colony by Spain during the sixteenth century, applying the international law doctrine of discovery.[2] Prior to this time a number of culturally and linguistically distinct indigenous peoples exercised full sovereignty over their territories. For more than 300 years after the arrival of, and settlement by, the Spanish, the Mapuche maintained control over the region of Araucania south of the Bio Bio River until 1833.[3] Today there remain nine indigenous ethnicities recognised by the *Indigenous Law*, all of which culturally and linguistically distinct, and which make up approximately 12.8 per cent of the total population.[4] The 2,185,792 people that self-identify as indigenous are the Mapuche, Aymara, Rapa Nui, Likan Antai (otherwise known as Atacameño), Quechua, Colla, Diaguita, Kawesqar, Yagan or Yamana.

Upon colonisation, the Spanish Crown declared for itself both pre-emption rights and underlying title to all lands and resources in Chile. The Crown acknowledged that the indigenous peoples held only limited residual rights amounting to temporary rights to occupy land.[5] It even relied occasionally on the doctrine of 'terra nullius' to claim title to vast areas of what it viewed as vacant lands, in a similar manner as occurred in Australia.[6] Early laws passed by the colonisers show the disparaging attitudes held towards the indigenous by Spanish settlers. For example, the preamble to the *Law of 1 June 1813* (Chile) refers to

[2] Robert J Miller, 'The International Law of Discovery, Indigenous Peoples, and Chile' (2010) 89 *Nebraska Law Review* 819, 850–3.
[3] Robert J Miller, 'The International Law of Colonialism: A Comparative Analysis' (2011) 15(4) *Lewis & Clark Law Review* 847, 896.
[4] 'Instituto Nacional de Estadísticas Chile, "Síntesis de Resultados Censo 2017"' [Synthesis of Census Results 2017] 16 www.censo2017.cl/descargas/home/sintesis-de-resultados-censo2017.pdf. A total of 2,185,792 people declare themselves as belonging to an indigenous group which is 12.8 per cent of the total population. The dominant groups are Mapuche with 1,745,147, Aymara with 156,754, Diaguita with 88,474.
[5] Miller, above n. 2, 861–2.
[6] Ibid. 850; Kuppe, Chapter 2 above n. 46, 108.

the 'indians' as being of 'extreme misery, inertia, incivility, lack of morals and education in which they live abandoned in their camps, with the supposed name of townships'.[7]

The cultural difference of the Chilean indigenous peoples was also ignored by the state until the end of the twentieth century. Government policy towards indigenous Chileans, for most of Chile's history until the early 1990s, sought to assimilate indigenous groups into the general population (a policy referred to as '*Chilenización*').[8] The *Chilenización* policy was characterised by a denial of ethnic diversity, and favoured the individuation of communal indigenous territories.[9]

Over time, the colonisers recognised that some indigenous groups held limited rights to possess or occupy land in certain parts of Chile, although such rights did not equate to 'ownership'. The recognition of indigenous land rights was different in the north than in the south. As mentioned, south of the Bio Bio River, the Mapuche had managed to maintain control over their lands for 300 years after the arrival of the Spanish conquistadors. The Mapuche were moved to '*reducciones*' (reservations) from the mid nineteenth century,[10] to which they were given possessory titles ('*titulos de merced*').[11]

---

[7] *Ley de 1 de Junio de 1813* [Law of 1 June 1813] (Chile) in Mylene Valenzuela Reyes and Sergio Oliva Fuentealba, *Recopilación de Legislación del Estado Chileno Para Los Pueblos Indígenas, 1813–2006* [Compilation of Legislation of the State of Chile for Indigenous Peoples 1813–2006] (Librotecnia, 2007) 15.

[8] Comisionado Presidencial para Asuntos Indígenas [Presidential Comission for Indigenous Issues], 'Informe de La Comisión Verdad Histórica y Nuevo Trato Con Los Pueblos Indígenas [Report of the Historical Trust Comission and New Agreement]' (First Edition, October 2008) 51.

[9] Chile, 'Primer Informe Comisión Especial Pueblos Indígenas' [First Report of the Special Commission for Indigenous Peoples], Cámara de Diputados [House of Representatives], 10 November 1992, in Biblioteca del Congresso Nacional de Chile [Library of National Congress of Chile], 'Historia de La Ley No 19.253 Establece Normas Sobre Protección, Fomento y Desarrollo de Los Indígenas, y Crea la Corporación Nacional de Desarrollo Indígena [History of Law No 19.253 to Establish Norms for the Protection, Creation and Development of the Indigenous, and to Create the National Corporation of Indigenous Development]' (5 October 1993) 67; interview with Hernando Silva (Temuco, 11 November 2011).

[10] See Jorge Contesse, 'The Rebel Democracy: A Look into the Relationship between the Mapuche People and the Chilean State' (2006) 26 *Chicano-Latino Law Review* 131, 140.

[11] *Ley de 4 de Diciembre de 1866 (Dispone la fundación de poblaciones en el territorio de los Indígenas y da normas para la enajenacion de estos)* [Law of 4 December 1886 (To provide for the creation of settlements in Indigenous territories and create norms for their alienation)] (Chile) in Valenzuela Reyes and Oliva Fuentealba, above n. 7, 32; *Decreto de 19 de Mayo de 1910 (Da normas para el ortogamiento de Titulos de Merced por la Comisión Radicadora de Indígenas)* [Decree of 19 May 1910 (To create norms for granting Titulos de Merced by the Indigenous Settlement Commission) (Chile)] in ibid. 63.

The situation for indigenous groups in northern Chile, including the Aymara and Atacameña people, was much less favourable. From the annexation of the north from Peru and Bolivia in 1879 and 1883 respectively at the conclusion of the War of the Pacific until the late nineteenth century, there was no indigenous communal title to land in Chile.[12] The lands on which the northern groups lived were declared *'tierra fiscal"* (State land) after they were annexed to Chile.[13] The pre-sovereignty land rights of the northern indigenous groups were simply not recognised. They were not given land titles to the fiscal lands they occupied, although they were allowed 'usufructuary' rights for their use.[14] Those indigenous people that came within the Republic of Peru from 1854 to 1879 had, in some situations, been granted individual titles to the lands they occupied by the Peruvian Government. By giving the indigenous occupants a title resembling ownership (subject only to a twenty-five-year prohibition on alienation), the government was able to collect taxes from the indigenous titleholders.[15] However, when these individual land titles were subsumed within the Republic of Chile after their annexation in 1879, the Chilean Government lifted any restriction on their alienation. This allowed the individual titles to be freely disposed of in Chile's general land market, leading to their widespread dispossession.[16]

The relatively more favourable treatment of indigenous land rights in Chile's south can arguably be explained by the existence of treaty-like documents, called *'parlamentos'*. The Mapuche independence and border at the Bio Bio was recognised in a number of *parlamentos*

---

[12] Héctor González Cortez and Hans Gundermann Kröll, 'Acceso a La Propiedad de La Tierra, Comunidad e Identidades Colectivas Entre Los Aymaras del Norte de Chile (1821–1930) [Land Property Right Access, Community and Collective Identities Among Aymara Communities in Northern Chile (1821–1930)]' (2009) 41(1) *Chungara: Revista de Antropología Chilena* 51, 51; Manuel Prieto, *Privatizing Water and Articulating Indigeneity: The Chilean Water Reforms and the Atacameño People (Likan Antai)* (The University of Arizona, 2014) 126–41.

[13] Chile, 'Primer Informe Comisión Especial Pueblos Indígenas' [First Report of the Special Commission for Indigenous Peoples], Cámara de Diputados [House of Representatives], 10 November 1992, in Biblioteca del Congreso Nacional de Chile [Library of National Congress of Chile], above n. 9, 62; see Aylwin, Chapter above n. 7, 5; Instituto de Estudios Indígenas, 'Los Derechos de los Pueblos Indígenas en Chile [Indigenous Peoples' Rights in Chile]' (Programa de Derechos Indígenas, Universidad de la Frontera, 2003) 38.

[14] Chile, 'Primer Informe Comisión Especial Pueblos Indígenas' [First Report of the Special Commission for Indigenous Peoples], Cámara de Diputados [House of Representatives], 10 November 1992, Biblioteca del Congresso Nacional de Chile [Library of National Congress of Chile], above n. 9, 62.

[15] Cortez and Kröll, above n. 12, 53–4 referring to legislation passed by the Peruvian Government between 1824–8.

[16] Ibid. 56.

between the Spanish Crown and Mapuche. The *parlamentos*, directly translating as 'talks', but really a sort of pact entered into by independent nations with treaty-like status, were reached between Mapuche groups and the Spanish from the seventeenth century.[17] The most famous of these was the *Parlamento de Quilin 1641*. The *parlamentos* recognised Mapuche territorial rights and statehood, and presumably played an important role in cementing Mapuche identity after colonisation. Nonetheless, even after the Mapuche were confined to the *reducciones*, they also experienced widespread dispossession of their *titulos de merced*. By the end of the twentieth century, the Mapuche, like other indigenous groups in Chile, claimed to have suffered largescale land loss.[18]

In an attempt to restore communal indigenous land titles, the Frei Montalva and Allende left-wing governments of the 1960s and 70s established a Directorate of Indian Affairs and Institute for Indigenous Development and commenced projects for restitution of land to legally recognised indigenous communities,[19] although the projects focused on the Mapuche.[20] Unfortunately, these indigenous land rights initiatives were overturned during the military dictatorship of 1973–1990, principally via a legal decree signed by General Augusto Pinochet in 1979.[21]

---

[17] Chile, 'Primer Informe Comisión Especial Pueblos Indígenas' [First Report of the Special Commission for Indigenous Peoples], Cámara de Diputados [House of Representatives], 10 November 1992, in Biblioteca del Congresso Nacional de Chile [Library of National Congress of Chile], above n. 9, 62.

[18] Aylwin, Chapter 2 above n. 7, 4; Patricia Richards, 'Of Indians and Terrorists: How the State and Local Elites Construct the Mapuche in Neoliberal and Multicultural Chile' (2010) 42 *Journal of Latin American Studies* 59, 82. According to Richards, the Mapuche now hold only 40 per cent of the original lands held under *titulos de merced*.

[19] *Decreto No. 60 1964 (Aprueba e Reglamento Organico de la Dirección de Asuntos Indígenas)* [Decree No. 60/1964 (Approve the Regulation of the Directorate for Indigenous Issues)] (Chile) created the Directorate for Indigenous Issues and charged it with supporting and promoting indigenous development; *Ley No. 17.729 1972 (Establece Normas Sobre Indígenas y Tierras de Indígenas, Transforma la Dirección de Asuntos Indígenas en Instituto de Desarrollo Indígena, Establece Disposiciones Judiciales, Administrativas y de Desarrollo Educacional en la Materia y Modifica o Deroga los Textos Legales que Señala)* [Law No 17.729 1971 (Establish Norms about Indigenous People and Lands, Transform the Directorate of Indigenous Issues into an Institute of Indigenous Development, Establish Legal, Administrative and Educational Development Dispositions on the Issue and Modify or Repeal Specified Legal Texts)] (Chile) 17 contained a provision on land restitutions (Part 1, Paragraph 3), and converted the Directorate for Indigenous Issues into the Institute for Indigenous Development (Part 2).

[20] Manuel Prieto, above n. 12, 133.

[21] *Decreto Ley No 2.568 1979 (Modifica Ley N° 17.729, Sobre Proteccion de Indígenas, y Radica Funciones del Instituto de Desarrollo Indígena en el Instituto de Desarrollo Agropecuario)* [Decree Law No 2.568 1979 (Modifies Law No 17.729, about Protection of the Indigenous, and Transfers the Functions of the Institute of Indigenous Development to the Institute of Agricultural and Fishing Policy (Chile)].

It was not until the dictatorship ended in the 1990s that indigenous communities began to reassert their rights to land and resources. From taking office in 1990, President Patricio Aylwin Azocar and his transition government embarked on a social justice project.[22] The project included responding to the demands made by indigenous groups (including claims for water rights).[23] Prior to his election, at the end of 1989, Aylwin signed an agreement with the Mapuche, Huilliche, Aymara and Rapanui indigenous peoples at Nueva Imperial. In the Nueva Imperial agreement, Aylwin promised attention to the 'grave indigenous problem' in exchange for the support and defence of his future government.[24] Representatives of the Aymara and Atacameña peoples of Chile's north attended the negotiations at Nueva Imperial,[25] and took the opportunity to raise their grievances with respect to water.[26] Araya, a former lawyer with the Indigenous Development Corporation, explains that the proposal for article 64 was in fact prepared by the Aymara and Atacameña indigenous communities themselves during these consultations.[27] Aylwin promised indigenous Chileans three policies under the Nueva Imperial agreement: constitutional recognition of indigenous peoples and their economic, social and cultural rights (still before Parliament); the establishment of a national indigenous development corporation and fund for indigenous ethnodevelopment to promote the economic, social and cultural development of indigenous people; and the creation of a special commission for indigenous peoples to advise the government on indigenous issues.[28]

Soon after his election, Aylwin set up the *Comisión Especial de Pueblos Indígenas* (Special Commission for Indigenous Peoples).[29]

[22] Phillip E Berryman, 'Report of the Chilean National Commission on Truth and Reconciliation: English Translation' (University of Notre Dame Centre for Civil and Human Rights, 1993) 7–9.

[23] Comisión Especial de Pueblos Indígenas [Special Commission for Indigenous Peoples], 'Nueva Ley Indígena: Borrador de Discusión' ['New Indigenous Law : Discussion Paper']' (Discussion Paper, 1990) 17; Comisión Especial de Pueblos Indígenas, 'Nueva Ley Indígena: Borrador de Discusión [New Indigenous Law : Discussion Paper]' (La Comisión, 1990) 5.

[24] Encuentro Nacional de Pueblos Indígenas [National Alliance of Indigenous Peoples] and Don Patricio Aylwin Azocar, 'Acta de Compromiso [Agreement], Nueva Imperial' (1 December 1989).

[25] José Bengoa, *Historia de un Conflicto: el Estado y los Mapuches en Siglo XX* [History of a Conflict: The State and Mapuches in the 20th Century] (Planeta, 1999) 183.

[26] Interview with Milka Castro (Santiago, 9 November 2011); interview with Fransisco Huenchumilla (Temuco, 22 January 2014).

[27] Interview with Juan Carlos Araya (Santiago, 15 November 2011).

[28] Encuentro Nacional de Pueblos Indígenas [National Alliance of Indigenous Peoples] and Don Patricio Aylwin Azocar, above n. 24.

[29] *Decreto 30 Crea Comisión Especial de Pueblos Indígenas* [Creating the Special Commission of Indigenous Peoples] 1990 (Chile) 30. See paragraph 6 of the preamble explaining that the

The principal task given to the Special Commission was to draft a 'new indigenous law'.[30] At the same time, the transition government presented the International Labour Organization's *Convention Concerning Indigenous and Tribal Peoples in Independent Countries* (*'Convention 169'*)[31] to Congress for ratification.[32] Although Chile did not ratify *Convention 169* until 2008,[33] it was on the political agenda during the preparation of the *Indigenous Law*, and influenced the Special Commission in developing its conceptual basis.[34] The drafting of the Law followed a nationwide consultation with indigenous groups culminating in the National Congress of Indigenous Peoples in Temuco in January 1991. Representatives of indigenous groups throughout Chile participated directly in the drafting of the Law's first discussion draft, including its water provisions.

7.2.2 Water Law Frameworks and Indigenous Rights

As in the other countries considered in this book, indigenous Chileans were excluded from water law frameworks for much of Chile's history, because they did not typically hold recognised land rights. Until the 1980s, rights to use water were attached to land title in Chile.

Chile originally operated pursuant to inherited Spanish law concerning water, which established a system of *'regalias'*, as did other Latin American countries including Colombia.[35] The *regalias* were a type of *'merced'* (concession), granted by kings or lords to authorise the use of water upon the payment of a fee. When the Republic of Chile gained independence from Spain in 1818,[36] it inherited the *regalias* from the Spaniards as property of the State.[37] However, the Chilean

---

Special Commission was made up of a mixture of persons appointed by the government and indigenous organisations.

[30] *Decreto 30 Crea Comisión Especial de Pueblos Indígenas* [Creating the Special Comission of Indigenous Peoples] 1990 (Chile), art. 3.

[31] *Convention Concerning Indigenous and Tribal Peoples in Independent Countries (No. 169)* [1989] 28 ILM 1382 (Entered into force 5 September 1991) ('*Convention 169*').

[32] D Solís and A Luis, 'Memoria: Comisión Especial de Pueblos Indígenas [Memoir: Special Commission for Indigenous Peoples]' (Comisión Especial de Pueblos Indígenas, 1993) 9.

[33] *Decreto Supremo 236* [Supreme Decree 236] (Chile) 2 October 2008 (Diario Oficial [Official Gazette]14 October 2008).

[34] Wolfram Heise, 'Indigenous Rights in Chile: Elaboration and Application of the New Indigenous Law (Ley No. 19.253) of 1993', in Rene Kuppe and Richard Potz (eds.), *Law and Anthropology: International Yearbook for Legal Anthropology* (Kluwer Law International, 2001) 41.

[35] See, generally, Alejandro Vergara Blanco, *Derecho de Aguas* [Water Law] (*Editorial Jurídica de Chile*, 1998) Vol. II, 309; Carl J Bauer, *Against the Current: Privatization, Water Markets, and the State in Chile* (Springer, 1998) 36.

[36] Bernardo O'Higgins declared Chilean independence from the Spanish Empire in 1818.

[37] Vergara Blanco, above n. 35, Vol. II, 310.

government soon began to legislate concerning water rights, in recognition of the importance of irrigation for productive land use in Chile.[38]

Water use rights were not comprehensively provided for in Chilean law until the passage of Chile's first Civil Code in 1855.[39] The *Civil Code 1855* (Chile), drafted by Andes Bello under the influence of Roman law,[40] declared that rivers and all natural water courses are *'bienes nacionales de uso publico'* (national property for public use), other than waters that 'are born, flow, and die within the same property'.[41] This conceptualisation of water carries through into Chilean water law frameworks today.[42] The characterisation of waters as national property for public use means that waters are a common good belonging to the public rather than the property of the Chilean State.[43] This is similar to Australia and New Zealand, where states or governments are vested with the right to the use, flow and control of water, but water itself is a common or public resource, incapable of ownership.

The *Civil Code 1855* also stipulated that water use rights would be provided for by means of *mercedes* 'granted by the competent authority'.[44] Further, from colonial times, the owner of land abutting a watercourse was entitled to take water according to their needs, in a rule comparable to riparian rights in the British tradition.[45] However, the Chilean government eventually recognised the risk of riparian owners over utilising waterways flowing through their land and abolished these riparian-type rights in 1951, in Chile's first comprehensive Water Code.[46]

---

[38] *Decreto Supremo de 18 noviembre 1819* [Supreme Decree of 18 November 1819] (Chile) reproduced in Alejandro Vergara Blanco, 'Contribución a la Historia del Derecho de Aguas: Fuentes y Principios del Derecho de Aguas Chileno Contemporáneo (1818–1981)' [Contribution to the History of Water Law: Sources and Principles of Contemporary Chilean Water Law (1818–1981)]' (1989) 1 *Revista de Derecho de Minas y Aguas* 118, 120. The Supreme Decree of 1819 was the first Chilean water law, providing rules for irrigation, water extraction and water sales.

[39] *Código Civil de la República de Chile* [Civil Code of the Republic of Chile] 1855 ('*Civil Code*').

[40] Bauer, above n. 35, 36.

[41] *Civil Code 1855* (Chile), art. 595.

[42] *Codigo de Aguas 1981* ['Water Code'] (Chile), art. 5.

[43] See Alejandro Vergara Blanco, 'Las Aguas Como Bien Público (No Estatal) y lo Privado en el Derecho Chileno: Evolución Legislativa y su Proyecto de Reforma' [Water as a Public (Non-State) and Private Good in Chilean Law: Legislative Evolution and Reform]' (2002) 1 *Revista De Derecho Administrativo Económico* 66.

[44] *Civil Code 1855* (Chile), art. 860. These *mercedes* were governed by public administrative law and could be cancelled by the state if desired.

[45] Ibid. art. 834.

[46] *Código de Aguas 1951* ['Water Code 1951 (Chile)'].

The *Water Code 1951* (Chile) provided that the right to use water (which it called a *'derecho de aprovechamiento'*) could be acquired by virtue of an administrative grant from the President of the Republic. Article 12 of the 1951 Code provided:

> A water use right is a real property right that attaches to waters in the public domain and consists of the right to use, enjoy and dispose of the water with the requirements and in conformity with the rules prescribed.

In the context of Chilean civil law, a *'derecho real'*[47] is an enforceable right in relation to a thing (similar to the Latin common law concept *in rem*), as opposed to a right enforceable against a particular person (similar to the Latin *in personam*).[48]

Derechos de aprovechamiento under the *Water Code 1951* were governed by private civil law and could technically be traded with the approval of canal users' associations.[49] However, the use and location could not change without requesting a new concession, which undermined the transferability of *derechos de aprovechamiento* separate from landholding. In a practical sense, and like the riparian right-type *mercedes* that preceded them, *derechos de aprovechamiento* under the 1951 Code were contingent on land title.

The *Water Code 1951* was amended substantially in 1967, as part of Chile's agrarian reform. The agrarian reform 'aimed to expropriate and redistribute large landholdings with the twin purposes of expanding the class of small landowners and modernising agricultural production'.[50] The 1967 amendment declared all waters to be *'bienes nacionales de uso publico'* (national property for public use), which was relied upon to expropriate private *derechos de aprovechamiento* for the purpose of redistribution. In order to provide for the expropriation of *derechos de aprovechamiento*, the government also had to amend the constitutional provision protecting private property rights.[51]

---

[47] *Civil Code 1855*, art. 577, which provides that a 'real right is one had over a thing without respect to a particular person'. Article 577 sets out the various different types of *derechos reales*, which are: *'dominio'* (otherwise called *'propiedad'*) (ownership), *'herencia'* (legacy), *'usufructo'* (usufruct), *'uso o habitación'* (use or habitation), *'servidumbres activas'* (easement), *'prenda'* (pledge), and *'hipoteca'* (mortgage).
[48] See generally Samantha Hepburn, *Principles of Property Law* (Cavendish Publishing, 1998) 2, 20.
[49] See generally Bauer, above n. 35, 38.
[50] *Ley No 16.615 Modifica la Constitución Política del Estado 1967* [Law No 16.615 to Modify the Political Constitution of State] 1967 (Chile).
[51] Ibid. art. 10.

The redistribution of *derechos de aprovechamiento* in the interests of 'rational and beneficial use' was tied to patterns of agricultural land use as determined by state planning agencies.[52] As a result of the 1967 amendment, *derechos de aprovechamiento* returned to their earlier status as mere administrative concessions. Public administrative rather than private civil law now governed their administration, and their transfer separate from landholding was prohibited in the absence of a particular administrative approval.[53]

Chilean water law frameworks were overhauled during the military dictatorship of the 1970s and 80s, introducing an 'integrated-market' approach to water regulation. The new approach to water regulation, similar to contemporary water regulation in Australia, combined centralised water regulation with trade in unbundled *derechos de aprovechamiento* in water markets. Chilean water law reform was part of a wider project of neoliberal reform implemented by the military dictatorship across a range of sectors, and was accompanied by rapid growth in water-related development such as mining and hydroelectricity.[54]

The unbundling of *derechos de aprovechamiento* from land titles was initiated in a Decree signed by the Military Junta in 1979. *Decree Law 2.603 1979* (Chile) was intended to support the *'normalizacion'*[55] (normalisation) of the different types of *derechos de aprovechamiento* in operation in Chile in order to make way for the new *Water Code*. In order to facilitate the transferability of *derechos de aprovechamiento* in water markets, there was (and is) little regulatory control over the purposes for which they may be exercised,[56] merely being classified as 'consumptive' or 'non-consumptive'.[57] A 'consumptive' *derecho de aprovechamiento* 'allows its holder to totally consume the waters in any activity'.[58] Until 2005, applicants for *derechos de aprovechamiento* were not required to specify what they would use the water for. Since the 2005 amendment to the *Water Code* there has been a requirement in

---

[52] Ibid. art. 10; see generally Bauer, above n. 35, 39.
[53] *Ley No 16.615 Modifica la Constitución Política del Estado 1967* [Law No 16.615 to Modify the Political Constitution of State] *1967* (Chile), art. 10.
[54] See generally Bauer, *Siren Song: Chilean Water Law as a Model for International Reform*, above n. 12, 4.
[55] *Decreto Ley 2.603 Modifica y Complementa Acta Constitucional N° 3; y Establece Normas Sobre Derechos de Aprovechamiento de Aguas y Facultades para el Establecimiento del Regimen General de las Aguas* [Decree Law 2.603 to Modify and Complement Constitutional Act 3; and Establish Rules about Water Rights and Arrangements for the Establishment of a General Water Regime] *1979* ('*Decree Law 2.603*') (Chile), art. 3.
[56] *Water Code 1981*, arts. 131–2, 140(6).
[57] Ibid. art. 12.
[58] Ibid. art. 13.

articles 131–2 to explain proposed use for applications above a certain flow, although the proposed use is not binding. Nor is there any restriction on the holders of *derechos de aprovechamiento* changing the use of water, say from agriculture to mining or tourism to domestic consumption, in the interests of facilitating the free transfer of *derechos de aprovechamiento* within the market.

The *Water Code*, together with the *Constitution 1980* (Chile),[59] retained the provision that waters are *bienes nacionales de uso publico* but reinstated the governmental power to grant private water use rights.[60] The *Water Code* also authorises the use of water in two ways, both of which attract protection as '*propiedad*' under the Constitution.[61] First, as authorised by a *derecho de aprovechamiento*, which can be translated as a 'use and enjoyment right' and is effectively an administrative concession.[62] Secondly, where authorised by legislation recognising a right to use water, in which case there is no need for a *derecho de aprovechamiento*.[63] An example of such an 'as of right' entitlement is provided for in article 10 of the *Water Code*, which provides that, '[t]he use of rainwater that falls or is collected on private property corresponds to the owner of the property, provided that they flow within the property and do not fall into natural courses of public use. Accordingly, the owner may store them within its property using adequate measures, provided this does not prejudice the rights of other rightholders'. These statutory use rights are similar to 'basic landholder rights' to use water 'as of right' in the common law context, typically for domestic and stock purposes.

The characterisation of *derechos de aprovechamiento* as *propiedad* (or *dominio*) under Chilean civil law means that the rights are equivalent to rights of ownership. Article 6 of the *Water Code* provides:[64]

---

[59] *Constitución Política de la República de Chile* [Political Constitution of the Republic of Chile] *1980* ('Constitution'), art. 19(24).

[60] *Water Code 1981* (Chile), art. 5.

[61] *Constitution 1980*, art 19(24), which provides that '[p]rivate water rights, recognised or created in accordance with the law, give a right of ownership to their holders'.

[62] *Water Code 1981*, art 20; see generally Vergara Blanco, above n. 35 Vol. II, 321, characterising a *derecho de aprovechamiento* as a 'concession' created by administrative act; but see Carolina de Lourdes Riquelme Salazar, '*El Derecho al Uso Privativo de las Aguas en España y Chile: Un Estudio de Derecho Comparado*' [Exclusive Water Rights in Spain and Chile: A Comparative Law Study] (Universitat Rovira i Virgili, 2013) 356. Riquelme Salazar argues that a *derecho de aprovechamiento*, is not in fact an administrative concession but a *sui generis* '*acto de autoridad*' or 'administrative act'.

[63] See also *Water Code 1981*, arts. 11, 20, 56.

[64] *Civil Code 1855* (Chile), art. 589(2).

A water use right is a real property right that attaches to water and consists of its use and enjoyment, in accordance with the rules established by this code. A water use right is the domain of its holder, who can use, enjoy and dispose of it in accordance with the law.

In fact, the right of *propiedad* is the strongest form of property right available under Chilean law.[65] It is described as an absolute, exclusive, and perpetual right to use, enjoy and dispose of a thing.[66] The right of *propiedad*, therefore, aligns with the common law estate of fee simple, which entails a right to possession to the exclusion of all others.[67]

Under the *Water Code*, indigenous groups would need to hold a *derecho de aprovechamiento* in order to lawfully access and use water, unless they held a legislative entitlement to use water as of right. The *Water Code* provides three ways in which a person can acquire a *derecho de aprovechamiento*:[68]

1. via the constitution of a new *derecho de aprovechamiento* as an administrative act pursuant to article 20;
2. via the judicial 'regularisation' of unregistered, or 'customary', water use that is recognised by legislation without need for an administrative act under transitional article 2 of the *Water Code*; or
3. by private bargaining in water markets.

However, few indigenous people sought to access *derechos de aprovechamiento* (via constitution, regularisation or market processes) after the passage of the *Water Code* in 1981. Indigenous Chileans were typically unaware of its passage until many years later, and did not know about the general ability to create new water use rights free of charge, or regularise their unregistered water use.[69] The *Water Code* was prepared and passed privately during the military dictatorship and was not subject to usual democratic debate and publicity. Even if indigenous people were aware of the new methods for acquiring water rights under the Code, many would be unable to satisfy the requirement to prove productive water use necessary to regularise their traditional

---

[65] Ibid. art. 577.
[66] Ibid. art. 582.
[67] *Yanner v. Eaton* (1999) 201 CLR 351 [25]. See also *Wurridjal v. The Commonwealth* 237 CLR 309 [295] Kirby J.
[68] See generally Carolina de Lourdes Riquelme Salazar, above n. 62, 214. Riquelme Salazar classifies the first two avenues as 'originating' while the third is 'derivative'.
[69] Jessica Budds, 'The 1981 Water Code: The Impacts of Private Tradeable Water Rights on Peasant and Indigenous Communities in Northern Chile' in William L Alexander (ed.), *Lost in the Long Transition: Struggles for Social Justice in Neoliberal Chile* (Lexington Books, 2009) 54; interview with Daniela Rivera (Santiago, 22 November 2011).

interests. In order to regularise a *derecho de aprovechamiento*, an applicant must prove continuous *'uso efectivo'* ('productive use')[70] for a period of five years prior to the passage of the *Water Code* (i.e. 1976).[71] Because indigenous Chileans had been historically excluded from both land title and water use rights, their access to water resources was often obstructed by other users.

Until the passage of the *Indigenous Law* in 1993, it was not possible for indigenous Chileans to obtain communal water use rights under the *Water Code*. Before the *Indigenous Law*, indigenous groups did not enjoy legal personality,[72] separate from that of their individual members, necessary to create or regularise *derechos de aprovechamiento* as a group under the *Water Code*.[73] The only option left to indigenous Chileans would have been to acquire water use rights as individuals, unless they incorporated as a 'water association' under the Code.[74] However, incorporating as a water association was not an attractive prospect for indigenous groups because, despite acting in common, the members of water associations exercise their rights to use, enjoy and dispose of their *derechos de aprovechamiento* in an individual manner.[75] This means that any one individual could alienate their 'shares' in a water association without the consent of the other members.[76] Prieto, in empirical research on indigenous water holdings in northern Chile, found that indigenous groups in the region of Antofagasta were not comfortable with holding individual water use rights under the universal provisions of the *Water Code* because any individual holder or member could alienate their water use rights without the consent of the rest of the group.[77] This was at odds with the group-based nature of indigenous social organisation. Some indigenous groups in the region of

---

[70] *Decree Law 2.603 1979*, art. 3.

[71] *Water Code 1981* trans, art. 2.

[72] Solís and Luis, above n. 32, 32.

[73] *Water Code 1981*, art. 5, which provides that 'particulares' [individuals] can hold *derechos de aprovechamiento*; see also Carolina de Lourdes Riquelme Salazar, above n. 62 240, who explains that the concept of *'particulares'* encompasses natural and legal persons acting in the private realm.

[74] *Water Code 1981*, arts. 187–282. The various water associations are *'comunidades de aguas'*, *'asociaciones de canalistas'* and *'juntas de vigilancia'*.

[75] Ibid. art. 193, which provides, '[t]he right of each one of the members in the common flow comes from their respective titles'.

[76] See also Chile, 'Primer Informe Comisión Especial Pueblos Indígenas' [First Report of the Special Commission for Indigenous Peoples], Cámara de Diputados [House of Representatives], 10 November 1992, in Biblioteca del Congresso Nacional de Chile [Library of National Congress of Chile], above n. 9, 64.

[77] Manuel Prieto, above n. 12, 201–4; interview with Manuel Prieto (Santiago, 4 September 2013).

Antofagasta acquired *derechos de aprovechamiento* as water associations using the general regularisation provisions of the *Water Code* prior to 1993, although it was by no means common.[78]

Further, the complicated administrative and judicial processes for obtaining *derechos de aprovechamiento* by constitution or regularisation usually required the applicant to obtain legal representation and pay court costs. These legal and court costs meant that the processes were inaccessible to most indigenous groups who were (and are) the poorest sector of Chilean society. Finally, some indigenous groups viewed acquiring *derechos de aprovechamiento* under the *Water Code* as unnecessary. Why would they take the time and expense of registering the use of water that was rightfully theirs?[79] If indigenous groups wanted to purchase *derechos de aprovechamiento* in the market, they would have to compete for water use rights at market prices. This was often impossible for indigenous groups with limited financial means.

Meanwhile, the unbundling of *derechos de aprovechamiento* from land titles under the 1981 *Water Code* enabled other water users to accumulate water rights at the expense of indigenous Chileans. Because *derechos de aprovechamiento* were now held separately from landholding, and were available for purchase in water markets, other users could acquire rights to use water resources on or affecting indigenous owned or occupied lands. Chile's rapid growth in water-related development such as mining and hydroelectricity during the military dictatorship, and increased competition for water in water markets, assisted encroachment on indigenous water resources. In the north, mining projects acquired large numbers of *derechos de aprovechamiento* in the 1980s.[80] After the exhaustion of surface waters, many mining projects turned to subsurface water resources, which placed further strain on aquifers supporting 'customary' indigenous water use.[81]

Where other users sought water use rights via the mechanisms of constitution or regularisation, the *Water Code* did not account for water use by indigenous people. Indigenous water use was not recorded or

---

[78] See, e.g., *No 13 Comunidad de Aguas 'Canal dos de Quillagua'* (Regularisation decision 619/155, Conservador de Bienes Raises y Comercio de Tocopilla [Real Estate and Business Office of Tocopilla], 10 December 1986); interview with Manuel Prieto (Santiago, 4 September 2013), above n. 77.

[79] Interview with Daniela Rivera (Santiago, 22 November 2011), above n. 69; Interview with Milka Castro (Santiago, 9 November 2011), above n. 26.

[80] Interview with Gonzalo Arevalo (Santiago, 18 November 2011); interview with Maria Angelica Alegria (Santiago, 17 November 2011).

[81] Interview with Milka Castro (Santiago, 9 November 2011), above n. 26; Interview with Nancy Yanez (Santiago, 22 November 2011).

registered anywhere, and was therefore not considered by the administrative and judicial bodies allocating *derechos de aprovechamiento*.[82] The *Water Code* did provide for public notification and objection processes where others sought to create or regularise water use rights.[83] However, few indigenous groups had access to the Official Gazette, radio or even local newspapers in order to be notified of new applications for *derechos de aprovechamiento*.[84] Thus, by the end of the 1980s, it was uncommon for indigenous Chileans to hold *derechos de aprovechamiento* and other users held almost all of the water use rights in Chile.[85] This distributive injustice left some indigenous groups with land rights but no right to use the water on the land. For those groups, the lack of water rights seriously undermined the productive potential of their lands, and was a practical barrier to their economic development.

7.3 RECOGNISING INDIGENOUS WATER RIGHTS IN CHILE

When Chile returned to democracy in 1990, Aylwin's transition government embarked on an ambitious project of social justice reform, within which he included addressing the rights of indigenous Chileans.[86] The Special Commission charged with drafting the *Indigenous Law* recommended the protection of *ancestral* water rights of indigenous groups in northern Chile together with finance for acquisition and allocation of water use rights for groups throughout the country.[87]

The Special Commission's recommendations culminated in article 64 of the *Indigenous Law*, which protects the *ancestral* water rights of the Aymara and Atacameña indigenous communities of northern Chile,

---

[82] It was only after 1992 that the General Water Directorate began to keep track of water rights that were 'regularisable'. See Vergara Blanco, above n. 35, Vol. II, 348.

[83] *Water Code 1981*, arts. 131–3.

[84] Interview with Gonzalo Arevalo (Santiago, 18 November 2011), above n. 80.

[85] Chile, 'Primer Informe Comisión Especial Pueblos Indígenas' [First Report of the Special Commission for Indigenous Peoples], Cámara de Diputados [House of Representatives], 10 November 1992, in Biblioteca del Congreso Nacional de Chile [Library of National Congress of Chile], above n. 9, 9.

[86] Interview with Fransisco Huenchumilla (Temuco, 22 January 2014), above n. 26. Huenchumilla was the President's representative on the Comisión Especial de Pueblos Indígenas, responsible for developing the draft *Indigenous Law*.

[87] Chile, 'Primer Informe Comisión Especial Pueblos Indígenas' [First Report of the Special Commission for Indigenous Peoples], Cámara de Diputados [House of Representatives], 10 November 1992, in Biblioteca del Congreso Nacional de Chile [Library of National Congress of Chile], above n. 9, 64.

and article 20(c), which funds the allocation of water use rights to indigenous groups throughout Chile. Article 64 has been implemented almost exclusively as part of the judicial process of regularisation of unregistered water rights into *derechos de aprovechamiento* provided for in transitional article 2 of the *Water Code*.

Article 20 establishes the *Fondo de Tierras y Aguas Indígenas* (Indigenous Land and Water Fund). The main task of the Indigenous Land and Water Fund is to 'grant subsidies for the acquisition of lands for indigenous people or communities (or part thereof) where the surface area of the respective community's lands is insufficient ... '.[88] However, its functions also include, at 20(c), 'to finance the constitution, regularisation or purchase of water rights or finance works destined to obtain the resource'. The Fund has been used to finance the constitution, purchase and regularisation of *derechos de aprovechamiento* for indigenous groups throughout Chile (including those relying on article 64).

Aside from these two main water provisions, the *Indigenous Law* now includes other provisions aimed at addressing water injustices. The first of these is article 65, which requires the *Corporación Nacional de Desarollo Indígena* (National Indigenous Development Corporation) to create special programmes for the recuperation and repopulation of communities and sectors currently abandoned by the Aymara and Atacameña ethnicities. Article 65 serves as an acknowledgement of indigenous migration away from their traditional territories as a consequence of water dispossession. A 1992 amendment to the *Water Code*[89] also inserted two new provisions prohibiting (without express permission) explorations and extractions of subterranean waters (or 'groundwater') from aquifers that supply certain wetlands in the first and second regions.[90] These amendments to the *Water Code* responded to concerns that wetlands on which indigenous groups depended for agriculture had dried up as a consequence of over-extraction of groundwater by mining companies in Chile's north.[91]

---

[88] *Ley No 19.253 (Establece Normas Sobre Protección, Fomento y Desarollo de los Indígenas, y Crea la Corporación Nacional de Desarrollo Indígena)* [Law No 19.253 (Establish Norms for the Protection, Creation and Development of the Indigenous, and to Create the National Corporation of Indigenous Development)] 1993 (*'Indigenous Law'*) (Chile), art. 20(a).

[89] *Ley 19.145 Modifica Articulos 58 and 63 Codigo de Aguas* [Law 19.145 to Modify Articles 58 and 63 of the Water Code] 1992 (Chile), arts. 1–2.

[90] *Water Code 1981*, arts. 58(5), 63(3) respectively.

[91] See generally Yañez and Molina, Chapter 2 above n. 7, 106.

7.3.1 The Rationale for Recognition

The Special Commission reports and minutes, and Parliamentary debates for the *Indigenous Law*, provide much detail about the cultural, reparative and distributive aims of the new water provisions. Following the global indigenous rights movement that started in the 1960s a wave of '*panindianismo*' swept across in Latin America during the 1970s and into the 80s,[92] seeking the recognition of indigenous cultural diversity and identity in response to assimilationist policies directed at indigenous groups.[93] Lawmakers preparing Chile's indigenous water rights provisions were concerned with recognising indigenous cultural difference,[94] and reversing the negative impacts of the assimilationist *Chilenizacion* policies.[95] Accordingly, the *Indigenous Law* observes the particular cultural relationships indigenous groups have with land and resources, as follows:[96]

> The State recognises that the indigenous of Chile are the descendants of the human groupings that exist on national territory since pre-Columbian times, who maintain their own ethnic and cultural manifestations and that for them the land is the principal fundament of their existence and culture.

After listing by name the nine indigenous ethnicities recognised by the State (Mapuche, Aymara, Rapa Nui, Atacameña, Quechua, Colla, Diaguita, Kawashkar and Yámana), article 1 goes on to 'value their existence as an essential part of the races of the Chilean nation, in addition to their integrity and development, in accordance with their customs and values'.[97]

---

[92] Engle, Chapter 2 above n. 16, 55, 59; Jorge Iván Vergara, Hans Gundermann and Rolf Foerster, 'Legalidad y Legitimidad: Ley Indígena, Estado Chileno y Pueblos Originarios (1989–2004) [Legality and Legitimacy: Indigenous Law, the State, and Native Peoples in Chile (1989–2004)]' (2006) 24(71) *Estudios Sociologicos* 331, 344–5.

[93] See generally Engle, Chapter 2 above n. 16, 55.

[94] Chile, 'Discusion Sala' [Parliamentary Debates], Cámara de Diputados [House of Representatives], 21 January 1993 (Octavio Jara) in Biblioteca del Congresso Nacional de Chile [Library of National Congress of Chile], above n. 9, 150–1.

[95] See Chile, 'Primer Informe Comisión Especial Pueblos Indígenas' [First Report of the Special Commission for Indigenous Peoples], Cámara de Diputados [House of Representatives], 10 November 1992 Biblioteca del Congresso Nacional de Chile [Library of National Congress of Chile], above n. 9 62.

[96] Indigenous Law 1993, art. 1.

[97] But see Chile, 'Mensaje Presidencial' [Presidential Message], Cámara de Diputados [House of Representatives], 15 October 1991 (Patricio Aylwin Azocar) in Biblioteca del Congresso Nacional de Chile [Library of National Congress of Chile], above n. 9, 9. In the first draft of the Bill presented to the House of Representatives, article 1 went further to acknowledge that their ethnic and cultural manifestations are 'distinct from the other habitants of the Republic, such as systems of life, ways of living, customs, work methods, language, religion or any other

The *Indigenous Law* emerged from a political climate of reparative or transitional justice, during which the transition government sought to repair the relationship between the State and its constituents, which had deteriorated during Chile's long and painful dictatorship. This is evident in the foreword to the Report of the Chilean National Commission on Truth and Reconciliation:[98]

> The lesson for the Aylwin administration was that it should stake out a policy it could sustain. Reparation and prevention were defined as the objectives of the policy. Truth and justice would be the primary means to achieve such objectives. The result, it was expected, would be to achieve a genuine reconciliation of the divided Chilean family and a lasting social peace.

The *Indigenous Law* was the core of the transition government's effort to repair its relationship with indigenous Chileans;[99] a 'historical debt', which had been characterised by 'a history of domination, of usurpation ... and of political, legal, ideological and cultural subordination.'[100] The President explained:[101]

> We believe that this is an opportune moment to reflect upon ourselves, about our history, about the relationship between us, the relationship between the mixed, creole and indigenous societies, aboriginal of our nation. It is a propitious moment to rethink our culture, to turn our eyes back to ourselves and ask ourselves who we are and, using our past as a starting point, look into the future.

Paying the 'historical debt' to the indigenous would require redress for the loss of land and resource rights.[102] As part of this, according to the Special Commission, the failure of Chilean water law frameworks to recognise '*ancestral*' water rights disadvantaged indigenous groups (particularly in the north).[103]

form of indigenous cultural manifestation', although this wording did not make it into the final law.

[98] Berryman, above n. 22, 14.

[99] Chile, 'Primer Informe Comisión Especial Pueblos Indígenas' [First Report of the Special Commission for Indigenous Peoples], Cámara de Diputados [House of Representatives], 10 November 1992, in Biblioteca del Congresso Nacional de Chile [Library of National Congress of Chile], above n. 9, 67.

[100] Chile, 'Discusión Sala' [Parliamentary Debates], Cámara de Diputados [House of Representatives], 21 January 1993 (Kumicic) in ibid. 150.

[101] Chile, 'Mensaje Presidencial' [Presidential Message], Cámara de Diputados [House of Representatives], 15 October 1991 (Patricio Aylwin Azocar) in ibid. 8.

[102] Comisión Especial de Pueblos Indígenas, above n. 23, 7.

[103] Chile, 'Primer Informe Comisión Especial Pueblos Indígenas' [First Report of the Special Commission for Indigenous Peoples], Cámara de Diputados [House of Representatives],

In the original discussion paper prepared for the *Indigenous Law* in 1990, the Special Commission proposed that indigenous peoples would be guaranteed the right to use water on their lands.[104] Such a provision would be an exception to the unbundled status of *derechos de aprovechamiento*, preventing other users from obtaining water use rights on indigenous lands. Presumably, the provision would also override the *derechos de aprovechamiento* already held by others in water resources on indigenous lands, although it was not suggested how the provision might be implemented, or whether third-party compensation would be provided. According to Huenchumilla, a former Chilean politician and member of the Special Commission preparing the Law, the project for the *Indigenous Law* never considered compulsory expropriation of water rights from other holders, the intention was only ever that rights would be purchased from other users.[105] However, at least one senator expressed concern about the inconsistency of such a provision with the market-based regulatory framework for water use rights under the *Water Code 1981*, pursuant to which *derechos de aprovechamiento* were held independent of land title.[106] In any event, in the first draft of the legislation presented to Congress the proposal had been pared back to prohibiting new water extractions affecting those waters that supply indigenous communities except in the case of express declaration.[107] As will be discussed below, the final version of article 64 only allows for the recognition of Aymara and Atacameña water rights to the extent that this does not prejudice other holders of *derechos de aprovechamiento*.

When embarking on the project of indigenous reform, the Special Commission was mindful of the fact that indigenous Chileans were 'the poorest of the poor'.[108] The overarching objective of the Law is therefore to 'promote indigenous development'.[109] While being rooted in a distinct culture, the recognition of indigenous land and water rights in

---

10 November 1992, in Biblioteca del Congresso Nacional de Chile [Library of National Congress of Chile], above n. 9, 64.

[104] Comisión Especial de Pueblos Indígenas, above n. 23.

[105] Interview with Fransisco Huenchumilla (Temuco, 22 January 2014), above n. 26.

[106] This was Senator Alessandri. See Chile, 'Primer Informe Comisión Especial Pueblos Indígenas' [First Report of the Special Commission for Indigenous Peoples], Cámara de Diputados [House of Representatives], 10 November 1992, in Biblioteca del Congresso Nacional de Chile [Library of National Congress of Chile], above n. 9, 347.

[107] Chile, 'Mensaje Presidencial' [Presidential Message], Cámara de Diputados [House of Representatives], 15 October 1991 (Patricio Aylwin Azocar) in ibid. 36. This draft also included (as original article 18) a provision giving indigenous landholders preferential rights for the constitution of water rights in areas of indigenous development.

[108] Comisión Especial de Pueblos Indígenas, 'Memoria: Comisión Especial de Pueblos Indígenas' [Memoir: Special Commission for Indigenous Peoples] (1993) 29.

[109] *Indigenous Law 1993*, art 1.

the *Indigenous Law* was intended to support indigenous economic progress, in an approach known as 'culture plus development'. This
approach was a reinterpretation of the 'ethnodevelopment' movement
influential Latin America in the late twentieth century as a means of
addressing structural economic inequality.[110] At the presentation of
the *Indigenous Law* before Parliament, President Aylwin emphasised
that, 'defence of culture and development are not two opposed situations. On the contrary, they are complementary'.[111] The Special
Commission further explained:[112]

> The spirit of this new indigenous law is totally different. It establishes the
> diversity of cultures existing in Chilean society and promotes their
> development. Accordingly, it recognises the character of indigenous
> groups and the right they have to develop according to their own
> cultural criteria and custom.

Those preparing the *Indigenous Law* probably had a particular form
of indigenous water use in mind when they drafted its water provisions, namely, the use of water for small-scale irrigated agriculture
using traditional methods within rural communities. Indeed, the
President remarked when presenting the law to Parliament that indigenous people have a 'particular cultural relationship in harmony
with nature, calling into question the type of progress to which
Chileans aspire'.[113] However, it was never assumed that indigenous
groups, and their resource use, must align with traditional, or pre-
sovereignty, practices or technology. The Special Commission made
this clear when it said that, 'this valorisation [of indigenous culture]
does not imply a level of folkloric conservatism which seeks to maintain these cultures in living laboratories or museums'.[114] President
Aylwin referred to indigenous Chileans', 'strong desire to conserve
their culture, their identity, and each of their idiosyncrasies, and

---

[110] See generally Engle, Chapter 2 above n. 16.

[111] Chile, 'Mensaje Presidencial' [Presidential Message], Cámara de Diputados [House of
Representatives], 15 October 1991 (Patricio Aylwin Azocar) in Biblioteca del Congresso
Nacional de Chile [Library of National Congress of Chile], above n. 9, 8.

[112] Comisión Especial de Pueblos Indígenas, above n. 23, 5. It is in the agreement at Nueva
Imperial that we first see a reference to 'ethnodevelopment'. Encuentro Nacional de Pueblos
Indígenas [National Alliance of Indigenous Peoples] and Don Patricio Aylwin Azocar, above
n. 24, cl. 2.

[113] See, e.g., Chile, 'Mensaje Presidencial' [Presidential Message], Cámara de Diputados [House of
Representatives], 15 October 1991 (Patricio Aylwin Azocar) in Biblioteca del Congresso
Nacional de Chile [Library of National Congress of Chile], above n. 9, 8.

[114] Comisión Especial de Pueblos Indígenas, above n. 108, 30.

further to progress, develop and incorporate their knowledge into the modern world'.[115]

The Special Commission warned that indigenous groups want 'to conserve their values and culture but do not want to pay the cost of remaining in extreme poverty'.[116] The re-titling of indigenous lands and development finance were presented in Chilean debates on the *Indigenous Law* as key to indigenous economic development.[117] Indigenous water rights were positioned as a necessary precondition for the development of indigenous lands, noting that in some situations indigenous groups held rights to land (at least to occupy land) without corresponding rights to use the water on that land.[118]

The main cause of encroachment, according to the Special Commission, was the unbundling of *derechos de aprovechamiento* from land title and the emergence of water markets under the *Water Code*.[119] Unbundling, and markets, had allowed others to accumulate water use rights at the expense of indigenous peoples. At the time of drafting the *Indigenous Law* the pendulum had swung away from the extreme market-centred reforms of the dictatorship and Latin America had embraced the 'post-Washington Consensus'; the idea that governments should intervene in markets to support social outcomes, particularly for poor and vulnerable groups to secure essential services such as water. Ruckert describes this policy shift as a more socially 'inclusive' form of neoliberalism. While this inclusive neoliberalism still promotes privatisation, liberalisation and deregulation initiatives along free-market lines, it also recognises that the poor cannot afford basic social services (or public goods) such as healthcare, education and housing at market rates and require financial

---

[115] Chile, 'Mensaje Presidencial' [Presidential Message], Cámara de Diputados [House of Representatives], 15 October 1991 (Patricio Aylwin Azocar) in Biblioteca del Congreso Nacional de Chile [Library of National Congress of Chile], above n. 9, 6.

[116] Comisión Especial de Pueblos Indígenas, above n. 108, 30.

[117] See, e.g., Chile, 'Discusion Sala' [Parliamentary Debates], Cámara de Diputados [House of Representatives], 21 January 1993 (Garcia) in Biblioteca del Congreso Nacional de Chile [Library of National Congress of Chile], above n. 9, 158.

[118] Chile, 'Discusion Sala' [Parliamentary Debates], Cámara de Diputados [House of Representatives], 21 January 1993 (Garcia) in ibid. in which Garcia discusses the need to 'promote the adequate exploitation of indigenous lands'; Chile, 'Discusion Sala' [Parliamentary Debates], Senado [Senate], 30 June 1993 (Alessandri) in ibid. 453. Alessandri discusses Aymara and Atacameña irrigated agriculture as an economic activity.

[119] Chile, 'Primer Informe Comisión Especial Pueblos Indígenas' [First Report of the Special Commission for Indigenous Peoples], Cámara de Diputados [House of Representatives], 10 November 1992, in Biblioteca del Congreso Nacional de Chile [Library of National Congress of Chile], above n. 9, 64.

help.[120] Thus, water markets were perceived as being hostile to indigenous interests, and government intervention would be required to provide a fair share of *derechos de aprovechamiento* to indigenous groups. According to the former Director of the General Water Directorate, Rodrigo Weisner, the necessity for recognition of particular indigenous water rights in Chile arose because the market is not designed to function for non-commercial interests.[121]

Nevertheless, the government was concerned that this process should not undermine the certainty of other water use rights. The fact that the *Indigenous Law*, and its water provisions, enjoyed majority support in the Chilean Parliament may perhaps be explained by the uncertainty posed to other water use rights, and associated investment, by indigenous political demands for water rights.[122] The proper functioning of a water market requires water use rights to be sufficiently clear and certain to encourage water trade and investment.[123] The neoliberal reforms of the 1980s had repositioned the Chilean economy as one that depended on foreign investment.[124] Chilean politicians in the early 1990s were concerned about the certainty of water use rights in order to attract investment in development projects that depended on water, such as hydroelectric and mining ventures. Although indigenous groups in Chile were excluded from water law frameworks allocating formal water use rights from the time of the Spanish acquisition of sovereignty, some indigenous groups continued to take and use water despite holding no state-sanctioned rights. This was a common occurrence in formerly remote areas of Chile's north, which were becoming increasingly subject to competition for water use rights from mining developments.[125] The ongoing customary use of water by indigenous groups, and their increasing political demands for water

---

[120] See generally Arne Ruckert, 'Towards an Inclusive-Neoliberal Regime of Development: From the Washington to the Post-Washington Consensus' (2006) 39(1) *International Development Studies* 34, 36, 42.

[121] Interview with Carlos Herrera Inzunza (Temuco, 11 November 2011); Interview with Rodrigo Weisner (Santiago, 18 November 2011).

[122] See Biblioteca del Congreso Nacional de Chile [Library of National Congress of Chile], above n. 9, 576–647, which breaks the voting down by provision, showing a strong majority in favour of each provision.

[123] Lin Crase, *Water Policy in Australia: The Impact of Change and Uncertainty*, Issues in Water (Resources for the Future, 2011).

[124] Carl J Bauer, 'In the Image of the Market: The Chilean Model of Water Resource Management' (2005) 3(2) *International Journal of Water* 146, 150.

[125] Interview with Milka Castro (Santiago, 9 November 2011), above n. 26; Interview with Nancy Yanez (Santiago, 22 November 2011), above n. 81.

rights, presented ongoing uncertainty for other water users and water markets more generally.

7.3.2 The Recognition of 'Ancestral Rights'

Article 64 of the *Indigenous Law* recognises ancestral water rights for certain Chilean indigenous groups as follows:

> The waters of the Aimara and Atacameña communities must be especially protected. Waters, including rivers, canals, streams and springs, found on the lands of the Indigenous communities established by this law will be considered property of ownership and use of the Indigenous communities, without prejudice to the rights that other rightholders have registered in accordance with the Water Code.
>
> New water rights must not be granted over lakes, ponds, springs, rivers and other aquifers that supply waters owned by the various Indigenous communities established by this law without first guaranteeing normal water supply to the affected communities.

Article 64 refers only to the water rights of the Aymara and Atacameña indigenous communities.[126] However, article 62 of the *Indigenous Law* provides that the specific provisions for the Aymara and Atacameña apply to other ethnicities from the north, such as the Quechua and Colla. Additionally, a number of indigenous communities have relied upon ILO *Convention 169* to recognise their water rights under article 64 despite not belonging to Aymara or Atacameña ethnicities.[127]

Like indigenous rights to water in the other countries in this book, the rights are recognised as 'communal' rights. Article 64 recognises the water rights of 'indigenous communities', which are 'constituted' and registered under the *Indigenous Law*.[128] The *Indigenous Law* provides the following definition of 'indigenous community':[129]

---

[126] Article 62 classifies as 'Aymara' those indigenous persons who belong to Andean communities located in the I Region and as 'Atacameño' those indigenous that belong to villages in the interior of the II Region, which also only refers to these two ethnicities.

[127] Interview with Juan Carlos Araya (Santiago, 15 November 2011), above n. 27. See also David Espinoza Quezada, 'Regularizaciones Remitidas por DGA Region de Antofagasta a Tribunales Competentes' [Regularisations Remitted by the DGA in the Region of Antofagasta to Competent Courts]' showing a number of regularisations to Quechua communities.

[128] Indigenous communities are constituted in accordance with articles 9 and 10 of the *Indigenous Law*. According to the National Indigenous Development Corporation's website, as at 25 September 2015 there were 3,213 registered communities with 125,033 members and 1,843 registered associations with 69,660 members listed in the Register of Indigenous Communities and Associations. See Interview with Carlos Herrera Inzunza (Temuco, 11 November 2011), above n. 121.

[129] *Indigenous Law 1993*, art. 3.

> Indigenous Community, for the purpose of this law, means the entire group of persons belonging to the same indigenous ethnicity and that come within one or more of the following situations: (a) They come from the same family tree; (b) They recognise a traditional leadership; (c) They possess or have possessed indigenous communal lands; and (d) They come from the same historic people.

The recognition of indigenous water rights under article 64 only applies on indigenous lands. Article 64 provides that '[w]aters ... found in the lands of the Indigenous communities established by this law will be considered property of ownership and use of the Indigenous communities'.[130] Article 64, therefore, seems inconsistent with the unbundled system of *derechos de aprovechamiento* in Chile, in which water use rights are held and may be transferred independently of land title. In fact, the proposal to recognise *ancestral* water rights in Chile's north, from inception, reflected a desire to reverse the impact of unbundling and secure to indigenous landholders the right to use water on their lands. The Chilean Supreme Court has confirmed that 'lands of the community' in article 64 means not only the lands for which the community holds registered title, but also the lands that they 'occupy or otherwise use'.[131] To reach this result, the Court relied on *Convention 169*, which provides at article 15 that '[t]he rights of the peoples concerned to the natural resources pertaining to their lands shall be specially safeguarded', and at article 13 that "*lands*' in articles 15 and 16 shall include the concept of territories, which covers the total environment of the areas which the peoples concerned occupy or otherwise use'.[132] The Court's reasoning reflects the approach typically taken by the Inter-American Court of Human Rights on ancestral land rights in Latin America.

Article 64 also recognises a right of '*propiedad*', which at Chilean civil law entails an 'absolute, exclusive, and perpetual right to use, enjoy and dispose of waters'.[133] Article 64 has been described by the Chilean Supreme Court as establishing 'a presumption of ownership and use of the waters of the Aimara and Atacameña Indigenous

---

[130] Ibid. art. 64(1).

[131] *Alejandro Papic Dominguez con Comunidad Indígena Aymara Chusmiza y Usmagama* [2009] Corte Suprema [Supreme Court] No. 2840–2008 (Chile) (25 November 2009) ('Chusmiza (Supreme Court Decision)') [7].

[132] *Convention Concerning Indigenous and Tribal Peoples in Independent Countries (No. 169)* [1989] 28 ILM 1382 (Entered into Force 5 September 1991) ('*Convention 169*'), art. 13, 15.

[133] *Civil Code 1855*, art. 582.

Communities'.[134] As mentioned above, the right of *propiedad* in water, is protected by article 19(24) of the *Constitution 1980* (Chile), which provides that '[p]rivate water rights, recognised or created in accordance with the law, give a right of ownership to their holders'. The Supreme Court has affirmed that the word 'law' in article 19(24) would include article 64 of the *Indigenous Law*.[135] *Ancestral* water rights under article 64 are, accordingly, much stronger than indigenous rights to water recognised in the other countries considered in this book, which cannot amount to rights of ownership, and do not convey the right to exclude.

Article 64 of the *Indigenous Law* has been applied by the Chilean courts to recognise *ancestral* water rights in a way similar to native title rights to water in Australia. Despite the fact that the word 'recognise' is not used in article 64, the Chilean courts have treated article 64 as a recognition mechanism. As explained by the Supreme Court in the recent *Atacameña Community* case, 'the law recognises the existence of ancestral uses carried out by communities of indigenous peoples, who collectively exercise a customary right recognised by law'.[136] The Chilean Courts are responsible for determining when, and on what conditions, indigenous water rights can be recognised under article 64. Waters in Chile are vested in the public as '*bienes nacionales de uso publico*' (national property for public use), giving the State the power to regulate water.[137] However, the Chilean courts have not found the public vesting of water to be an obstacle to allocating private rights of *propiedad* in water. They have reconciled the public nature of water with private rights of *propiedad* by providing that the right of *propiedad* attaches to the right to use the water and not to the water itself.[138]

The water rights provided for in article 64 do not need a registered title (a *derecho de aprovechamiento*) in order to be protected.[139] An

---

[134] *Chusmiza* (Supreme Court Decision) [2009] Corte Suprema [Supreme Court] No. 2840–2008 (Chile) (25 November 2009) [7]. See also *Comunidad Indígena Atacameña con Sociedad Química y Minera SA y otros* [2018] Corte Suprema de Chile [Supreme Court of Chile] No 44.255–2017 (Chile) (22 August 2018).

[135] *Chusmiza* (Supreme Court Decision) [2009] Corte Suprema [Supreme Court] No. 2840–2008 (Chile) (25 November 2009) [4].

[136] *Comunidad Indígena Atacameña con Sociedad Química y Minera SA y otros* [2018] Corte Suprema de Chile [Supreme Court of Chile] No 44.255–2017 (Chile) (22 August 2018) 14.

[137] *Water Code 1981*, art. 5.

[138] *Chusmiza* (Supreme Court Decision) [2009] Corte Suprema [Supreme Court] No. 2840–2008 (Chile) (25 November 2009) [4].

[139] *Water Code 1981*, trans art. 2, art. 122. See generally Vergara Blanco, above n. 830 Vol. II, 327: 'Accordingly, and there exist water rights that are not registered, but are, nonetheless, recognised by law'.

indigenous community could, for example, rely on article 64 to enforce its water rights against prejudicial state action using administrative law writs such as the *recurso de protección* (action for protection of constitutional rights).[140] However, According to Sotomayor, a former lawyer for the National Indigenous Development Corporation, indigenous communities have not sought protection of the article 64 rights via judicial review because, in the absence of a registered title, it would be difficult to prove their rights.[141] By convention, *ancestral* water rights recognised in article 64 have been provided for in the process of 'regularisation' under the *Water Code*. The Supreme Court in *Chusmiza* summarised the interrelationship of article 64 and the regularisation process, as follows:[142]

> ... in these rules the ancestral rights of the aimara and atacameña communities over their waters are legislatively recognised, recognition which also guarantees their constitutional protection, and allows their regularisation and subsequent registration.

Transitional article 3 of the *Indigenous Law* required the Development Corporation and the General Water Directorate (Directorate) to enter into an agreement about the implementation of article 64. The agreement must provide, specifically, for the 'protection, creation and re-establishment of the Aimara and Atacameña communities' ancestral water property rights'.[143] Soon after the commencement of the *Indigenous Law* the Development Corporation and the Directorate met to discuss the implementation of article 64.[144] They decided to use the existing procedures of the *Water Code* to provide for the indigenous water rights recognised and allocated pursuant to the *Indigenous Law*.[145] As previously explained, there are three ways in which a person can acquire a *derecho de aprovechamiento* under the *Water Code*. They are: 'constitution' of a new right as an administrative act; 'regularisation' of an unregistered water use recognised by

---

[140] *Constitution 1980*, art. 181.
[141] Interview with Diego Sotomayor (Santiago, 23 December 2013). See generally Bauer, *Siren Song: Chilean Water Law as a Model for International Reform*, above n. 12, 92. Bauer explains that in Chile those with few resources tend to avoid the formal legal system: judicial review being a costly and lengthy process, requiring the applicant to retain legal representation and expert evidence.
[142] *Chusmiza* [Supreme Court Decision] [2009] Corte Suprema [Supreme Court] No. 2840–2008 (Chile) (25 November 2009) [8].
[143] *Indigenous Law 1993* trans art. 3(2).
[144] Interview with Diego Sotomayor (Santiago, 23 December 2013), above n. 141.
[145] Ibid.

legislation; or purchase in the market. The Development Corporation and the Directorate agreed to use the regularisation process to recognise and provide for Aymara and Atacameña water rights pursuant to article 64.[146] The Indigenous Land and Water Fund would finance the regularisation process,[147] covering the cost of legal representation, court fees and expert evidence (hydrological evidence, anthropological evidence and engineering evidence).[148] Where indigenous landholders had been unable to continue to exercise their *ancestral* rights, including where other users had been allocated inconsistent rights, the intragovernmental agreement between the Development Corporation and the Directorate stated that compensation would be provided, or subsidy or purchase programmes would be used to reallocate *derechos de aprovechamiento*.[149]

The regularisation process was originally devised to formalise the vast numbers of registered and unregistered water rights existing under different laws at the time of passing the *Water Code*. According to Vergara Blanco, at the commencement of the *Water Code* around 70 per cent of water use in Chile was customary in nature, being carried out without any registered title.[150] Normalising, and titling, the distinct types of water rights was considered necessary to clarify property rights to water in order for water markets to emerge.[151] Regularisation adopted the logic of 'prescription' from Chilean civil law. The doctrine of prescription is based on the idea that the possessor of a thing, or a right in a thing, for a determined period of time without title or ownership can acquire ownership in that thing, or right in a thing, on the general understanding

---

[146] Corporación Nacional de Desarollo Indígena [National Indigenous Development Corporation] and Dirección General de Aguas [General Water Directorate], *Convenio Marco Para La Proteccion, Constitución y Reestablecimiento de Los Derechos de Agua de Propiedad Ancestral de Las Comunidades Aymaras y Atacamenas* [Convention for the Protection, Constitution and Reestablishment of the Ancestral Water Property Rights of the Aymara and Atacamena Communities] (on File with the Author) (Gobierno de Chile) (Intergovernmental Agreement), 1997) cl. 2.

[147] *Indigenous Law* 1993, art 20(c); interview with Diego Sotomayor (Santiago, 23 December 2013), above n. 141. According to Sotomayor, the regularisations have exclusively been carried out with resources from the Indigenous Land and Water Fund and there have been no 'privately funded' applications for regularisation.

[148] Interview with Waldo Contreras (Santiago, 16 November 2011). Contreras was an official at the General Water Directorate at the time of the interview.

[149] Corporación Nacional de Desarollo Indígena [National Indigenous Development Corporation] and Dirección General de Aguas [General Water Directorate], above n. 146 cl. 2.

[150] Alejandro Vergara Blanco, 'Comentario: Regularización de Derechos de Aguas y Publicidad en el Uso de las Mismas [Commentary: Regularisation of Water Rights and Publicity of Their Use]' (1996) VII *Revista de Derecho de Aguas* 254333.

[151] Dinko Tomislav Rendic Veliz, *Derechos de Agua y Pueblos Indígenas* [Water Rights and Indigenous Peoples] (Librotecnia, 2009) 148–9.

that the owner or title holder has lost possession and done nothing to recuperate it. The basis for prescription, and therefore regularisation, is 'longstanding possession'.[152] The regularisation process is provided for in transitional article 2 of the *Water Code*:

> Registered water use rights that are being used by people distinct from their titleholders at the date that this code entered into effect, can be regularised when those users have achieved five years of uninterrupted use, counted from the date on which they started to do so, in accordance with the following rules:
>
> a) The use must have been carried out without force or secrecy and without recognising third party rights;
> b) The application is made to the General Water Directorate in accordance with the form, time frame and process prescribed in paragraph 1, title I, part II of this code;
> c) Affected other rightholders may oppose the application by way of a presentation that accords with the rules set out in the previous paragraph; and
> d) After the legal time frames have expired, the General Water Directorate will remit the application and all its details with the opposition, if applicable, to the Local Civil Court Judge, who will consider and determine the case in accordance with the procedure established by article 177 and following of this code.
>
> The same procedure applies in the case of people who, provided they satisfy all the requirements set out in the previous paragraph, apply to register unregistered use rights, which are extracted individually from a natural source.

The Chilean Courts have applied the process of regularisation in conjunction with article 7 of *Decree Law 2.603 1979*, which (as explained earlier in the chapter) deemed the person making '*uso efectivo*', or 'productive use', of a water right to be its owner.[153] According to Rendic Veliz, *Decree Law 2.603 1979* 'is the origin of "recognition" of customary water uses, and is central to understanding the spirit of the regularisation process'.[154]

Indigenous groups have relied on the final paragraph of transitory article 2, together with *Decree Law 2.603 1979*, to apply for regularisation of their unregistered water use, protected by article 64 of the *Indigenous*

---

[152] Vergara Blanco, above n. 35 Vol. II, 347–9.
[153] *Comunidad Atacamena Toconce con Essan SA* [2004] Corte Suprema [Supreme Court] No. 4064–2004 (Chile) (22 March 2004) 6 (*'Toconce'*).
[154] Dinko Tomislav Rendic Veliz, above n. 151, 148, 150.

Law.[155] An application for regularisation must be made to the Directorate, which refers the application to the local civil court to determine whether or not it meets the statutory requirements.[156] This means that the civil (or appeal) courts are responsible for determining in which situations indigenous water users enjoy recognition and protection of *ancestral* water rights under article 64, as well as deciding which water uses may be regularised. Unfortunately, adopting the regularisation process has meant that further threshold requirements must be satisfied before *ancestral* water rights can be recognised, aside from the requirements of article 64. This is the case despite the fact that the judicial process of regularisation under the *Water Code* predates the *Indigenous Law*, and was not designed with indigenous water rights in mind.

The first of the additional requirements is that an applicant must prove that it has made uninterrupted use of the water since five years before the commencement of the *Water Code* (i.e. 1976).[157] This time frame simply adopted the standard rule for prescription of real estate in the *Civil Code 1855* (Chile), which is five years' uninterrupted possession.[158] Secondly, the use must have been conducted 'without force or secrecy', and 'without recognising the rights of others'. These requirements are similar to the requirements to establish adverse possession at common law, of 'physical control that is open rather than secret, peaceful rather than forceful, and without the consent of the actual true owner'.[159] Finally, applicants must show that they have made '*uso efectivo*' ('productive use') of water, in order to meet the requirements of *Decree Law 2.603 1979*.[160] As will be discussed in the following part, article 64 of the *Indigenous Law* is a very limited response to the historical exclusion of indigenous Chileans from water law frameworks and, much like the native title model in Australia discussed in Chapter 4, problems of continuity and priority undermine the potential application of the recognition mechanism.

---

[155] Interview with Diego Sotomayor (Santiago, 23 December 2013), above n. 141.
[156] *Water Code 1981* trans art. 2(d).
[157] Ibid. trans art. 2.
[158] *Civil Code 1855*, art. 2508.
[159] Hepburn, above n. 48, 64. Hepburn refers to the test from *Mulcahy v. Curramore Pty Ltd* [1974] 2 NSWLR 464, 475. See also *Riley v. Pentilla* [1974] VR 547: it must also be established that the person intended to possess the land adversely. But Hepburn at 64 makes a distinction between 'adverse possession' (a right based on limitation) and 'long standing use' (a right based on prescription). The distinction between prescription and limitation appears not to apply in the same way in Chile where prescription appears to be the equivalent to adverse possession.
[160] Although once indigenous water rights have been regularised as a *derecho de aprovechamiento*, their holders can in principle use the water for any sort of use from commercial extractive use to in-stream cultural or environmental use.

(a) The Problem of Continuity

As mentioned, the Chilean courts have characterised water rights under article 64 as being '*ancestral*' in nature.[161] Transitional article 3 of the *Indigenous Law* is the source of the idea that they are *ancestral*:[162]

> Equally, the [Indigenous Development] Corporation and the Directorate will establish an agreement for the protection, creation and re-establishment of the Aimara and Atacameña communities' ancestral water property rights under article 64 of this law.

However, the *Indigenous Law* provides no definition for *ancestral*. The word '*ancestral*' appears only twice in the *Indigenous Law*, once in transitional article 3. Its other appearance is in article 26, which provides that the Minister for Planning and Cooperation can establish areas of indigenous development in territorial spaces where indigenous ethnicities have lived '*ancestralmente*' ('ancestrally'). According to the *Spanish Language Dictionary*, the Spanish word '*ancestral*' can refer to both 'belonging and relative to ancestors' or 'traditional and of remote origin'.[163] In debates leading up to the *Indigenous Law*, *ancestral* water rights were taken to mean rights originating before Spanish colonisation. For example, the discussion draft of the *Indigenous Law* prepared by the Special Commission uses the word *ancestral* when describing the Atacameña community's pre-Columbian water management systems, with canal cleaning ceremonies as a principal traditional festival.[164]

When the Chilean courts characterise water rights under article 64 as *ancestral*, they too appear to be referring to rights that originate prior to the acquisition of sovereignty.[165] The courts refer to *ancestral* water rights as existing since 'time immemorial',[166] reflecting terminology

[161] *Chusmiza* (Supreme Court Decision) [2009] Corte Suprema [Supreme Court] No. 2840–2008 (Chile) (25 November 2009) [8]; *Toconce* [2004] Corte Suprema [Supreme Court] No. 4064–2004 (Chile) (22 March 2004) 7.

[162] *Indigenous Law 1993* trans art. 3(2).

[163] *Diccionario de la Lengua Española* [Spanish Language Dictionary] (*Real Academia Española, 2011*).

[164] Biblioteca del Congresso Nacional de Chile [Library of National Congress of Chile], above n. 9, 154.

[165] See generally Manuel Cuadra, 'Teoría Práctica de Los Derechos Ancestrales de Agua de las Comunidades Atacameñas [Practical Theory of the Ancestral Water Rights of the Atacamena Communties]' [2000] (19) *Estudios Atacameños* 93, 101.

[166] *Chusmiza* (Supreme Court Decision) [2009] Corte Suprema [Supreme Court] No. 2840–2008 (Chile) (25 November 2009) [5], [8]; *Toconce* [2004] Corte Suprema [Supreme Court] No. 4064–2004 (Chile) (22 March 2004).

from early US indigenous land rights jurisprudence.[167] Sotomayor explains that while the courts do not define 'time immemorial' it is used to mean since a long time ago, certainly long prior to 1976, which is the requirement for regularisation under the *Water Code*.[168] They also construe the rights recognised by article 64 as arising out of the 'historical use' of the resources by the indigenous group. This approach resembles United States[169] and Canadian[170] indigenous rights jurisprudence rather than the Australian native title conceptualisation, discussed in Chapter 4, of rights arising out of the 'traditional laws and customs' of the native title group.

The *ancestral* quality of indigenous water rights comes from customary use of waters, carried out since time immemorial. This 'historical use' approach is consistent with the Chilean approach to the basis of indigenous title more generally as 'immemorial occupation and use'.[171] It also reflects the characterisation of ancestral rights to land and resources in Latin American jurisprudence, based on 'historical possession'.[172] The historical use is a 'customary' use, because it does not derive from a historical title or *derecho de aprovechamiento*.[173] According to the Court of Appeal of Iquique, because article 64 recognises rights that derive from customary use rather than formal titles, this article represents 'a recognition of the customary law of these

---

[167] See, e.g., *Coos Bay, Lower Umpqua and Siuslaw Indian Tribes* v. *United States* (1938) 87 Ct Cl 143; see generally *Mabo* (1992) 175 CLR 1189 (Toohey J); *Milirrpum (1971)* 17 FLR 141152 (Blackburn J).

[168] Interview with Diego Sotomayor (Santiago, 23 December 2013), above n. 141.

[169] The approach taken in the United States jurisprudence is that the source of 'Indian title' is the indigenous group's exclusive use and occupation of land over a long period of time. See, e.g., *United States* v. *Santa Fe Public Railroad Company* (1941) 314 US 339, 62 S Ct 248. See generally Chapter 4 Young, above n. 115, 45, 85–90; *Mabo* (1992) 175 CLR 1114–5, 189 (Toohey J).

[170] The Canadian Aboriginal title cases also emphasise occupation of land prior to the acquisition of sovereignty as the source of a *sui generis* title. *Calder* v. *Attorney-General of British Colombia* (1973) 34 DLR (3rd) 145 (1973) 34 DLR 3rd 145185, 187–90; *Delgamuukw* v. *British Colombia (1997)* 153 DLR 4th 193, 241–53; *Tsilhqot'in Nation* v. *British Colombia* [2014] SCC 44 [14].

[171] Gonzalo Aguilar Cavallo, 'El Titulo Indígena y su Aplicabilidad en el Derecho Chileno' [Indigenous Title and Its Application in Chilean Law] (2005) 11(1) *Revista Ius et Praxis* 269, 271–2.

[172] See, e.g., *Mayagna (Sumo) Awas Tingni Community* v. *Nicaragua (Judgment)* [2001] Inter-American Court of Human Rights (Ser C) Case No. 79 (31 August 2001) ('*Mayagna (Sumo) Awas Tingni Community* v. *Nicaragua (Judgment)* (Inter-American Court of Human Rights (Ser C) Case No. 79, 31 August 2001)') [127–8]; *Case of the Saramaka People* v. *Suriname* Inter-Am. Court Hum. Rights Ser. C No 18 17 Sept. 2003 [96]: '[t]he foundation of territorial property lies in the historical use and occupation which gave rise to customary land tenure systems'.

[173] See *Chusmiza* (Supreme Court Decision) [2009] Corte Suprema [Supreme Court] No. 2840–2008 (Chile) (25 November 2009) [4], referring to the water rights as customary water rights.

aboriginal ethnicities'.[174] Some commentators have argued that the recognition of indigenous water rights in article 64 represents recognition of indigenous law-making systems.[175]

Ancestral water rights may be recognised pursuant to article 64 where indigenous landholders can show that they have carried out a, 'historical use' of water. The Development Corporation routinely commissions evidence on behalf of indigenous communities to accredit *ancestral* use, since time immemorial.[176] The evidence commonly refers to the antiquity of water infrastructure and agricultural land (terraces).[177] For example, the Civil Court in *Associación Atacameña de Regantes y Agricultores Aguas Blancas* recognised *ancestral* water rights based on evidence of:[178]

> ... the existence of antique water capture works, and agricultural irrigation systems of immemorial origin, based on terrace structures and stone canals.

In *Toconce* the Supreme Court accepted:

> ... with the testimony of the applicant it has been accredited that since time immemorial the inhabitants of Toconce have made an uninterrupted use of the waters from the river for human and animal consumption, as owners and in sight of the whole world, with the consequence that this is taken as established the use of the waters in the terms indicated. This was corroborated in addition during the site inspection made by the tribunal.

An applicant for regularisation must prove uninterrupted water use since five years prior to the commencement of the *Water Code* (i.e. 1976). That water use must be an *uso efectivo* (productive use), applying the standard in *Decree Law 2.603 1979*. Proof of uninterrupted

---

[174] *Corporación Movimiento Unitario Campesino y Etnias de Chile con Dirección General de Aguas* [2014] Corte Suprema [Supreme Court] (Chile) No. 7899–2013 (Chile) (5 May 2014) 4 ('*Corporación Movimiento Unitario Campesino y Etnias de Chile Con Dirección General de Aguas*, No. 7899–2013, Corte Suprema [Supreme Court] (Chile) (5 May 2014)') [9].

[175] Interview with Nancy Yanez (Santiago, 22 November 2011), above n. 81; Interview with Maria Angelica Alegria (Santiago, 17 November 2011), above n. 80.

[176] Interview with Diego Sotomayor (Santiago, 23 December 2013), above n. 141.

[177] *Toconce* [2004] Corte Suprema [Supreme Court] No. 4064–2004 (Chile) (22 March 2004) [2]; *Alejandro Papic con Comunidad Indígena Aymara Chuzmira y Usmagama* [2008] Corte de Apelaciones de Iquique [Iquique Court of Appeal] No. 817–2006 (9 April 2008) (Chile) [10].

[178] *Inscripción Sentencia Derechos de Approvechamiento de Aguas Associacion Atacamena de Regantes y Agricultores de Aguas Blancas* [1997] Segundo Juzgado Civil de Calama, Chile [Second Civil Court of Calama, Chile] NR-II-1381 (19 November 1997) [2].

productive use, for the purposes of the *Water Code*, is usually provided in a technical report prepared by the Directorate. Before preparing its report, the Directorate carries out an inspection of construction and maintenance of physical water infrastructure such as canals and wells suggestive of productive use.[179] It also consults with the Development Corporation, which in turn retains anthropologists to consult the relevant indigenous communities.[180] As an example, the Court in the *Associacion Atacameña de Regantes y Agricultores Aguas Blancas* accepted evidence from the Directorate's technical report in satisfaction of the requirement for uninterrupted use:[181]

> The report referred to highlights that it was satisfied that the antiquity of the stone works found in the majority of water sources, as well as the rustic irrigation works (terraces), accredit an immemorial use of the resource.

Unfortunately, many indigenous groups in Chile are unable to prove that they have made uninterrupted historical use of particular water resources. After *derechos de aprovechamiento* were unbundled from land titles in the early 1980s, other users acquired water use rights in water resources on or affecting indigenous lands, meaning that in many situations indigenous groups have not continued to use water resources since pre-sovereignty times. Others do not have the finance needed to construct or maintain water infrastructure to accredit productive use.[182]

There is also uncertainty as to the extent to which *ancestral* water rights can adapt or evolve over time or whether they must be practised in the same way as the communities did prior to the arrival of the

---

[179] Interview with Diego Sotomayor (Santiago, 23 December 2013), above n. 936; interview with Manuel Cuadra (Antofagasta, 23 November 2011); interview with Hernando Silva (Temuco, 11 November 2011), above n. 9; interview with Waldo Contreras (Santiago, 16 November 2011), above n. 148.

[180] Interview with Carlos Herrera Inzunza (Temuco, 11 November 2011), above n. 121. Herrera was a water lawyer working for the National Indigenous Development Corporation at the time of interview.

[181] *Inscripción Sentencia Derechos de Approvechamiento de Aguas Associación Atacamena de Regantes y Agricultores de Aguas Blancas* (Unreported, Segundo Juzgado Civil de Calama, Chile [Second Civil Court of Calama, Chile], NR-II-1381, 19 November 1997) [2].

[182] Interview with Carlos Herrera Inzunza (Temuco, 11 November 2011). But see Interview with Juan Carlos Araya (Santiago, 15 November 2011). Araya, a lawyer for the National Indigenous Development Corporation at the time of the interview, argues that it is not necessary for a community to have canals to show use and has in cases managed to regularise rights without productive use, but admits that the approach is usually opposed.

Spanish. In the *Chusmiza* case, the use of water for human and animal consumption and irrigation was considered to be an *ancestral* use of water, because the communities had carried out that use, in an unin-terrupted manner, since 'time immemorial'.[183] In *Toconce* the Supreme Court accepted evidence about significant terraced cultivation.[184] Some cases have relied on evidence of sustaining wetlands together with agriculture and grazing.[185] However, there is still an expectation by Chilean courts and government officials that indigenous water use will be consistent with pre-sovereignty uses. Cuadra explains, 'not just any use of water enjoys legal recognition, rather only those that satisfy certain conditions' to which he includes that it must be an 'antique use' of the water resource, carried out continuously since pre-Columbian times and evidenced by antique water infrastructure.[186]

The requirement to prove productive use in the process of regular-isation also means that *ancestral* water rights have typically been recog-nised in reliance on article 64 for the consumptive use of surface waters only.[187] It would be difficult for an applicant for regularisation to prove productive use that is non-consumptive or subterranean in the absence of water infrastructure.[188] This is despite the fact that clause 5 of the interdepartmental agreement between the Development Corporation and the Directorate provides that non-consumptive and subterranean water rights are also contemplated within the concept of *ancestral*

---

[183] *Chusmiza* (Supreme Court Decision) [2009] Corte Suprema [Supreme Court] No. 2840–2008 (Chile) (25 November 2009) [5].

[184] *Toconce* [2004] Corte Suprema [Supreme Court] No. 4064–2004 (Chile) (22 March 2004) [2].

[185] *Inscripción Sentencia Derechos de Approvechamiento de Aguas Comunidad Atacamena de Peine* [1997] Segundo Juzgado Civil de Calama, Chile [Second Civil Court of Calama, Chile], NR-II-1383 (19 November 1997) *Inscripción Sentencia Derechos de Approvechamiento de Aguas Comunidad Atacamena de Cupo* [1997] Segundo Juzgado Civil de Calama, Chile [Second Civil Court of Calama, Chile] NR-II-1387 (19 November 1997).

[186] Cuadra, above n. 179, 101–2.

[187] David Espinoza Quezada, above n. 127. In the second region of Antofagasta almost all of the water rights regularised to indigenous communities pursuant to article 64 up until 2012 have been permanent consumptive surface water rights. In six cases they were eventual and continuous, meaning that water could only be taken once there was a surplus after satisfying other, permanent rights.

[188] Interview with Nancy Yanez (Santiago, 22 November 2011), above n. 81. See also *Modifica la Ley No 19.253, Relativa a la Proteccion, Fomento y Desarollo de los Pueblos Indígenas, Estableciendo la Regularizacion de Derechos de Agua Potable Rural* [Modification of Law No 19.253, Relating to the Protection, Promotion and Development of Indigenous Peoples, Establishing the Regularisation of Rights for Rural Drinking Water] 2012 (Chile).

rights.[189] Only those groups who have continued to maintain their water rights since pre-Colombian times have had water rights recognised, and regularised, in reliance on article 64. This has led some commentators to observe that article 64 added little to the procedures already available to rural agricultural communities under the general provisions of the *Water Code*.[190]

(b)   The Problem of Priority

Another reason why it is difficult to make out a claim for *ancestral* water rights in Chile is that such rights cannot be recognised if to do so would be inconsistent with the rights of other users. Chilean law, at least, provides for *ancestral* water rights to take 'priority' as against the water use rights sought by other users in the future. Article 64(2) provides:

> New water rights must not be granted over lakes, ponds, springs, rivers and other aquifers which supply waters that are property of the various Indigenous communities established by this law without first guaranteeing normal water supply to the affected communities.

This means that other users must not be allocated new *derechos de aprovechamiento* if this would prevent normal water supply to the indigenous communities.[191] Ensuring 'normal water supply' to indigenous communities is dealt with in the intragovernmental agreement between the Directorate and Development Corporation, mentioned above. The agreement requires the Directorate to request a technical report from the Development Corporation evaluating the impact on indigenous communities of all new applications for *derechos de aprovechamiento* in indigenous areas of Chile's first and second regions.[192] The report must cover any existing water use by an indigenous community, the type

---

[189]   Corporación Nacional de Desarollo Indígena [National Indigenous Development Corporation] and Dirección General de Aguas [General Water Directorate], above n. 146 cl. 5.

[190]   Interview with Manuel Cuadra (Antofagasta, 23 November 2011). The formalisation of unregistered water use was already provided for in transitional article 2 of the *Water Code*. Where indigenous groups owned land they also already enjoyed an entitlement to take and use water sources which are 'born, flow and die' within a parcel of land as of right without the need for a *derecho de aprovechamiento* under article 20 of the Code, although, prior to the *Indigenous Law* indigenous groups could not hold water rights in the corporate sense.

[191]   See *Codelco Chile División Chuquicamata* v. *Dirección General de Aguas y otra* (Unreported, Corte de Apelaciones de Antofagasta, Chile [Court of Appeal of Antofagasta, Chile], Rol: 14003–2013, 15 May 2014) in which the Court of Appeal upheld a decision of the General Water Directorate refusing an application for an authorisation to explore and extract subterranean waters on fiscal lands on the basis that this would cause prejudice to the indigenous occupiers of the land, whose water rights were protected under article 64.

[192]   Corporación Nacional de Desarollo Indígena [National Indigenous Development Corporation] and Dirección General de Aguas [General Water Directorate], above n. 146 part 2, para 6.

of water use carried out, and the flow claimed. The Development Corporation contracts lawyers, anthropologists and geographers to check whether a community will be affected by a new application for a *derecho de aprovechamiento* and if necessary recommends that the application is refused. This is difficult, Araya explains, in the case of new applications with respect to subterranean waters, as there are few studies of subterranean waters meaning that it is hard to anticipate the effect on surface waters.[193] On the question of what is meant by 'normal water supply' there is little guidance, being a case-by-case assessment made by the Development Corporation when preparing its report. However, the protection in article 64(2) may be of less consequence as might first appear, because rights to use surface water resources in Chile's north were already largely allocated to other rightholders at the commencement of the *Indigenous Law*.[194]

In terms of *derechos de aprovechamiento* already held by others, article 64(1) prioritises the other rights ahead of the *ancestral* water rights of indigenous communities, as follows:

> Waters, including rivers, canals, streams and springs, found on the lands of the Indigenous communities established by this law will be considered property of ownership and use of the Indigenous communities, without prejudice to the rights that other titleholders have registered in accordance with the Water Code.

The priority mechanism in article 64(1) significantly undermines the potential for recognising indigenous water rights. Surface *derechos de aprovechamiento* in northern Chile were largely fully allocated to other titleholders around the time the *Indigenous Law* came into effect. Accordingly, recognising indigenous water rights would prejudice other water use rights, and be incapable of recognition, in many situations. Notwithstanding, the courts have continued to recognise indigenous water rights in reliance on article 64 in Chile's north as part of the regularisation process.[195] However, the regularisations have mostly concerned minimal flows still being customarily used by indigenous communities, which were still available within the water system.

---

[193] Interview with Juan Carlos Araya (Santiago, 15 November 2011).
[194] Jessica Budds, 'Power, Nature and Neoliberalism: The Political Ecology of Water in Chile' (2004) 25(3) *Singapore Journal of Tropical Geography* 322, 326.
[195] See David Espinoza Quezada, above n. 927. According to General Water Directorate records provided by Espinoza Quezada, for example, in the second region of Antofagasta a total flow of 2,729.6 litres per second of water was allocated to indigenous communities between 1995 and 2012 in reliance on article 64.

According to former Development Corporation lawyer, Diego Sotomayor, the *ancestral* water rights regularised in reliance on article 64 are not always 'viable economically'.[196]

The landmark Supreme Court decision in *Chusmiza* is emblematic of the tension between indigenous water rights and the registered rights of non-indigenous third parties. The indigenous communities of Chusmiza and Usmagama in the north of Chile claimed to have used the thermal mineral waters emanating from the Chusmiza spring for irrigation and human and animal consumption, distributed via an array of wells, canals and other infrastructure, since time immemorial.[197] However, a water bottling company held *derechos de aprovechamiento* authorising it to use the water from the mineral spring, as well as the title to the land on which it was found, undermining the continued use of the waters by the communities.[198]

Despite the company's rights, which the company claimed would be prejudiced by the decision, the Supreme Court applied article 64 to recognise the communities' water rights.[199] The Supreme Court made its decision with reference to what it saw as the objective of article 64. That was, the repopulation, subsistence and development of rural indigenous communities in Chile's north, which depend on adequate water supply:[200]

> Accordingly, it is without doubt that the central axis of the protection given to indigenous waters resides in the idea of repopulation of Andean communities, for which it seems essential that they may have access to the water resources necessary for their subsistence and development.

---

[196] Interview with Diego Sotomayor.

[197] *Agua Mineral Chusmiza SAIC con Comunidad Indígena Aymara Chusmiza* [2006] Juzgado de Letras de Pozo Almonte [Pozo Almonte Civil Court] No. 1194–1996 (Chile) (31 August 2006) 11 (*'Chusmiza – First Instance Decision'*).

[198] *Chusmiza* (Supreme Court Decision) [2009] Corte Suprema [Supreme Court] No. 2840–2008 (Chile) (25 November 2009) [1].

[199] *Chusmiza – First Instance Decision* 1 referring to *Decreto No 1540*, 3 August 1948, Pozo Almonte 4 February 1983, Resolution no 406, 29 September 1983; *Chusmiza*, No. 2840–2008, Corte Suprema de Chile [Supreme Court of Chile] [25 November 2009] [3] referring to judgement [3695–05]. The company inherited a historical water allocation for 10 litres per second and 10,000 litres per day granted by Supreme Decree by the Ministry of Public Works in 1948, which was regularised by the Civil Court under the transitional provisions to the *Water Code* in 1983. It also held a mixture of consumptive and non-consumptive registered rights to 50 cubic metres per day and 5 litres per second originating in 1996/1997 the validity of which was confirmed by a separate decision of the Constitutional Division of the Supreme Court in 2005.

[200] *Chusmiza* (Supreme Court Decision) [2009] Corte Suprema [Supreme Court] No. 2840–2008 (Chile) (25 November 2009) [7].

With this objective in mind, the Supreme Court emphasised that the communities' water rights arose prior to the registered *derechos de aprovechamiento* held by the company:[201]

> Of course, it is worth remembering that in this case that which is regularised is the ancestral right of the applicant indigenous community, whose members from time immemorial have made uninterrupted use of the waters for human and animal consumption and irrigation. It follows that the water right recognised to the Aymara community is therefore prior to the constitution of water use rights created in favour of other titleholders and as a corollary; it is prior to the origin of the registered rights of the Company.

Presumably, because the communities were already using the water there could be no prejudice to others by recognising the communities' right. According to Sotomayor, the courts have been able to do this because they have focused on the regularisation procedure established in the *Water Code*, rather than the requirements of article 64. In the process of regularisation, provided that the applicant can prove it is currently using the water the courts have no option but to approve the regularisation. It follows from that line of reasoning, the use is already being made of the water, so there can be no prejudice to other users.[202] Accordingly, the Court reasoned, the communities' water right already exists; it is simply being registered in order to provide certainty as to the amount of water being used, at what location and by whom:[203]

> On this topic it is useful to make clear that any lack of registration of customary water rights does not entail their absence, but only the lack of formalisation and registration and, precisely because the right exists, is recognized by law and only for the purpose of having certainty about its existence, location of point of capture of the waters and accuracy of water resource use, a regularization system has been created which permits their eventual registration.

The Court also remarked at one point that the company's water use rights and the communities' water rights could co-exist.[204] How exactly the rights could co-exist is unclear, given that the Court of Appeal had found (relying on the hydrological report of the Directorate) that the

---

[201] Ibid. [5].
[202] Interview with Diego Sotomayor (Santiago, 23 December 2013), above n. 141.
[203] *Chusmiza* (Supreme Court Decision) [2009] Corte Suprema [Supreme Court] No. 2840–2008 (Chile) (25 November 2009) [4].
[204] See also *Toconce* [2004] Corte Suprema [Supreme Court] No. 4064–2004 (Chile) (22 March 2004).

supply from the spring was no more than eight to ten litres per second,[205] and a full ten litres per second was regularised to the communities.

Some have described the *Chusmiza* case as a 'triumph' of indigenous water rights over the registered water use rights of a non-indigenous party.[206] Others have criticised the decision for causing prejudice to the holder of legally valid water use rights,[207] and producing legal uncertainty around water rights ownership.[208] After the decision there existed two separate sets of enforceable rights over waters from the Chusmiza spring, one held by the company and one held by the communities, and not enough water to satisfy both. Each set of rights was judicially recognised, protected by the Constitution, and recorded in the system of titles in Chile's Real Estate Office. Weisner, the former director of the General Water Directorate, believes that if the Supreme Court was inclined to recognise the communities' rights to the waters from the spring it should have cancelled the company's *derechos de aprovechamiento* so as to prevent an over allocation,[209] although the Court may not have had the jurisdiction to do so as part of the cassation appeal. The lawyer for the water bottling company has described the result in *Chusmiza* as a form of compulsory redistribution, without compensation.[210]

Many years after the decision of the Court in *Chusmiza* the indigenous communities' water rights are still uncertain. The communities took their ongoing claims for water rights to the Inter-American Commission on Human Rights in 2013.[211] They argued that they have been unable to exercise their water rights recognised by the Supreme Court, which have not been registered as *derechos de aprovechamiento* by the Directorate. They maintained that the Chilean courts should have declared the company's registered *derechos de aprovechamiento* to be void. In 2017, the Rapporteur on the Rights of Migrants of the Inter-American Commission convened a meeting to try and secure

---

[205] *Alejandro Papic Con Comunidad Indígena Aymara Chuzmira y Usmagama* [2008] Corte de Apelaciones de Iquique [Iquique Court of Appeal] *No. 817–2006* (9 April 2008) (Chile), 3.
[206] See, e.g., 'Conflictos por el agua en Chile: entre los derechos humanos y las reglas del mercado [Water conflicts in Chile: Between Human Rights and Market Rules]' (Programa Chile Sustentable, 2010) 105; Yañez and Molina, Chapter 1 above n. 7, 144–5.
[207] Interview with Gonzalo Arevalo (Santiago, 18 November 2011), above n. 80.
[208] Interview with Rodrigo Weisner (Santiago, 18 November 2011), above n. 121.
[209] Ibid.
[210] Interview with Gonzalo Arevalo (Santiago, 18 November 2011), above n. 80.
[211] *Admissibility Aymara Indigenous Community of Chusmiza-Usmagama and its Members, Chile* [2013] Inter American Commission on Human Rights 1288–06, 29/13 (2013).

a settlement of the *Chusmiza* case.[212] The result of the meeting was the signing of a Memorandum on the Rationale for a Friendly Settlement Agreement.[213] In 2018, the Executive Secretariat of the Inter-American Commission visited Chile to continue negotiations towards the settlement. However, the outcome of this process remains uncertain, as the decisions of the Inter-American system are not directly enforceable in Chile's internal legal framework.[214]

The *Chusmiza* dispute remains unresolved and illustrates the difficulty inherent in recognising pre-sovereignty water rights many years after colonisation, in the presence of other rights. The case shows that, in situations of full resource allocation, it is impossible to simply recognise indigenous water rights without impacting on the water rights held by third parties. In such situations, some form of redistribution is inevitable.

7.4 ALLOCATING INDIGENOUS WATER RIGHTS IN CHILE

When preparing the *Indigenous Law*, the Special Commission also recommended the creation of the Indigenous Land and Water Fund to 'amplify' indigenous territories, via the purchase of land and water rights from other users. This recommendation reflected the reality that almost all the *derechos de aprovechamiento* in Chile had already been allocated to other water users and indigenous peoples had been dispossessed of their traditional territories. The Fund would address these more difficult indigenous land and water issues, where indigenous groups could not simply have their *ancestral* territories 'regularised' because they no longer occupied the lands, and other users now held

---

[212] Gil Botero and subsequently Luis Vargas oversaw the negotiations. Commissioner Gil Botero also facilitated a working meeting on Case 12,904 (Chusmiza Usagama Aymara Community) to promote negotiations for a friendly settlement agreement to uphold the community's rights regarding sources of water in its ancestral territory. At the meeting, the parties signed a memorandum on the rationale for a friendly settlement agreement in the case. Germana Aguiar Ribeiro do Nacimiento, 'El Derecho al Agua y su Protección en el Contexto de La Corte Interamericana de Derechos Humanos' [The Right to Water and Its Protection in the Context of the Interamerican Court of Human Rights] (2018) 16(1) *Estudios Constitucionales* 245, 272.

[213] OAS, IACHR, *Progress on Friendly Settlements in Petitions and Ongoing Cases before the IACHR Concerning Chile* (23 June 2016) www.oas.org/en/iachr/media_center/PReleases/2016/085 .asp.

[214] Judith Schönsteiner and Javier Couso, '*La Implementación de las Decisions de los Órganos del Sistema Interamericano de Derechos Humanos en Chile: Ensayo de un Balance*' [The Implementation of Decisions of the Interamerican Human Rights System in Chile: The Study of a Balance] (2015) 22(2) *Revista de Derecho Universidad Católica del Norte Sección: Estudios* 315.

registered title.[215] It also responded to the uncertainty such indigenous groups posed for the market, by continuing to use water resources without formal rights and by making political claims for water use rights held by other users;[216] charging the Fund with redistributing *derechos de aprovechamiento* on market terms.

The functions of the Fund are set out in article 20 of the *Indigenous Law*. They include: (a) granting subsidies for the acquisition of land, and (b) financing solutions to land problems like land transfers, exchanges or court action. Indigenous water rights are provided for in the function at 20(c), 'to finance the constitution, regularisation or purchase of water use rights or finance works destined to obtain the resource'. As mentioned above, the Fund has been used in northern Chile in conjunction with article 64 to finance the regularisation of *ancestral* water rights. It has also been used to finance the regularisation of indigenous water rights in other parts of Chile (without article 64) using the process provided for in transitional article 2 of the *Water Code*.[217] However, the Fund is also used to finance the constitution and purchase of *derechos de aprovechamiento* for indigenous landholders throughout Chile, often in conjunction with government-sponsored indigenous economic development projects.[218] *Derechos de aprovechamiento* acquired for indigenous landholders with the assistance of the Fund are the same as any other *derechos de aprovechamiento* in Chilean water law frameworks. They are held independent of land title, and are exercisable for any purpose (including commercial purposes).

If anything, indigenous Chileans have been dissuaded from exercising their *derechos de aprovechamiento* for non-commercial purposes. This

---

[215] Chile, 'Primer Informe Comisión Especial Pueblos Indígenas' [First Report of the Special Commission for Indigenous Peoples], Cámara de Diputados [House of Representatives], 10 November 1992, in Biblioteca del Congreso Nacional de Chile [Library of National Congress of Chile], above n. 9, 64. In fact, 'regularisation' was not included in the original version of article 20(c) in the Bill presented to Parliament, which provided only for the 'creation' or 'purchase' of water rights in article 20(c). See also Chile, 'Discusión Sala' [Parliamentary Debates], Cámara de Diputados [House of Representatives], 27 January 1993 (Huenchumilla) in ibid., 251, '[i]n article 20, between "constitution" and "or purchase", the word "regularisation" is inserted. This indication was formulated by the executive and is intended to signify another outcome for finance granted by the Indigenous Land and Water Fund. It is approved unanimously'.

[216] Interview with Diego Sotomayor (Santiago, 23 December 2013), above n. 936.

[217] Corporación Nacional de Desarollo Indígena [National Indigenous Development Corporation] and Dirección General de Aguas [General Water Directorate], 'Convenio Dirección General de Aguas y Corporación Nacional de Desarollo Indígena' [Convention between the National Indigenous Development Corporation and General Water Directorate] (Interdepartmental Convention, on file with the author 2000). See also Interview with Nancy Yanez (Santiago, 22 November 2011), above n. 81.

[218] Interview with Diego Sotomayor (Santiago, 23 December 2013), above n. 141.

has occurred as a result of the levying of 'fees for non-use' on the holders of *derechos de aprovechamiento* who do not use the water for 'productive purposes', introduced into the *Water Code* in 2005,[219] although, recent judicial and political developments suggest a disinclination to charge 'fees for non-use' to indigenous users.[220]

The *derechos de aprovechamiento* acquired with support from the Fund are constitutionally protected rights of *propiedad*. This means, as above, that they are roughly equivalent to a right of full ownership, or the estate of fee simple in the common law sense, including the right to exclude others. However, *derechos de aprovechamiento* acquired with the support of the Fund are subject to a restriction on their alienation for twenty-five years.[221] This means that *derechos de aprovechamiento* acquired with finance from the Fund typically remain outside of water markets for twenty-five years, unless administrative approval is obtained or the Fund is repaid.[222] The rights can still, however, be transferred within and between indigenous communities of the same ethnicity. The ability to trade *derechos de aprovechamiento* within indigenous communities was intended to encourage an 'indigenous water market'.[223]

The Development Corporation administers the Indigenous Land and Water Fund. It receives *derechos de aprovechamiento* from the State or private holders for allocation to indigenous communities.[224] Regulations dealing with the operation of the Indigenous Land and Water Fund provide that the finance envisaged in article 20(c) is

---

[219] *Water Code 1981*, art. 129 bis 4.
[220] See *Corporación Movimiento Unitario Campesino y Etnias de Chile con Dirección General de Aguas*, No. 7899–2013, Corte Suprema (5 May 2014) *[2014]* Corte Suprema [Supreme Court] (*Chile*) *No. 7899–2013 (Chile)* (5 May 2014), which found that fees for non-use could not be levied against indigenous communities holding *derechos de aprovechamiento* acquired with finance from the Indigenous Land and Water Fund, because to do so would contravene the restriction on alienation of such rights under section 22 of the Indigenous Law. See also *'Reforma el Codigo de Aguas, Examinando del Pago de Patente a Pequenos Productores Agricolas y Campesinos, a Comunidades Agricolas y a Indígenas y Comunidades Indígenas que se Señalan'* [Reform of the Water Code, Examining the Payment of Tax by Small Agricultural Producers, to Agricultural and Indigenous Communities Included] (Boletin No 8315–01), which proposes to exempt indigenous people and communities under the indigenous law from fees for non-use.
[221] The National Indigenous Development Corporation can authorise the alienation of *derechos de aprovechamiento* if the value of the subsidy provided by the Fund is repaid. *Indigenous Law 1993*, art. 22(2).
[222] Ibid. art. 22(1).
[223] Chile, 'Discusion Sala' [Parliamentary Debates], Senado [Senate], 30 June 1993 (Navarrete) in Biblioteca del Congresso Nacional de Chile [Library of National Congress of Chile], above n. 9, 417.
[224] *Indigenous Law 1993*, art. 21.

a subsidy used to acquire water use rights, which indigenous communities or individuals can apply for in accordance with a number of conditions.[225] Where rights to use particular water resources are not already fully allocated, *derechos de aprovechamiento* can simply be constituted or regularised in the name of indigenous communities. The Fund finances the processes for acquiring the rights established in the *Water Code* in each of these processes.[226]

Derechos de aprovechamiento purchased for indigenous communities with assistance from the Fund are bought in the open market, and their title is transferred at the local Real Estate Office.[227] The Regulations also set out the factors the Development Corporation must consider before granting subsidies for water rights acquisition. They are: the number of persons or size of the community; the deterioration or degradation of lands affected by a lack of water; the sanitation conditions of families located on the property affected by a lack of water; and agricultural benefits from irrigation for the lands affected.[228]

The allocation of *derechos de aprovechamiento* to indigenous landholders with the support of the Fund is not limited by the problem of continuity in the same way as the recognition of *ancestral* rights pursuant to article 64. There is no express requirement in either the *Indigenous Law* or Regulations to prove prior water use in order to access the Fund.[229] The Fund can, therefore, be used to finance water rights acquisition in situations where the indigenous community has no historical relationship with the particular water, or where the community's water use has been interrupted at some point since sovereignty.

The government's intention has always been that the Fund will support the economic development of indigenous lands. In an earlier draft of article 20(c) the link between indigenous water rights and productive land development was explicit. This version stated the Fund's objective as to '[p]urchase water rights where indigenous communities [sic] do not have them, or carry out works in order to put or

---

[225] *Decreto 395 Que Aprueba el Reglamento sobre el Fondo de Tierras y Aguas 1994* [Decree 395 Approving the Land and Water Fund Regulations 1994] (Chile) 395.

[226] In order to 'constitute' new water rights, indigenous communities must apply to the General Water Directorate for new water rights in accordance with the process set out in *Water Code*, art.140.

[227] *Water Code 1981*, arts. 112–3 (Pt. II, Div. 1).

[228] *Decreto 395 Que Aprueba el Reglamento sobre el Fondo de Tierras y Aguas 1994* [Decree 395 Approving the Land and Water Fund Regulations 1994] (Chile) cl. 8.

[229] Interview with Diego Sotomayor (Santiago, 23 December 2013), above n. 936.

restore the resource for indigenous land production'.[230] The Development Corporation continues to assume that water rights purchases will be financed for the general productive benefit of land.[231] There is also an assumption in the Regulations that *derechos de aprovechamiento* will be provided specifically for irrigation.[232]

The Fund deals with the problem of priority by funding the acquisition of *derechos de aprovechamiento* from others for allocation to indigenous landholders. This redistribution is made possible by the unbundled status of *derechos de aprovechamiento*, and their availability for purchase in water markets. Significantly, water use rights may even be provided to indigenous landholders in situations where rights to use water resources are already fully allocated. Furthermore, other users are not adversely impacted, as they are willing sellers and receive market price. The Fund is, therefore, a 'creative mechanism' to deal with the problem of priority, as the Special Commission explained:[233]

> In Chile's political conditions, as in many other countries, it is necessary to design ways to resolve the territorial conflict that respects historical indigenous rights and also respects historical rights acquired by non-indigenous. To this end laws must take creative mechanisms of arbitration and financial mechanisms that allow for fair solutions.

The Fund does not necessarily allow indigenous communities to recoup their particular *ancestral* lands and waters. However, there is no doubt that Chile's Indigenous Land and Water Fund is an important supplement to the recognition of *ancestral* water rights in article 64. Unlike the *ancestral* rights recognition model, the Fund responds to the reality that indigenous peoples have, in most cases, been dispossessed of their *ancestral* interests, and rights to use water resources historically used by indigenous landholders are now held by others. Instead of focusing on historical rights it responds to the *ongoing* injustice experienced by indigenous groups.

---

[230] Chile, 'Mensaje Presidencial' [Presidential Message], Cámara de Diputados [House of Representatives], 15 October 1991 (Patricio Aylwin Azocar) in Biblioteca del Congreso Nacional de Chile [Library of National Congress of Chile], above n. 9, 14.

[231] Interview with Diego Sotomayor (Santiago, 23 December 2013), above n. 141. According to Sotomayor, of the 200 regularisation cases he worked on in the south of Chile, the vast majority were for agriculture while only a few were for grazing.

[232] *Decreto Supremo N°1.220 Que Aprueba el Reglamento del Catastro Público de Aguas* [Supreme Decree No 1.220 Approving the Public Water Cadastre Regulations] (Diario Oficial [Official Gazette] 25 July 1998) 1997 ('*Decreto Supremo N°1.220*') (Chile) cl. 8.

[233] Solís and Luis, above n. 31, 33.

Because water resources in many parts of Chile were approaching full allocation by the time the *Indigenous Law* was passed, with no share of *derechos de aprovechamiento* having previously been set aside for allocation to indigenous groups, purchases would be instrumental. Yet, the Chilean Government has been criticised for providing the Fund with inadequate finance.[234] Accordingly, the Chilean experience illustrates the importance of setting aside a share of water resources for future allocation to indigenous landholders prior to water resources reaching full allocation, in order to reduce the cost of buying back water use rights in the future.

7.5 THE FUTURE OF INDIGENOUS WATER RIGHTS IN CHILE

The future of indigenous rights to water in Chile is uncertain. Successive Chilean governments have attempted to reform the *Water Code*, and overhaul its market-based conceptual underpinnings, since the transition to democracy in the 1990s. As in the other countries considered in this book, indigenous claims have often been eclipsed within this complicated political bargaining. The Chilean government managed to effect a minimal reform of the *Water Code* in 2005, which, as explained above, included some protections of wetlands and corresponding indigenous uses in Chile's north. However, large-scale reform has been impossible to secure.

Meanwhile, public concern about the adequate protection and fair distribution of water continues, among increasing pressure on water resources from industry (mining and hydroelectric development), irrigated agriculture, urbanisation and climate change. It was estimated in 2017 that only 10 per cent of surface water rights and 40 per cent of ground water rights remain unallocated in Chile.[235] In terms of

---

[234] Chile, 'Discusion Sala' [Parliamentary Debates], Cámara de Diputados [House of Representatives], 21 January 1993 (Ribera) in Biblioteca del Congresso Nacional de Chile [Library of National Congress of Chile], above n. 9, 190. See also interview with Carlos Esteves (Santiago, September 2017); interview with Carlos Herrera Inzunza (Temuco, 11 November 2011). Inzunza, a lawyer for the National Indigenous Development Corporation explained that the Fund is financed significantly better for land acquisitions than for water.

[235] Hernando Silva, '*El Derecho Humano al Agua de los Pueblos Indígenas y el Proyecto de Reforma al Código de Aguas*' [Indigenous Peoples' Human Right to Water and the Reform of the Water Code] in *Derechos humanos y pueblos indígenas en Chile hoy: las amenazas al agua, a la biodiversidad y a la protesta social* [Human Rights and Indigenous Peoples in Chile today: threats to water, biodiversity and social protest] (Observatorio Ciudadano, 2017) 15–30.

indigenous water access, water rights acquisitions at market prices are costly and the Indigenous land and Water Fund continues to receive limited governmental support. The failure of Chilean governments to set aside a portion of water use rights for indigenous use in the future, prior to water resources reaching full allocation, has undermined the effectiveness of the Fund.

In 2011 the Chilean Parliament introduced a Parliamentary motion for a comprehensive amendment to the *Water Code*. Yet it wasn't until 2014, during Bachelet's Government, that the motion proceeded after a number of water related conflicts broke out in Chile,[236] including major social protest about the construction of the Hydroaysen dam complex which led Chileans onto the streets in the tens of thousands. Following this, the second Bachelet administration of 2014 to 2018, developed the *Proyecto de Reforma al Código de Aguas* (Chile) ('*Water Code Amendment Bill*').[237] The Bill was presented as a move away from *Decree Law 2.603*, which separated land and water holdings and made way for the water market. *Decree Law 2.603*, discussed above, was passed during the military dictatorship, as part of a neoliberal reform implemented across a range of sectors promoting a growth in water-related development; a context adding to its political unpalatability today. This Bill responded to public concern about the fair distribution of water among uses and the hoarding of private water rights.[238] It also responded to broader public concern about foreign investment in natural resource exploitation in Chile, and a perceived neoliberal or propertised approach to water allocation in Chilean water law frameworks, to the benefit of the wealthy, combined with inadequate protection of the human rights of the vulnerable.

The Water Bill also includes some provisions dealing with indigenous rights to water. At clause 5 the Bill provides:[239]

> In the case of indigenous territories, the State will ensure the integrity between land and water, and protect waters for the benefit of indigenous communities, in accordance with laws and international treaties ratified by Chile that are currently in force.

---

[236] Interview with Carlos Estevez (Santiago, 8 September 2017); Andrei Jourvalev.

[237] *Reforma al Código de Aguas* [Reform of the Water Code]: *Primer Trámite Constitucional* [First Constitutional Stage] 2018.

[238] Interview with Carlos Estevez (Santiago, 8 September 2017), above n. 236; interview with Sara Larrain (Temuco, 6 September 2017).;

[239] *Proyecto de Reforma al Código de Aguas* [Water Code Amendment Bill], art. 5 bis.

This amendment attempts to reverse the separation of water and land rights in current Chilean water law frameworks,[240] instead adopting the logic of ILO *Convention 169* in relation to the integrity of indigenous territory.[241] The territorial nature of indigenous water rights in Chile, in line with international and regional law and jurisprudence, has been emphasised since the development of the *Indigenous Law*. The Special Commission explained that the concept of 'territory' in international indigenous rights law extends beyond the land itself to 'water, air lakes foreshore, surface and subsurface land and flora and fauna'.[242] Of course, the real strength of clause 5 is the explicit reference to international law,[243] given that human rights treaties are 'self-executing' in Chile's monist constitutional law framework.[244] Commentators, however, express concern that Chile has not in the past afforded effective protection of human rights, despite the domestic weight of international law.[245]

The second protection of indigenous water rights in the Bill is the prioritisation of water use for human consumption and sanitation,[246] in line with a shift back towards emphasising water as a *bien nacional de uso public* (national property for public use) and ongoing concerns about priority of use and unfair distribution of water. The Bill recognises the use of water for subsistence purposes, carried out by many indigenous communities throughout Chile.[247] However, there is little clarity in the Bill as to how the priority mechanism will work in practice.

The Bill also includes a number of exceptions to general principles for indigenous communities, as a sort of 'positive discrimination'.[248] Indigenous communities that use water for traditional subsistence agriculture will be exempted from the requirement to set aside an ecological flow of water for environmental purposes.[249] Other exemptions include: exempting indigenous communities from the five-year limit on regularisations of water use rights under transitory article 2 to

---

[240] Interview with Sara Larrain (Temuco, 6 September 2017), above n. 238. Interview with Carlos Estevez (Santiago, 8 September 2017), above n. 236.
[241] International Labour Organization, Chapter 6 above n. 26 arts 6, 7.
[242] Comisión Especial de Pueblos Indígenas, above n. 23, 7.
[243] *Proyecto de Reforma al Código de Aguas* [Water Code Amendment Bill] art. 5.
[244] *Constitution 1980*, art. 5(2).
[245] Schönsteiner and Couso, above n. 214.
[246] *Proyecto de Reforma al Código de Aguas* [Water Code Amendment Bill], art. 5.
[247] Interview with Sara Larrain (Temuco, 6 September 2017), above n. 238. Larrain was on the committee responsible for preparing the Bill.
[248] Ibid.
[249] *Proyecto de Reforma al Código de Aguas* [Water Code Amendment Bill], art. 307 bis.

the *Water Code*; exempting indigenous communities from the payment of fees for non-use of water; and exempting indigenous communities from restrictions on exercising their water rights once a basin has been declared 'exhausted'.[250]

Although these protections look promising, controversy has surrounded the reform process and the Chilean Government has been accused of failing to properly consult with indigenous peoples. According to the then Director of the Chilean Water Directorate, the Parliamentary committees charged with developing the law reform proposal decided to leave the indigenous protections out of early development of the reform project in order to avoid consulting with indigenous peoples.[251] This occurred despite the fact that the need to consult indigenous peoples about matters that affect them is underscored in Chile's commitment to ILO *Convention 169*,[252] as well as domestic legislation around consultation with indigenous peoples.[253] The Committees did so in order to avoid delays in consultation,[254] given that the Parliament does not have procedures and mechanisms in place for consulting indigenous peoples.[255] Instead, the water provisions affecting indigenous water rights were inserted at a later stage of the Parliamentary process, avoiding the need for consultation.

In any event, the Bachelet administration was unable to pass the *Water Bill* prior to the change of government, and as at 2019 it remains in the last phase of discussion in the Senate.[256] The incumbent right-wing Piñera Government (returned for the second time in 2017) has little appetite for reforming water law away from a market-based logic. In fact, President Piñera has announced his opposition to the *Water Code Amendment Bill*, as prepared, on the basis that it presents uncertainty and probable loss for the Chilean agricultural sector.[257] His new government would provide security to water rights by re-establishing

---

[250] Ibid. transitory, art. 2.
[251] Interview with Carlos Estevez (Santiago, 8 September 2017), above n. 236.
[252] International Labour Organization, Chapter 6 above n. 26 arts 6, 7.
[253] *Indigenous Law 1993*, art 34.
[254] Interview with Sara Larrain (Temuco, 6 September 2017), above n. 238. Accordingly issues concerning indigenous peoples' rights that were separated for later discussion and further consultation in an urgent attempt to push through the discussion of the draft Bill.
[255] Ibid.; interview with Alvaro Duran (Santiago, 29 August 2017); interview with Carlos Estevez (Santiago, 8 September 2017), above n. 236.
[256] As at February 2019 the Bill was before the Agricultural Committee of the Senate.
[257] Sebastian Piñera, *En Nuestro Gobierno Vamos a Asegurar la Disponibilidad de Agua* [In Our Government We Are Going to Secure Water Availability] (12 October 2018) www.sebastianpinera.cl/sebastian-pinera-en-nuestro-gobierno-vamos-a-asegurar-la-disponibilidad-de-agua.

the legal certainty of new and ancestral water rights as *propiedad*. Meanwhile, indigenous water reform remains unresolved.

7.6 CONCLUSION

Indigenous Chileans have clearly been excluded from water law frameworks since colonisation and settlement. Their continuing political demands for recognition and allocation of water rights have stressed the need for more equitable distribution of access and use rights. When the Chilean government finally responded to these demands at the end of the twentieth century, giving indigenous landholders the right to use water on their lands for productive purposes, this was not only considered necessary in the interests of justice and fairness; it was intended to support indigenous economic development. The outcomes were the recognition of the *ancestral* water rights of indigenous peoples in article 64 of the *Indigenous Law* and the creation of an Indigenous Land and Water Fund for the acquisition of rights in the market.

The effectiveness of the recognition of *ancestral* water rights has been limited. Article 64 of the *Indigenous Law* only serves to formalise water interests that have continued to be practiced by indigenous groups since 'time immemorial', and where inconsistent rights are not held by other users. It is, therefore, an incomplete response to the ongoing exclusion indigenous peoples experience from rights allocated within water law frameworks. The Fund, by contrast, specifically responds to the situation where indigenous peoples have been unable to continue to exercise their water rights, because their lands and waters have been acquired by others. Indigenous communities are not required to prove continuous use of water resources since 'time immemorial' in order to acquire water use rights via the Fund. In the case of water resources already fully allocated to others, the Fund finances the purchase of water use rights in markets for redistribution to indigenous landholders. An interesting lesson from the Chilean experience, therefore, is that market mechanisms, like funds, may in some situations be a 'creative' response to the injustice in water rights distribution. However, by setting aside a share of water use rights before water resources are already fully allocated, governments reduce the cost of buying-back water use rights for allocation to indigenous peoples in the future.

Indigenous demands for water justice remain unresolved in Chile. In 2019 the Chilean Government has a reform proposal before the Senate, which attempts to deal with the continuing exclusion of indigenous peoples from an equitable share of the water market. The proposal seeks to overcome the separation of water from indigenous lands within regulatory frameworks, and support subsistence water uses by indigenous peoples. However the reform proposal has little likelihood of passage in its current form, at least during the current administration, revealing that market-based solutions are no panacea, especially without adequate resourcing and commitment from governments.

PART III

LESSONS LEARNT

CHAPTER EIGHT

INDIGENOUS WATER RIGHTS IN COMPARATIVE LAW: JURISDICTION AND DISTRIBUTION

8.1 INTRODUCTION

What lessons can be drawn from the study of four historically, legally, politically, culturally and socially distinct countries about indigenous water rights in law and regulation? The country studies offer new perspectives on both the reasons why states should respond to indigenous water injustice and support indigenous water jurisdiction and distribution, as well as providing some broad guidance as to how this might (or should not) be done.

In the first part of this chapter, I discuss the imperative for governments to finally address historical water injustice, and respond to the exclusion indigenous peoples have experienced, and continue to experience, from water law frameworks. In the second part of the chapter, I consider how a more complete response to indigenous water exclusion might be achieved. This cannot be done, the country studies tell us, if indigenous peoples lack either the jurisdiction to exercise authority and influence over water management and governance in their territories, or a fair distribution of substantive rights to use water under legal and policy frameworks. I follow with a reflection on how a more complete response to indigenous water injustice might look.

8.2 INDIGENOUS WATER INJUSTICE

There remains no consensus in the indigenous rights literature on whether indigenous groups should be entitled to specific land and resource rights, nor is there consensus on the reason why indigenous land and resource rights might be needed. In Chapter 2 I explained how indigenous land and resource rights, and especially indigenous rights to water, have been conceived of in a particularly limited way in law, policy and scholarship. Indigenous water interests are presented as being traditional and cultural in nature, a framing that often limits the rights that flow from them. This portrayal is inconsistent with indigenous demands for water rights for a full spectrum of purposes from cultural, spiritual or environmental uses through to commercial purposes like agriculture, tourism or industry.

In Australia, the *Native Title Act 1993* (Cth) restricts native title rights and interests in water to traditional, cultural purposes only, which may remove the potential to recognise or allocate indigenous water rights for modern or commercial water interests. In New Zealand, although governments have often been willing to consult with Māori about decision making about water under the *Resource Management Act 1991* (NZ), the Crown has maintained its 'bottom line' that no one, not even Māori, can own water and has actively resisted Māori making any sort of substantive, commercial water claim. The Crown has been more prepared to involve Māori in collaborative governance arrangements for rivers, even via the use of legal person models such as that for the Whanganui River. The Colombian courts have adopted a similar approach, inspired by the New Zealand developments to design a complicated institutional model for the Atrato River as a legal subject under the guardianship of indigenous and afrodescendent communities. Yet the Colombian Government has made no broader attempts to provide indigenous Colombians with the substantive right to use water in their territories. In Chile, the Government seems better disposed to allocate water rights to indigenous landholders for any purpose under the *Indigenous Law 1993* (Chile), although the surrounding policy framework suggests that such rights are intended for subsistence farming purposes at the expense of cultural or conservation uses. The reasons why the Chilean Government took a broader view of indigenous water interests will be explored further below. In any event, in all four countries, indigenous water interests are perceived by governments in particular restrictive ways. This perception flows on to the legal frameworks that provide for the rights, and continues to undermine their potential to resolve indigenous water injustice.

I have argued in this book that the limited cultural conception of indigenous water rights stems from an over-reliance on the idea of reparative justice for historical wrongs. Reparative justice does help to explain why indigenous peoples should be provided with water rights. However, focusing only on the need to recognise indigenous water rights that were lost as a result of colonisation is problematic. If the object of indigenous water rights is to recognise pre-sovereignty water rights, must the rights be constrained by pre-sovereignty notions of resource use? What happens where rights to the resources traditionally used by that indigenous group have (after colonisation) been allocated to other users? The focus, instead, should be on enduring indigenous water injustices, by which indigenous communities continue to be shut out of allocation frameworks and laws and policies that regulate resource use.

As shown in Chapter 7, the Chilean indigenous water rights provisions were intended, not only as reparation for historical injustices against indigenous Chileans including land and water dispossession. The Chilean lawmakers hoped to redress the concentration of water use rights in the hands of private water users, and provide some distributive justice for indigenous Chileans, where necessary by redistributing water use rights to indigenous landholders. Instead of focusing just on traditional indigenous culture, concerns about the relationship between productive land use and indigenous poverty highlighted a need for 'culture plus development'. Providing indigenous people, the 'poorest of the poor', with the right to use water on their lands was intended to support the productive use of that land, in order to combat indigenous economic disadvantage. Because indigenous water rights in Chile were designed to address indigenous disadvantage, and support the development of indigenous lands, they logically needed to extend beyond limited traditional, cultural purposes.

Indigenous peoples in all the countries studied in this book have undeniably suffered historical injustices. However, indigenous peoples continue to be excluded from the right to use water on their lands and to experience economic and social disadvantage. We learned in Chapter 4 that, 'Aboriginal peoples and Torres Strait Islanders have become, as a group, the most disadvantaged in Australian society'.[1] Indigenous groups in Australia now hold more than 30 per cent of the land in Australia, under native title and titles granted under land rights

---

[1] *Native Title Act 1993* preamble.

legislation but less than 0.01 per cent of water diversions go to indigenous people.[2]

In Chapter 5 I explained how the Māori peoples of Aotearoa New Zealand have been excluded from legal frameworks allocating rights to manage and use their water resources since the British acquisition of sovereignty. Since 1903 the New Zealand government vested all water resources in the Crown, and gave itself the sole right to regulate and allocate water, without any acknowledgement of the ongoing water use by, and water relationships of, Māori. The Crown continues to resist Māori proprietary claims for water rights on the basis that 'no one can own water', because that would be inconsistent with the public vesting of all waters in the Crown on behalf of the New Zealand public. However, the denial of any substantive rights to water for Māori, and the right for Māori to benefit from the development of water, undermines the potential for the economic development of indigenous peoples and territories in New Zealand.

In Chapter 6 I introduced the indigenous peoples of Colombia, who have suffered widespread dispossession and disenfranchisement since the time of colonisation, including the loss of traditional territories and disrupted access to water resources. Despite the limited historical recognition of indigenous rights to territory and self-government in Colombia, nowhere in Colombian law is any specific provision made for indigenous peoples to use water on their territories. Despite there being no specific recognition of an indigenous right to water in the Constitution, the Constitutional Court has developed a line of jurisprudence attempting to protect the human right to water, including for indigenous communities. The *Tierra Digna* decision on the Atrato River presents new hope for how governments might reconstitute relationships with indigenous groups around access to and management of water. However, the Atrato decision is about the management of water rather than water use, and in no way impacts on water allocation frameworks, which continue to ignore and disadvantage the impoverished indigenous peoples.

In Chapter 7, I discussed the case of Chile, where indigenous peoples have suffered a long history of physical, social and economic disadvantage including the loss of, or failure by the state to recognise their rights to own and govern their traditional territories, including their water

---

[2] Jon Altman and Francis Markham, 'Burgeoning Indigenous Land Ownership: Diverse Values and Strategic Potentialities' in Sean Brennan et al. (eds.), *Native Title from Mabo to Akiba: A Vehicle for Change and Empowerment?* (Federation Press, 2015) 126.

resources. Their continuing political demands for recognition and allocation of water rights have stressed the need for more equitable distribution of access and use rights. When the Chilean government finally responded to these demands at the end of the twentieth century, giving indigenous peoples the right to use water on their lands for productive purposes, this was not only considered necessary in the interests of fairness; it was intended to support indigenous economic development. However, against a background of ongoing concern about the separation of water rights from indigenous lands and the continuing exclusion of indigenous peoples from an equitable share of the water market in Chile, indigenous demands for water justice remain unresolved.

Where indigenous landholders do not enjoy substantive rights to use the water for any purpose this is an obvious restriction on the potential for the development of indigenous peoples and territories. There is increasing momentum around the indigenous right to development, both in international and domestic comparative law. In Aotearoa New Zealand, for example, the Waitangi Tribunal found that the Māori right to water includes a right to develop the resources and benefit from them commercially. The Tribunal reasoned:[3]

> The interested parties emphasised the Treaty right of development and the choice of Māori to work in two worlds: to resist assimilation and protect their mātauranga Māori and tikanga (knowledge and law) but also to benefit commercially from development, as guaranteed by the Treaty and affirmed by the United Nations Declaration on the Rights of Indigenous Peoples.

Where indigenous water claims are not settled they may also present uncertainly for regulatory regimes for water. The study of debates preceding Chile's *Indigenous Law* in Chapter 7 revealed that the water provisions were considered desirable because political demands for water rights by indigenous groups undermined the certainty of other water use rights, which overlapped or conflicted with indigenous interests. Both the government and industry saw the endurance of 'customary' water interests as a threat to the efficiency of water markets, which depended on clear and certain water use rights. Addressing indigenous water claims would contribute to a more complete and effective water

---

[3] Waitangi Tribunal, *Stage 1 Report on National Freshwater and Geothermal Resources*, Chapter 5 above n. 34, 35.

market, through the clarification of property rights and minimisation of transaction costs caused by legal disputes.

Certainty for water users, investors and stakeholders is a key pre-condition for markets to operate effectively,[4] and a key precondition for all regulatory regimes for water.[5] The Australian National Water Initiative, which consolidated Australia's contemporary approach to water regulation in the 1990s, was intended to provide greater certainty in Australia's water management regimes.[6] Concerns about certainly of title have also featured consistently in debates about native land reform in Australia.[7] They are reflected in the preamble of the *Native Title Act*, providing that the 'needs of the broader Australian community require certainty and the enforceability of acts potentially made invalid because of the existence of native title'. Certainty, too, has been seen as a key driver for indigenous land rights legislation in Australia, in light of the uncertainty ultimately provoked by native title, as McRae et al. explain:[8]

> Collectively these land rights laws have yielded far more of the contemporary Indigenous estate around Australia than native title has to date, often with a stronger title and associated set of rights, as well as greater certainty for Indigenous people and better protection against unwanted activity on their land. Statutory land rights also typically offer a clearer and more secure basis for Indigenous economic activity.

More recently in Northern Australia, the *White Paper* specifically underscores the importance of certainty in water use rights in order to attract investment and support economic water uses, including by indigenous users.[9]

---

[4] See, e.g., Crase, Chapter 7 above n. 123; Corbett A Grainger and Christopher Costello, 'The Value of Secure Property Rights: Evidence from Global Fisheries' [2011] (No. w17019) *NBER Working Paper*.

[5] Jon Altman, 'Indigenous Interests and Water Property Rights' (2004) 23(3) *Dialogue* 29, 30. See generally Rutgerd Boelens, 'The Politics of Disciplining Water Rights' (2009) 40(2) *Development and Change* 307, 319, 320; Altman; WD Nikolakis and R Grafton, WD Nikolakis and RQ Grafton, 'Assessment of the Potential Costs and Benefits of Water Trading Across Northern Australia' (Report, Tropical Rivers and Coastal Knowledge, March 2011)' 217, 3, 4.

[6] Commonwealth of Australia and the Governments of New South Wales, Victoria, Queensland, South Australia, the Australian Capital Territory and the Northern Territory, above n. 23, cl. 15.

[7] Short, Chapter 2 above n. 47, 47–8.

[8] Heather McRae, *Indigenous Legal Issues: Commentary and Materials* (Thomson Reuters (Professional) Australia, 4th ed, 2009) 222.

[9] Australian Government, 'Our North, Our Future: White Paper on Developing Northern Australia', Chapter 4 above n. 232, 21, 41.

Those calling for commercial water rights for Māori in New Zealand have also emphasised the benefits of clarifying indigenous water claims in order to reduce water-related disputes, improve the certainty of other water use rights, and market conditions for security and investment purposes.[10] In the context of commercial fisheries in Aotearoa New Zealand, Grainger and Costello discuss the impact of clear, certain tradeable fisheries quota rights, which are structured as a perpetual right to fish, applicable as collateral for credit.[11] Highlighting the benefits of settling indigenous water claims in terms of certainty is an appeal to a more liberal mindset than that usually relied on to support indigenous claims, yet as pointed out by Sunstein, 'people can often agree on constitutional practices, and even on constitutional rights, when they cannot agree on constitutional theories'.[12]

As discussed in Chapter 3, as governments respond to indigenous water exclusion they must make political and constitutional trade-offs about the scope of rights, and confront the contestation inherent in diverse regulatory approaches. For example, are indigenous water rights a basic human right? Are they a right to an economic good; a distribution of available water; applicable to commercial as well as subsistence purposes and potentially transferable in markets? Are indigenous water rights essentially environmental rights, where indigenous interests run concurrent with the rights of nature? The country studies in this book have shown how the need to respond to indigenous water exclusion puts the onus on lawmakers to come up with a more complete response. A more complete response would contemplate both: jurisdiction for indigenous peoples to exercise authority over water governance and management; and, broader (including commercial or economic) water interests for indigenous peoples via an equitable distribution of substantive water rights.

8.3 A MORE COMPLETE RESPONSE

8.3.1 Indigenous Water Jurisdiction
In Chapter 2 I argued that that states must allow indigenous groups to practice their cultural ways over, and exercise their cultural

---

[10] Sin, Murray and Wyatt, Chapter 5 above n. 179, 6.
[11] Grainger and Costello, above n. 4.
[12] Cass R Sunstein, 'Incompletely Theorized Agreements in Constitutional Law' (2007) 74(1) *Social Research* 1, 1.

relationships with respect to, their water resources, applying their own laws and customs in water management and governance, as an imperative of indigenous water jurisdiction. Indigenous peoples in all four countries studied have, among other things, asked their governments to enable them to participate in or control the management of their water resources. In many cases indigenous governance has been taken away from indigenous peoples with no negotiation, consent or compensation and governments are doing a poor job of managing waterways. Indigenous peoples are not 'just another stakeholder' in natural resource governance.[13] What indigenous peoples seek has been described as a 'space for indigenous governance to continue to "breathe"',[14] or 'establishing the necessary conditions under the law (access to water and autonomy for management), in order to stay out of the way of the law'.[15]

In New Zealand and Colombia, courts and legislatures have responded with innovative legal approaches to include indigenous peoples in water governance; declaring rivers to be legal persons and enabling traditional owners to 'voice' the river's concerns as their guardians. The New Zealand and Colombian cases do both, to an extent, recognise and provide room for indigenous water jurisdiction. They accommodate, to an extent, diverse legal and cultural interests in rivers, in order to establish a new relationship between the state and river communities in the interests of improved outcomes.

In Chapter 7 I discussed how the idea of extending legal personality to the Whanganui River was part of a political settlement with Whanganui Iwi intended to repair past wrongs and recognise Māori relationships with water. The Te Awa Tupua model reflects the conceptualisation of the relationship between humans and the natural world in the *tikanga* (law) of the Whanganui Iwi. According to their *tikanga*, the river has rights to which the *Iwi* belong and not the other way around, exemplified in the *Iwi's* idiom: 'I am the river, and the river is me'.

The Constitutional Court of Columbia's recognition of the 'biocultural' rights of the Atrato communities, considered in Chapter 6, also underscores the role of legal and cultural pluralism in the legal person idea. The Court frames the ethnic communities' rights as the right 'to administer and exercise trusteeship in an autonomous manner over

---

[13] O'Bryan, above n. 19.
[14] Strelein and Tran, Chapter 2 above n. 26, 47.
[15] Boelens et al., Chapter 2 above n. 27, 167.

their territories – in accordance with their own laws and customs'.[16] The Court points out that these biocultural rights 'are not new rights for the ethnic communities', but rather a category that unifies their interconnected rights in natural resources and to culture.[17] This approach resembles recognition models like native or aboriginal title, where state law recognises and gives effect to pre-existing indigenous laws and customs.

In Australia, the recognition of the Yarra River as a 'living and integrated natural entity' under the *Yarra River Protection (Wilip-gin Birrarung murron) Act 2017* (Vic) discussed in Chapter 4 takes an approach similar to the Whanganui and Atrato Rivers, although without declaring that the river is a legal person. This legislation 'recognises the intrinsic connection of the traditional owners to the Yarra River and its Country and further recognises them as the custodians of the land and waterway which they call Birrarung'.[18] The Yarra River model demonstrates how indigenous values may be used to reframe non-indigenous relationships to a river.[19] However, with only two traditional owner members on a council of twelve, the legislation does not go nearly as far as it should in terms of enabling indigenous jurisdiction.

The recognition or grant of legal rights for rivers in all these cases, although emerging from and existing within very different circumstances, is an attempt by the state to recognise and accommodate legal and cultural pluralism.[20] In each country, legal personality is a mechanism adopted by the state to recognise river interests and relationships existing in indigenous and tribal customs and laws. When states recognise indigenous rights and interests, they are using approximations from western law to attempt to match indigenous norms and cosmologies. States risk essentialising indigenous culture through a traditional, cultural lens; leaving open the question whether legal person models can be applied outside of such a cultural context.

As I argued in Chapter 2, the way in which indigenous peoples choose to 'use' natural resources may not, in fact, coincide with western notions of authentic indigenous culture. Indigenous peoples may freely determine, for example, to dispose of their natural wealth.[21] Where the

---

[16] *Tierra Digna [2016]* Corte Constitucional [Constitutional Court], Sala Sexta de Revision [Sixth Chamber] (Colombia) No T-622 of 2016 (10 November 2016) 43.
[17] Ibid. 44.
[18] *Yarra River Protection (Wilip-Gin Birrarung Murron) Act 2017* preamble.
[19] Interview with Erin O'Donnell (Melbourne, 9 July 2018), Chapter 4 above n. 217.
[20] Macpherson and Clavijo Ospina, Felipe, above n. 19.
[21] Tobin, Chapter 2 above n. 1, 121.

state determines the parameters of legal person or guardianship models for indigenous peoples they may undermine the potential for indigenous peoples to exercise full authority, or jurisdiction, over their territories. They may contradict indigenous peoples' development aspirations in line with their cultural norms and practices, such as the 'right to development' in international law, or in Latin America the idea of 'ethnodevelopment', promoting alternative but self-determined models of indigenous economic development.[22]

The settlement for the Whanganui River discussed in Chapter 5, although going a long way to accommodate the *tikanga* of the Whanganui Iwi with respect to the river, does not accommodate their position as river 'owners' as well as river 'guardians'. Māori interests in the river under *tikanga* Māori are proprietary and territorial rights to the entire catchment and all its resources based on familial and ancestral connections. Yet, in the political compromise which is the *Te Awa Tupua Act*, no one has the right to 'own' the river as a whole. Furthermore, the failure to give the river the legal right to its own water is contradictory. In this context can Te Awa Tupua really be described as an 'indivisible and living whole from the mountains to the sea incorporating physical and metaphysical elements'.[23] Nor is the indigenous jurisdiction in the Te Awa Tupua Model exclusive. While both recognising and limiting an indigenous jurisdiction, the model is an advanced collaborative governance approach, in which the interests of the river are emphasised but Māori are just one interest group in its management, along with other government, community and business interests.[24]

In Colombia, although the Atrato communities' biocultural rights are positioned as being territorial in nature, the Court does not recognise a right of property for the communities in the Atrato River, nor for the river to own itself. In the context of increasing completion for water use from a range of economic, social and environmental interests, not giving the communities substantive rights to use the water will surely undermine their potential to influence decision-making in the catchments.

The studies of Aotearoa New Zealand and Colombia in this book also show the tendency for indigenous peoples and states to adapt

---

[22] Engle, Chapter 2 above n. 16. The Constitutional Court in the Atrato case actually refers to the idea of ethnodevelopment in support of its approach.
[23] *Te Awa Tupua (Whanganui River Claims Settlement) Act 2017* (NZ), s. 12.
[24] Macpherson and Clavijo Ospina, Felipe, above n. 19.

constitutional norms and processes, and specifically human rights guarantees, in response to water concerns. The Constitutional Court of Colombia was able to recognise the Atrato River as a legal person because of the strong constitutional protection of indigenous and environmental rights, and a rich national and international jurisprudence regarding indigenous rights to natural resources in Latin America. The Court is developing rights for nature in the Colombian constitutional context as an extension of human rights law, as a 'third generation' of human rights, in a similar way to that in which Latin American cases have used the protection of indigenous 'territory' by expanding the right to 'property'.[25] In the New Zealand context, the Te Awa Tupua model does not rest on obvious constitutional human rights documents like the *New Zealand Bill of Rights Act 1990* (NZ), although as part of the Treaty of Waitangi constitutional reconciliation process, it certainly shows how human rights doctrine can develop outside of core human rights laws. As Sanders argues, the legal person model for the Whanganui River is a new 'constitutional' framework to 'regulate human relationships' and establish a forum for disagreement and compromise: recognising the competing claims of the Crown and Māori to political authority.[26]

Thus, both the Colombian and New Zealand models are an attempt to reset the relationship between the state and communities, in recognition of the culturally specific guardianship relationship between communities and nature. Legal personality is a mechanism used to recognise indigenous and tribal relationships and jurisdictions to manage the natural world. However, the jurisdiction is not recognised in its complete form, and may actually be limited via the process of recognition. Whether either model results in improved river outcomes, or increased indigenous or community jurisdiction to govern, will depend on the surrounding institutional framework, including ensuring that river institutions are adequately funded. Although in the New Zealand and Colombian context, the attempt at recognising cultural conceptions of a river as a person or ancestor under the guardianship of indigenous peoples is frustrated by the over prescription of the legal frameworks and institutions accompanying that recognition. Indigenous peoples need room to

---

[25] See, e.g., *Mayagna (Sumo) Awas Tingni Community v. Nicaragua (Judgment) (Inter-American Court of Human Rights (Ser C) Case* No. 79 (31 August 2001) [2001] Inter-American Court of Human Rights.
[26] Sanders, Chapter 3 above n. 64, 1.

breathe, to exercise their jurisdiction over water governance and use. Where governments are devising legal person models in other contexts, they should heed these warnings.

Ultimately, the use of legal person or subject models in settling indigenous water claims in Aotearoa New Zealand and Colombia has been *ad hoc* and reactive. Questions remain about the practical utility of such *ad hoc* responses. Why these rivers and not others? Could the same objectives be achieved with strong institutions and meaningful colla-boration, without the legal fiction of the 'person'? The legal person models discussed in this book are clearly attempts to settle disputes over water governance, and reconstitute relationships between states and peoples, in the context of indigenous concerns about poor resource management by central and local governments by giving more author-ity and control to indigenous users. At the same time, they are clearly political compromises, which internalise power imbalances between states and indigenous peoples. Perhaps more significantly though, they are interim solutions to indigenous territorial claims for more than governance, putting off the more difficult questions around water dis-tribution and 'ownership'.

8.3.2 Indigenous Water Distribution

Indigenous peoples in all four countries studied in this book certainly suffered historical injustices because of colonisation, as their land and resource rights were not recognised. This historical injustice of non-recognition sets indigenous experiences aside from other disadvantaged groups, and began a chain of events leading to the current injustice in the distribution of rights. If states are to correct this ongoing injustice in the distribution of water use rights, they must include indigenous people in water law frameworks in the way that might have occurred had indigenous land and water rights been recognised at the acquisition of sovereignty. Put simply, indigenous people should have the right to use water for both cultural and commercial purposes, as other water users enjoy its use. Where water use rights have now been allocated to other users, and as demand for water continues to grow for irrigation, sanitation and industry as well as ecological or environmental needs, there may logically need to be some sort of redistribution to enable indigenous water rights.

In Chapter 3 I discussed the trajectories of, and conflict inherent within, modern regulatory frameworks for water, in particular the ten-sion between public and private interests. How does such a distributive imperative respond to the reality that water catchments are already in many parts of the world fully or over allocated, and there are increasing pressures on supply and quality from water users and the changing climate, and existing debates about the use of water for development versus the protection of water as a human right? The country studies for Australia and Chile, in particular, provide important lessons about how to respond to the distributive imperative of indigenous water rights in this context.

As explained in Chapter 4, native title rights to water in Australia deliver an important function of recognising traditional-cultural interests, generally reflected in water planning frameworks as 'environmental and public benefit' outcomes. There is a growing consensus around including such interests in instream 'cultural flows', which may not require a water use right. Yet there remain conceptual and legal obstacles to establishing and exercising substantive native title rights to water for broader purposes, especially for economic development. Some remain hopeful that native title will evolve to authorise broader water uses, and the Australian Law Reform Commission recommends as such, even though the native title determination process is a long, difficult, expensive and *ad hoc* approach to resolving indigenous claims to land and resources.[27] The Australian experience serves as a warning to other countries, including Aotearoa New Zealand, of the risks involved in taking a native title-style recognition approach.

As mentioned in Chapter 4, some commentators argue that the 'traditional law and custom' approach to proving native title rights in Australia is responsible for Australia's particular predicament with 'continuity'.[28] Pearson has argued for a possessory title to land and resources for indigenous peoples in Australia, on the basis that indigenous inhabitants in possession of their lands as at colonisation are presumed to have a fee simple estate, unless someone else can claim a better right. There is no need to prove any pre-sovereignty laws and customs, or ongoing continuity, but merely to prove occupation at the time of sovereignty. The idea that native title could be based on historical possession rather than traditional laws and customs is sometimes put forward with reference to American and Canadian approaches to establishing native title, which appears to accommodate

---

[27] Australian Law Reform Commission, Chapter 4 above n. 65, 15–16.
[28] Pearson, 'The Concept of Native Title at Common Law', Chapter 4 above n. 45; Pearson, 'Land is Susceptible of Ownership', Chapter 4 above n. 118, 83; see also Pohle, Chapter 4 above n. 118; see also O'Connor, Chapter 4 above n. 118.

better the evolution or change of native title rights and interests over time.[29]

Yet, as discussed in Chapter 7 indigenous rights to land and resources in Chile can be best characterised as arising out of historical possession or use, consistent with the approach in Latin American jurisprudence that: '[t]he foundation of territorial property lies in the historical use and occupation which gave rise to customary land tenure systems'.[30] Nonetheless, the problem of continuity has limited the potential to recognise *ancestral* water rights in Chile, in much the same way as has occurred with respect to native title rights to water in Australia. As found in Chapter 7, those seeking recognition of *ancestral* water rights are expected to prove that they have continuously used a particular water resource for *ancestral* purposes since 'time immemorial'. This problem of continuity precludes recognition for those who no longer exercise their water rights, including because other people have acquired land and water rights in their *ancestral* territories. Historical possession and use of land and resources may be easier to prove than the content of pre-sovereignty laws and customs authorising resource use. Yet, recognition mechanisms based on historical possession cannot avoid the problem of continuity altogether. All recognition mechanisms are an attempt to recognise historical rights. Recognising pre-sovereignty rights raises inevitable tensions around continuity of connection, because of the time that has elapsed since colonisation.

Simply recognising historical rights via a native title model would inevitably raise issues of priority, given the fact that other users now hold state-sanctioned rights to the same resources. In the Australian context, discussed in Chapter 4, the *Native Title Act* includes a number of provisions that make native title rights to water ineffective if they are inconsistent with State water legislation and water use rights under it

---

[29] See, e.g., *Tsilhqot'in Nation v. British Colombia* [2014] SCC 44 [24], [45]. In that case the Supreme Court held that in terms of Canadian Aboriginal title continuity is only a requirement where current occupation is used to establish an inference of pre-sovereignty occupation. The Supreme Court explains at [45], 'Continuity simply means that for evidence of present occupation to establish an inference of pre-sovereignty occupation, the present occupation must be rooted in pre-sovereignty times. This is a question for the trier of fact in each case'.

[30] *Case of the Saramaka People v. Suriname (Preliminary Objections, Merits, Reparations and Costs)* [2007] *Inter-American Court of Human Rights Series* C No. 172 (28 November 2007) [96]. See also *Mayagna (Sumo) Awas Tingni Community v. Nicaragua (Judgment)* (Inter-American Court of Human Rights, (Ser C) Case No. 79, 31 August 2001) [2001] Case No. 79 (31 August 2001) [127–8], where the Court found that possession of land was all that was required to be proved in order to obtain state recognition. Where communities have lost possession since colonisation through no fault of their own they retain the right to property in their traditional lands.

(although limited procedural rights remain). This problem of priority means that third-party water rights override native title rights to water in the case of fully allocated water resources. A similar problem of priority, I showed in Chapter 7, undermines the potential to recognise *ancestral* water rights under article 64 of the *Indigenous Law*. Because surface water resources in northern Chile were largely fully allocated to others around the time the *Indigenous Law* came into effect, there was little scope left for the recognition of *ancestral* water rights.

The Chilean courts have attempted to recognise *ancestral* water rights despite the presence of other users, in an effort to support the repopulation of indigenous territories where indigenous landholders have been dispossessed of their water rights. In the *Chusmiza* case, the Supreme Court's decision sought to redistribute water use rights away from another user for allocation to an indigenous community. However, as I found in Chapter 7, the *Chusmiza* case illustrates the difficulty inherent in recognising pre-sovereignty water rights many years after the acquisition of sovereignty, in the presence of other rights. In fact, the legacy of *Chusmiza* is that in the case of fully allocated water resources, a recognition mechanism cannot provide for indigenous water rights without some form of redistribution.

The use of market mechanisms in the regulation of water resources, in both Chile and Australia, added urgency to demands for indigenous water rights, because the unbundling of water use rights from land-holding, and the introduction of water markets, has enabled other users to acquire the right to use water on or affecting indigenous lands. Furthermore, competitive water markets are designed to grow to encompass more traders and more water use rights, meaning that there will be less water available for responding to indigenous water demands in the future.

I make a hermeneutic distinction between recognition and allocation models in this book, although I acknowledge that the line between laws that recognise and allocate land and resource rights to indigenous groups is blurred. Allocation mechanisms may themselves be a type of recognition where they are devised by the state in recognition of indigenous interests in land and resources, and some actually refer to traditional land associations. However, it is possible to differentiate laws that 'recognise' rights that have their origins in and take their content from pre-sovereignty traditional laws and customs (such as those related to the Australian native title process); and laws that 'allocate' state-generated and state-defined rights to land and resources to indigenous groups. Ritter distinguishes statutory land rights from native title in Australia in terms of the source of the right arguing, '[s]tatutory land rights, though, do not represent "native title" in a technical sense: the former is created by parliament while the latter refers to an inherent common law right, the recognition of something already there, with origins not in the authority of the settler state but in pre-existing systems of law and custom'.[31] New Zealand Court of Appeal judge, Williams J, sees the difference as 'a choice between upholding existing rights, and making good on those that were unfairly taken away'.[32] I highlight that difference in the hope that, by shifting the focus away from recognition, alternative models will emerge.

There are already a few discrete examples where legislation has been used in Australia to allocate commercial water rights to indigenous groups, outside of the native title process. For example, Aboriginal 'access licences' may be allocated under the *Water Management Act 2000* (NSW), some of which are exercisable for commercial purposes, although they have enjoyed little take-up. One explanation for this is because the licences only authorise the use of small quantities of water, and are only available in well-watered coastal areas and not the over-allocated Murray Darling Basin. The New South Wales experience demonstrates the need for a statutory allocation mechanism to provide for the redistribution of water use rights where rights to use water resources are already fully allocated. In Queensland and the Northern Territory, there are examples of indigenous reserves being included in water plans to set aside a share of the consumptive pool of water to support 'indigenous economic aspirations'. These initiatives appear better able to anticipate the impact on indigenous water rights of other water use and development. However, as mentioned in Chapter 4, most Australian governments have distanced themselves from the indigenous water reserve policies.

The study of Chilean laws providing for indigenous water rights in Chapter 7 confirms the potential workability of an allocative model. Chile's Indigenous Land and Water Fund finances the allocation of water use rights to indigenous landholders who would not be able to make out *ancestral* water claims because they have been dispossessed of their historical rights. They are not required to prove continued

---

[31] David Laurence Ritter, *Contesting Native Title: From Controversy to Consensus in the Struggle over Indigenous Land Rights* (Allen & Unwin, 2009) 3.
[32] Williams, Chapter 2 above n. 44, 3.

ancestral water use since 'time immemorial' in order to acquire water use rights with the support of the Fund. The Fund also deals with the reality that other users now commonly hold water use rights in water resources claimed by indigenous groups, by providing for the redistribution of *derechos de aprovechamiento* on market terms.

Allocative models for indigenous land rights, like the *Aboriginal Land Rights (Northern Territory) Act 1976* (Cth), which provides for the grant of fee simple estates to Land Trusts 'for the benefit of Aboriginals entitled by Aboriginal tradition to the use or occupation of the land concerned', have a strong tradition in Australia. Ritter has warned that statutory land and resource rights, when compared to native title, amount to a 'favour' rather than a 'right'.[33] Statutory mechanisms for indigenous rights depend on the political will of governments and the political vulnerability of indigenous water rights policy is clear in recent acts of Australian governments. As mentioned, the Northern Territory Government has declared that it will not include strategic indigenous reserves in future water plans. There have been no developments on the Indigenous Economic Water Fund since the options paper released in 2012, and the First Peoples' Water Engagement Council is no longer active. However, at least in Victoria, there is renewed enthusiasm for supporting indigenous economic water demands via a potential allocative model.

An allocation model was adopted as part of a comprehensive response to indigenous commercial fishing claims in Aotearoa New Zealand. The 1992 'Sealords Deal' provided for the allocation of transferable quota rights to Māori iwi (tribes) in the context of a market-based allocation model for commercial fishing rights.[34] The Sealords deal was a pragmatic and comprehensive alternative to recognising indigenous commercial fishing claims on a claim-by-claim basis. As full and final settlement of Māori claims to commercial fishing rights, the deal gave Māori $150 million to be used for the development and involvement of Māori in the commercial fishing industry. This included participation in the acquisition of a joint share of Sealord Products Limited (a major fishing company) and 20 per cent allocation of all quota for species brought within the quota management system for commercial fisheries in New Zealand Aotearoa (including quota

---

[33] Ritter, above n. 31, 3.
[34] *Maori Fisheries Act 2004* (NZ), s. 3.

purchased by the government for allocation to Māori).[35] The Deed of Settlement for the Sealords Deal explains that it responds to:[36]

> ... uncertainty and dispute between the Crown and Maori as to the nature and extent of Maori fishing rights in the modern context and as to whether they derive from the Treaty or common law or both (such as by customary law or aboriginal title or otherwise).

While there has been criticism of the Sealords Deal,[37] it is generally accepted that the statutory allocation model has, at least to an extent, benefited Māori economic development and improved the certainty of commercial fishing rights.[38]

This book does not set out to detail the design or implementation of an allocative model for indigenous water rights, which is a matter for indigenous peoples to determine. The process of providing such rights should not be just the 'cut and dried incorporation of a discrete set of private rights', but involve a 'longer process of interaction, mutual adaptation and incitement to reflection and reform'.[39] In the following pages, I discuss some of the decisions that would need to be made by governments designing a distributive response.

(a)   An Indigenous Water Access Entitlement
One element of an allocative model might be an indigenous water access entitlement, to take and use water from the consumptive pool for any purpose, equivalent to the water rights of other users. Difficult decisions would need to be made about which groups would be entitled to access such an entitlement.[40] Vertongen acknowledges the difficulties of this in the New Zealand context, asking which level of *iwi* body would be entitled to a water right. Is it the local *marae* (community meeting place) or *hapū* who are actually in that area, in the same way as a farmer who is given an allocation, or is it a more global allocation to the *iwi* bodies in the catchment?[41] Such an entitlement could be made

---

[35] *Treaty of Waitangi (Fisheries Claims) Settlement Act 1992* (NZ) preamble *(1)*. There were other forms of compensation included in the Deal, set out in the preamble to the Act.
[36] Ibid. preamble (c).
[37] See, e.g., Steven Bourassa and Ann Louise Strong, 'Restitution of Fishing Rights to Maori: Representation, Social Justice and Community Development' (2000) 41(2) *Asia Pacific Viewpoint* 21; Toon Van Meijl, 'Changing Property Regimes in Maori Society: A Critical Assessment of the Settlement Process in New Zealand' (2012) 121(2) *The Journal of Polynesian Society* 181.
[38] See generally De Alessi, Chapter 2 above n. 57.
[39] Webber, Chapter 2 above n. 28, 60, 70.
[40] Interview with Riki Ellison (Wellington), above n. 69.
[41] Interview with Baden Vertongen (Wellington, 28 August 2018), Chapter 5 above n. 110.

available to existing indigenous landholders, with either legal or customary titles to use and possess their traditional lands. In the Australian context, for example, this might be those indigenous groups that hold title to land either under native title or land titles granted pursuant to land rights legislation like the *Native Title Act 1993* (Cth). In the New Zealand context, it might be the holders of Māori land titles under *Te Ture Whenua Māori Act 1993* (NZ) or the holders of Māori customary title. In Chile it may be the holders of land titles recognised or allocated under the *Indigenous Law 1993* (Chile), customary rights, or occupation rights. In Colombia it may be the occupants of indigenous *resguardos* or *consejos mayores* (reservations) or those recognised as having territorial rights under *ILO Convention 169*. A question remains about the rights of indigenous peoples no longer remaining on their traditional territories who, as a result of colonisation, have been pushed towards other regions and urban centres. Given the experience of historical injustice experienced by those groups, and their ongoing economic disadvantage, should not they too be able to access water rights for cultural and economic development purposes, especially if their experience of historical injustice means that they no longer hold land?[42]

Enshrining an indigenous water access entitlement in legislation would better protect the right from arbitrary change, noting that indigenous water rights policies have in the past been vulnerable to political uncertainty. The strategic indigenous reserves for economic development in the Northern Territory, for example, were promptly discontinued after a change of government. Such a legislative provision could be included in existing indigenous land rights legislation, like for example, the statutory native title rights to water under section 211 of the *Native Title Act*. Yet, such an approach may be problematic in the Australian context, because indigenous water rights could be read down by association with existing native title limitations, and a large indigenous estate remains under indigenous land rights legislation outside of native title. In Chile, indigenous water rights have been included within indigenous land rights legislation, in the *Indigenous Law*. An indigenous water access entitlement could alternatively be provided for in water legislation. As explained in Chapter 4, this has been done to varying extents in some state-based water legislation in Australia, although in the context of competing (often powerful) demand from other water users all have limited scope and content. A standalone

---

[42] Interview with Riki Ellison (Wellington), above n. 69.

indigenous water rights statute might reduce the possibility of indigenous water rights being read-down or prioritised after the water rights of other users.

What should be the nature and incidents of an indigenous water access entitlement prescribed by legislation? Native title rights to water in Australia, have been particularised in some depth, based on the courts' interpretation of traditional laws and customs, leaving little opportunity for them to evolve and develop for anything other than traditional, cultural purposes. In order to avoid translating and transforming indigenous interests to accommodate them within state law, and failing to account for the continually changing state of indigenous law, such an entitlement should be defined broadly avoiding overly prescribing the content of the right or its conditions, enabling indigenous peoples to administer the rights themselves within indigenous water jurisdictions. A positive aspect of Chilean law providing for indigenous water rights discussed in Chapter 5 is that indigenous landholders are simply allocated water use rights by the Indigenous Land and Water Fund, resisting the urge to further particularise the right in terms of the uses to which they may be put.

Because governments in all four countries studied in this book did not recognise indigenous land rights until the late twentieth century, the indigenous peoples were deprived of the opportunity to acquire water use rights as an incident of landholding. An indigenous water access entitlement should have the same broad characteristics as the water rights held by other users. In the Australian context, for instance, this would mean 'a perpetual or open-ended share of the consumptive pool of a specified water resource, as determined by the relevant water plan'.[43] The indigenous water access entitlement would take from the 'consumptive pool' and be exercisable for any use, including consumptive uses like irrigation, industry, urban and stock and domestic use.

The quantum of water allocated to an indigenous water access entitlement in any particular area is something that requires further consideration, together with indigenous peoples and water planners. In the context of New Zealand's Sealords Deal for commercial fisheries there is ongoing debate as to whether commercial fishing quota should be calculated based on tribal population or coastline landholding.[44]

---

[43] Commonwealth of Australia and the Governments of New South Wales, Victoria, Queensland, South Australia, the Australian Capital Territory and the Northern Territory, above n. 23, cll. 28–34.

[44] See generally De Alessi, Chapter 2 above n. 57, 403–4.

Michael O'Donnell has suggested in the Australian context that the quantum be determined with reference to a number of factors, including indigenous landholding and indigenous disadvantage.[45]

If an indigenous water access entitlement is to have the same characteristics as the water rights held by other users, would that mean that it might, depending on the regulatory frameworks of the country in question, be able to be traded as a commodity? In Chile, rights to use water have the status of *propiedad*; an interest very close to the western notion of 'ownership', including the exclusive right to, not only use, but freely alienate and encumber the interest. In Australia, water access entitlements must be: 'able to be traded, given, bequeathed or leased'; 'able to be subdivided or amalgamated'; 'be mortgageable' (and in this respect have similar status as freehold land when used as collateral for accessing finance); 'be enforceable and enforced'; and 'be recorded in publicly-accessible reliable water registers that foster public confidence and state unambiguously who owns the entitlement, and the nature of any encumbrances on it'.[46]

Whether indigenous rights to land and resources should be transferable is controversial in the literature on indigenous rights.[47] Legal mechanisms that recognise indigenous rights to land and resources often place limits on the alienability of such rights, typically justified on the basis that permanent alienation of property rights is not possible in traditional indigenous culture.[48] They are also considered necessary to maintain indigenous control of rights to land and resources for the economic and cultural benefit of future generations.

Indigenous land and resource rights in all four of the countries studied in this book are typically subject to restrictions on alienation, although the restrictions are not always absolute. Native title rights to water are inalienable, meaning that they cannot be transferred, leased or mortgaged. In Chile, *derechos de aprovechamiento* acquired with finance from the Indigenous Land and Water Fund are 'inalienable' for twenty-five years without administrative consent (including by transfer of ownership, embargo, tax, prescription or lease), although the restriction may be lifted if the finance is repaid, in order to

---

[45] See O'Donnell, 'Indigenous Rights in Water in Northern Australia', above n. 124.

[46] Commonwealth of Australia and the Governments of New South Wales, Victoria, Queensland, South Australia, the Australian Capital Territory and the Northern Territory, above n. 23, cl. 31.

[47] See, e.g., Tyron J Venn, 'Economic Implications of Inalienable and Communal Native Title: The Case of Wik Forestry in Australia' (2007) 64(1) *Ecological Economics* 131.

[48] See *Mabo* (1992) 175 CLR 1, 59 (Brennan J).

maintain resources in indigenous ownership for future economic development, while also supporting government return on investment. In Aotearoa New Zealand. Māori land under both customary law and *Te Ture Whenua Māori Act 1993* is inalienable, as 'a taonga tuku iho [treasure] of special significance to Maori people and, for that reason, to promote the retention of that land in the hands of its owners, their whanau [family], and their hapu [subtribe]'.[49] Indigenous territories in the Colombian *resguardos* can similarly not be alienated.

Nevertheless, restrictions on alienation are often criticised for their dampening effect on indigenous development, particularly by preventing the raising of finance.[50] The *White Paper on Developing Northern Australia*, for example, argues that 'Indigenous Australians should be able to use their exclusive native title to attract capital necessary for economic development'.[51] Similarly, the twenty-five-year restriction on alienation of indigenous water rights acquired with finance from Chile's Indigenous Land and Water Fund provoked prolonged debate in the Parliament, with some representatives arguing that such a 'market limitation' would only reduce the value of the resource.[52] As a compromise, the Chilean Parliament agreed to allow transfers of indigenous land and water rights 'within the same indigenous ethnicity'.[53] Senator Navarrete described the mechanism, as it applied to land, as creating an 'indigenous land market', whereby properties could be commercialised, alienated or transferred within indigenous communities.[54]

Whether the alienation of indigenous water rights should be restricted is equally controversial in Australia, although policy makers are experimenting with novel approaches and compromises. For example, restrictions on alienability have been relaxed to an extent under the *Water Management Act 2000* (NSW), with Aboriginal access

---

[49] See, e.g., *Te Ture Whenua Māori Act 1993* (NZ) preamble.

[50] See, e.g., Joshua Hitchcock, 'Financing Maori Land Development: The Difficulties Faced by Owners of Maori Land in Accessing Finance for Development and a Framework for the Solution' (2008) 14 *Te Mata Koi: Auckland University Law Review* 217.

[51] Australian Government, 'Our North, Our Future: White Paper on Developing Northern Australia', Chapter 4 above n. 232, 25.

[52] Chile, 'Discusion Sala' [Parliamentary Debates], Cámara de Diputados [House of Representatives], 21 January 1993 (Huepe) in Biblioteca del Congreso Nacional de Chile [Library of National Congress of Chile], Chapter 2 above n. 8, 143.

[53] *Indigenous Law 1993 253*, art. 13.

[54] Chile, 'Discusion Sala' [Parliamentary Debates], Senado [Senate], 30 June 1993 (Navarrete) in Biblioteca del Congreso Nacional de Chile [Library of National Congress of Chile], Chapter 2 above n. 18, 417.

licences being transferable among Aboriginal people.[55] The North Australian Indigenous Land and Sea Management Alliance has proposed that indigenous water rights be 'exclusive and able to be temporarily traded, subdivided or amalgamated, mortgageable, enforceable and registered'.[56] Some Māori groups considering the potential for commercial water rights in Aotearoa New Zealand require that indigenous water rights be inalienable.[57] However, they do contemplate that indigenous water rights can be leased, providing an important income stream for Māori through water rentals.

The extent to which indigenous rights to land and resources should be capable of alienation (if at all) is fraught. However, the potential for transfer of an indigenous water access entitlement within or between indigenous groups, or leasing of an indigenous water access entitlement to other water users, suggests future potential revenue streams for indigenous landholders, while protecting inter-generational equity. Ultimately, whether an indigenous water access entitlement should be subject to restrictions on alienation is a matter for further research and consultation. Indigenous people should lead these discussions, and by their own chosen representative structures control or influence the design of the entitlement more generally.

(b) An Indigenous Water Holder

An important lesson from the country studies in this book is that in the case of fully allocated water resources, indigenous water justice cannot be achieved without some form of redistribution. Any law providing indigenous peoples with a water access entitlement must confront the fact that in many situations, water resources are already fully allocated to other users. Thus, another element of an allocative model for indigenous water rights might be an 'indigenous water holder', to acquire, hold and distribute a portfolio of water use rights for the benefit of indigenous peoples.

Chile's Indigenous Land and Water Fund, discussed in Chapter 6, was intended to balance indigenous and non-indigenous historic rights in water as a 'fair solution'. The Fund was a 'creative mechanism' because it allowed the redistribution of water use rights from other

---

[55] See, e.g., *Water Sharing Plan for the Tweed River Area Unregulated and Alluvial Water Sources 2010* (NSW), cl. 38. Allocations under these licences are able to be traded to non-Aboriginal people; however, the licence itself can only be traded among Aboriginal people, and as such will remain in the Aboriginal community for the life of the licence.

[56] Northern Australian Land and Sea Management Alliance, Chapter 2 above n. 24, 11.

[57] Sin, Murray and Wyatt, Chapter 5 above n. 179.

users to indigenous communities in a way that did not impact adversely on other users, because water use rights would be voluntarily acquired in water markets. As discussed in Chapter 4, funding mechanisms have been used in Australia in the past to redistribute indigenous land rights, via the land market.

Aside from funding the redistribution of water rights, the study of Chile's Indigenous Land and Water Fund in Chapter 6 demonstrates the importance of setting aside a 'water reserve' in areas where water resources are not already fully allocated. The failure of the Chilean government to set aside a portion of water use rights for indigenous use in the future, prior to resources reaching full allocation, undermined the effectiveness of the Fund, because water rights acquisitions at market prices are costly and the Fund receives limited governmental support. Accordingly, as well as establishing a fund to finance the purchase of water use rights, a share of available rights to use specific water resources must be reserved for future granting to indigenous peoples. As noted in Chapter 4, some water reserves have already been set aside in Queensland and the Northern Territory, to help indigenous communities achieve economic and social aspirations.

An allocative model, comprising both a water fund and water reserve, has already been implemented in Australia with respect to environmental water rights.[58] As well as managing in stream environmental water interests, the Commonwealth Environmental Water Holder has the capacity both to hold water use rights and purchase (and sell) water use rights in the market, in the interests of the environment.[59] Its portfolio of water use rights was acquired through a combination of government purchases and savings of water via investment in water supply infrastructure that reduced water losses and incentivised reduced water use.[60]

As discussed in Chapter 5, the New Zealand government is considering how to go about allocating substantive water rights to Māori as water resources approach full allocation, and some have floated the idea of an allocative model within a market-based framework, perhaps like that adopted for commercial fisheries. The New Zealand Māori

---

[58] This was the Commonwealth Environmental Water Holder established by the *Water Act 2007* (Cth).
[59] Ibid. ss. 104–15.
[60] Australian Government, 'Water for the Future: Fact Sheet', Chapter 4 above n. 232. See generally O'Donnell and Macpherson, Elizabeth, Chapter 3 above n. 84, 32–3.

Council has suggested a need for some sort of water commission, which, among other functions, would oversee water allocation and management on behalf of Māori. However, tensions remain in Aotearoa New Zealand, and Australia, around who should control any sort of water holder, trust or commission, and whether local communities and tribal groups would best manage any such entity within their traditional areas.

The literature on indigenous water rights often portrays the unbundling of water use rights from landholding, and the presence of water markets, as a threat to continued water access by indigenous groups, enabling the accumulation of water use rights by other users. Boelens, Guevara-Gil and Panfichi argue instead for 'grassroots' approaches to water regulation which allow for the continuance of indigenous and peasant water use in customary or common property regimes.[61] Water markets encourage competition for water use rights, placing further strain on the availability of water use rights for attending to indigenous water demands in the future. Chapter 7 showed how unbundling intensified indigenous exclusion from water markets in Chile, and indigenous exclusion from water law frameworks also intensified as a result of unbundling in Australia. However, in Chile, other users began to accumulate water use rights long before the advent of water markets, using administrative processes that created property rights in water. Indigenous Australians have been excluded from laws providing for water use rights as an incident of landholding since the acquisition of sovereignty. Indigenous peoples in New Zealand and Colombia have also been excluded from legal and policy frameworks authorising the use and management of water, without the presence of water markets. In all cases, it is that historical exclusion, and not the advent of water markets, that allowed other users to accumulate water use rights.

Where governments, for better or worse, adopt the use of market mechanisms in water regulation there may be an opportunity for indigenous groups to gain access to legal rights to use water via the market. Where water use rights are available for purchase, separate from landholding in water markets, they may be redistributed to indigenous people, in a way that may not undermine the certainty of other water use rights. This has been done in Chile, via the Indigenous Land and

---

[61] See, e.g., Boelens, Armando Guevara-Gil and Aldo Panfichi, above n. 4.

Water Fund, which employs market mechanisms to enable the redistribution of *derechos de aprovechamiento*. And although there are many hard lessons to be learnt from the Chilean experience, ultimately, public investment in these statutory allocation models was warranted in the public interest.

CHAPTER NINE

CONCLUSION

Imagine a river catchment where indigenous peoples have their traditional territory. They have used and cared for the river for countless generations in accordance with their cultural water management and distribution practices. They will do so for future generations for many years to come. Often these normative frameworks are unwritten and uncodified, existing in common under customary law.

Imagine that the cultural, environmental and commercial interests those indigenous peoples have in the river are recognised and provided for in law. The obligation of the indigenous peoples to care for the river for future generations is reflected, perhaps, in a legal person model, where the river has its own rights of protection and survival separate from the indigenous peoples, but the peoples will defend the river's rights if necessary. The indigenous peoples as guardians have the jurisdiction to make decisions about the river's management according to their own laws and customs.

Imagine that there are other users of the catchment; agricultural, industry, recreational, domestic and urban, and environmental. These other people have legal rights to use the river, and their perspectives and needs are taken into account in managing the river. But both the river itself and the indigenous peoples also have legal rights to use the water in the river. The river holds or 'owns' a sufficient share of the water rights in it to maintain its own health and life under the guardianship of the indigenous peoples, and the indigenous peoples hold or own a fair share of the water rights in it, proportional to their territory or population, exercisable for any purpose from cultural to commercial use. If water rights are unfairly distributed, the government will secure a fair share for indigenous peoples. This is in the interests of all river users, as it clarifies rights and reduces disputes. The indigenous rights may be held by the peoples directly, or managed by some sort of indigenous water holder, trust or commission under their control, which has the power to distribute water rights in the interests of the peoples (and possibly even trade in those interests, should the peoples wish to).

Because the indigenous peoples are guardians of the river, they have the authority or jurisdiction to make or influence decisions about the river's governance and management. Because the indigenous peoples hold a fair share of the substantive rights in the river they have the autonomy to conduct practical and cultural livelihoods as they think fit in accordance with their laws and customs. Such a model does not yet exist. But we have seen glimpses of it in all of the four comparative country studies included in this book.

GLOSSARY

A MĀORI TERMS

Aotearoa	New Zealand
Hapū	Subtribe
Hui	Gathering/meeting
Iwi	Tribe
Kaitiaki	Guardian
Kaitiakitanga	Guardianship/stewardship
Kawanatanga	Government
Ki uta ki tai	From the mountains to sea
Mahinga kai	Food gathering area
Manaakitanga	Caring for others/hospitality
Mātauranga Māori	Māori knowledge
Mauri	The life-force which generates, regenerates and upholds creation
Pākehā	New Zealand European
Rūnanga	Tribal representative group
Te Awa Tupua	Whanganui River as recognised by the *Te Awa Tupua (Whanganui River Claims Settlement Act) 2017* (NZ)
Te Mana o Te Wai	The power of water, a policy approach for incorporating Māori water values in New Zealand water planning
Te Reo Māori	The Māori language

Te tino rangatiratanga	Highest chieftainship, authority
Te Tiriti o Waitangi	The Treaty of Waitangi 1840
Rohe	Area of influence/tribal area
Taonga	Treasures
Tikanga Māori	Māori laws and customs
Tupuna	Ancestor
Wāhi tapu	Sacred sites
Wairua	Spiritual existence
Whakapapa	Genealogies
Whakataukī	Proverb
Whanaungatanga	Kinship

B   COLOMBIAN SPANISH TERMS

Acción de tutela	Writ for protection of constitutional rights
Bien de uso publico	Public property
Buen vivir	Living well/the good life
Cabildos	Councils of *resguardos*
Comisión de Guardianes del Río Atrato	Commission of Guardians of the Atrato River
Consejos	Boards acting according with community uses and customs
Consejos Mayores	Reservations similar to *resguardos* Representative boards adopting the concept of councils or local authorities
Constitución Ecológica	Ecological Constitution
Despacho del Viceministro	Deputy Minister's Office
Dirección de Gestión Integral de Recurso Hídrico	Directorate for Integrated Water Management
Dominio	The right to use enjoy and dispose
Encomenderos	Colonial settlers
Encomienda	Colonial settlements
Entidad sujeto de derechos	Legal subject entity
Entradas	Incursions
Equipo asesor	Advisory group

Estado Social de Derecho	Social welfare state based on the rule law
Gerente del Río Atrato	Atrato River Manager
Gran Colombia	New Granada, Venezuela and Ecuador in the early 1800s
Mercedes	Colonial concession to use water upon the payment of a fee
Mercedes de tierra Corona	Land concessions granted by the Spanish Crown
Panel de expertos	Panel of experts
Personalidad jurídica	Legal personality
Resguardo	Reservation
Vida digna y bienestar general	Human dignity and common welfare

C CHILEAN SPANISH TERMS

Ancestral	Rights that originated prior to the acquisition of sovereignty
Bienes nacionales de uso público	National property for public use
Chilenización	Chilean government policy providing for assimilation of Chilean indigenous groups
Comisión Especial de Pueblos Indígenas	Special Commission for Indigenous Peoples
Corporación Nacional de Desarollo Indígena	National Indigenous Development Corporation
Derecho de aprovechamiento	Water use right
Derecho real	An enforceable right in relation to a thing
Propiedad/Dominio	Ownership
Fondo de Tierras y Aguas Indígenas	Indigenous Land and Water Fund
Normalización	Normalisation
Parlamento	treaty-like status between Mapuche and the Spanish Crown in the seventeenth century
Propiedad	Ownership

Recurso de protección	Action for protection of constitutional rights
Reducciones	Indigenous reservations
Regalias	Water concession granted by kings or lords in colonial Latin America
Tierra fiscal	State land
Títulos de merced	Possessory titles in the mid-nineteenth century

BIBLIOGRAPHY

A   ARTICLES/BOOKS/REPORTS

'Civil Law' [2013] *Columbia Electronic Encyclopedia, 6th Edition* 1

'HC Starts Delivering Verdict on Illegal Structures on Turag River's Bank' *The Daily Star*, 31 January 2019 www.thedailystar.net/city/illegal-structures-on-turag-rivers-bank-high-court-start-delivering-verdict-1695025

'Our Fresh Water 2017' (Ministry for the Environment & Statistics New Zealand, 2017)

'Universal Declaration of Rights of Mother Earth' (World People's Conference on Climate Change and the Rights of Mother Earth Cochabamba, Bolivia, 22 April 2010)

Aguiar Ribeiro do Nacimiento, Germana, 'El Derecho al Agua y su Protección en el Contexto de la Corte Interamericana de Derechos Humanos' [The Right to Water and Its Protection in the Context of the Inter-American Court of Human Rights] (2018) 16(1) *Estudios Constitucionales* 245

Aguilar Cavallo, Gonzalo, 'El Titulo Indígena y su Aplicabilidad en el Derecho Chileno' [Indigenous Title and Its Application in Chilean Law] (2005) 11 (1) *Revista Ius et Paxis* 269

Altman, JC and WS Arthur, 'Commercial Water and Indigenous Australians: A Scoping Study of Licence Allocations' (CAEPR Working Paper No. 57/2009, Centre for Aboriginal Economic Policy Research, September 2009)

Altman, Jon and Francis Markham, 'Burgeoning Indigenous Land Ownership: Diverse Values and Strategic Potentialities' in Sean Brennan et al. (eds.), *Native Title from Mabo to Akiba: A Vehicle for Change and Empowerment?* (Federation Press, 2015) 126

Altman, Jon and Francis Markham, *Value Mapping Indigenous Lands: An Exploration of Development Possibilities* (Centre for Aboriginal Economic Policy Research, The Australian National University, 2013)

Altman, Jon, 'Indigenous Interests and Water Property Rights' (2004) 23(3) *Dialogue* 29

Amparo Rodríguez, Gloria and Andrés Gómez Rey, 'La Participación como Mecanismo de Consenso para la Asignación de Nuevos Derechos' [Participation as a Consensus Mechanism for Assignment of New Rights] (2013) 37 *Pensamiento Jurídico* 71

Anker, Kirsten, 'Law in the Present Tense: Tradition and Cultural Continuity in Members of the Yorta Yorta Aboriginal Community v Victoria' (2004) 28-File Attachments (1) *Melbourne University Law Review* 1

Australian Government, 'Our North, Our Future: White Paper on Developing Northern Australia' (2015) https://northernaustralia.dpmc.gov.au

Australian Government, 'Position Statement: Indigenous Access to Water Resources' (National Water Commission, 2012) www.nwc.gov.au/__data/assets/pdf_file/0009/22869/indigenous-position-statement-june-2012.pdf

Australian Government, 'Water for the Future: Fact Sheet' (Department of Sustainability, Environment, Water, Population and Communities, 2010) www.environment.gov.au

Australian Law Reform Commission, '*Connection to Country: Review of the Native Title Act 1993 (Cth)*' (126, 2015)

Aylwin, Jose, *Pueblos Indígenas de Chile: Antecedentes Historicos y Situación Actual* [Indigenous Communities of Chile: History and Current Situation] (Instituto de Estudios Indígenas Universidad de la Frontera, 1994) Vol. 1

Ballara, Angela, *Iwi: The Dynamics of Māori Tribal Organisation from c. 1769 to c. 1945* (Victoria University Press, 1998)

Barcham, Manuhuia, 'The Limits of Recognition' in Benjamin Richard Smith and Frances Morphy (eds.), *The Social Effects of Native Title: Recognition, Translation, Coexistence* (ANU E Press, 2007)

Barlow, Maude, 'The World's Water: A Human Right or a Corporate Good? Whose Water Is It?' in B McDonald and D Jehl (eds.), *The Unquenchable Thirst of a Water-Hungry World* (National Geographic Society, 1st ed, 2003)

Barrera-Hernandez, Lila, 'Got Title Will Sell: Indigenous Rights to Land in Chile and Argentina' in Aileen McHarg et al. (eds.), *Property and the Law in Energy and Natural Resources* (Oxford University Press, 2010)

Barrera-Hernández, Lila, 'Indigenous Peoples, Human Rights and Natural Resource Development: Chile's Mapuche Peoples and the Right to Water' (2005) 11(1) *Annual Survey of International & Comparative Law*

Barry, Brian M, *Culture and Equality: An Egalitarian Critique of Multiculturalism* (Cambridge, Polity, 2001)

Bauer, Carl J, 'In the Image of the Market: The Chilean Model of Water Resource Management' (2005) 3(2) *International Journal of Water* 146

Bauer, Carl J, *Against the Current: Privatization, Water Markets, and the State in Chile* (Springer, 1998)

Bauer, Carl J, *Siren Song: Chilean Water Law as a Model for International Reform* (Resources for the Future, 2004)

Bavikatte, K and T Bennett, 'Community Stewardship: The Foundation of Biocultural Rights' (2015) 6(1)

Behrendt, Jason and Peter Thompson, 'The Recognition and Protection of Aboriginal Interests in NSW Rivers' (2004) 3 *Journal of Indigenous Policy* 37

Bello, Juan Carlos, *Atlas de la Biodiversidad de Colombia* (Instituto Alexander von Humboldt, 2000)

Bengoa, José, *Historia de un Conflicto: El Estado y los Mapuches en Siglo XX* [*History of a Conflict: The State and Mapuches in the 20th Century*] (Planeta, 1999)

Berry, Thomas, *The Great Work: Our Way into the Future* (Bell Tower, 1999)

Berryman, Phillip E, 'Report of the Chilean National Commission on Truth and Reconciliation: English Translation' (University of Notre Dame Centre for Civil and Human Rights, 1993)

Biblioteca del Congresso Nacional de Chile [Library of National Congress of Chile], 'Historia de la Ley No 19.253 Establece Normas Sobre Protección, Fomento y Desarrollo de Los Indígenas, y Crea la Corporación Nacional de Desarrollo Indígena' [History of Law No. 19.253 to Establish Norms for the Protection, Creation and Development of the Indigenous, and to Create the National Corporation of Indigenous Development] (5 October 1993)

Boast, Richard, *Buying the Land, Selling the Land: Governments and Maori Land in the North Island 1865–1921* (Victoria University Press, 2008)

Boast, Richard, *The Native Land Court 1862–1887: A Historical Study, Cases, and Commentary* (Thomson Reuters, 2013)

Boelens, Rutgerd et al., 'Contested Territories: Water Rights and the Struggles over Indigenous Livelihoods' (2012) 3(2) *International Indigenous Policy Journal* 1

Boelens, Rutgerd et al., 'Special Law' in Dik Roth, Rutgerd Boelens and Margreet Zwarteveen (eds.), *Liquid Relations: Contested Water Rights and Legal Complexity* (Rutgers University Press, 2005)

Boelens, Rutgerd, 'The Politics of Disciplining Water Rights' (2009) 40(2) *Development and Change* 307

Boelens, Rutgerd, Armando Guevara-Gil and Aldo Panfichi, 'Indigenous Water Rights in the Andes: Struggles Over Resources and Legitimacy' (2010) 20 *Water Law* 268

Bonet, Jaime, '¿Por Qué es Pobre El Chocó? Documentos de Trabajo Sobre Economía Regional' [Why Is Chocó Poor? Working Paper on Regional Economics] (90, Banco de la República, Bogotá, Centro de Estudios Económicos Regionales, 2007)

Bourassa, Steven and Ann Louise Strong, 'Restitution of Fishing Rights to Maori: Representation, Social Justice and Community Development' (2000) 41(2) *Asia Pacific Viewpoint* 21

Boyd, David R, 'The Constitutional Right to a Healthy Environment' (2012) 54(4) *Environment: Science and Policy for Sustainable Development* 3

Boyd, David R, *The Rights of Nature: A Legal Revolution That Could Save the World* (ECW Press, 2017)

Braithwaite, J, *Regulatory Capitalism: How It Works, Ideas for Making It Work Better* (Edward Elgar Publishing, 2008)

Brierley, Gary et al., 'A Geomorphic Perspective on the Rights of the River in Aotearoa New Zealand: Geomorphology and the Rights of the River' [2018] (Special Issue Paper) *River Research and Applications*

Budds, Jessica, 'Power, Nature and Neoliberalism: The Political Ecology of Water in Chile' (2004) 25(3) *Singapore Journal of Tropical Geography* 322

Budds, Jessica, 'The 1981 Water Code: The Impacts of Private Tradeable Water Rights on Peasant and Indigenous Communities in Northern Chile' in William L Alexander (ed.), *Lost in the Long Transition: Struggles for Social Justice in Neoliberal Chile* (Lexington Books, 2009)

Bulkan, Arif, 'Disentangling the Sources and Nature of Indigenous Rights: A Critical Examination of Common Law Jurisprudence' (2012) 61(4) *International & Comparative Law Quarterly* 823

Burns, Marcelle, 'Closing the Gap between Policy and "Law" – Indigenous Homelands and a Working Future' (2009) 27(2) *Law in Context* 114

Butterly, Lauren and Benjamin J Richardson, 'Indigenous Peoples and Saltwater/Freshwater Governance' (2016) 8(26) *Indigenous Law Bulletin* 3

Butterly, Lauren, 'Unfinished Business in the Straits: Akiba v Commonwealth of Australia [2013] HCA' (2013) 34(8) *Indigenous Law Bulletin* 3

Castro Lucic, Milka et al., *El Derecho Consuetudinario en la Gestión del Riego en Chiapa. Las Aguas del 'Tata Jachura'* [Customary Rights in Irrigation Management in Chiapa. The Waters of 'Tata Jachura'] (Konrad Adenauer Stiftung, 2017)

César Rodr; iguez Garavito, *Etnicidad.Gov – Los Recursos Naturales, Los Pueblos Indígenas y el Derecho a la Consulta Previa en los Campos Sociales Minados* [Natural Resources, Indigenous Peoples and the Right of Prior Consultation in Mining Territories] (Centro de Estudios de Derecho, Justicia y Sociedad, Dejusticia, 2012)

Clapcott, Joanne et al., 'Mātauranga Māori: Shaping Marine and Freshwater Futures' (2018) 52(4) *New Zealand Journal of Marine and Freshwater Research* 457

Comisión Especial de Pueblos Indígenas [Special Commission for Indigenous Peoples], 'Nueva Ley Indígena: Borrador de Discusión' ['New Indigenous Law: Discussion Paper'] (Discussion Paper, 1990)

Comisión Especial de Pueblos Indígenas, 'Memoria: Comisión Especial de Pueblos Indígenas' [Memoir: Special Commission for Indigenous Peoples] (1993)

Comisión Especial de Pueblos Indígenas, 'Nueva Ley Indígena: Borrador de Discusión' [New Indigenous Law: Discussion Paper] (La Comisión, 1990)

Comisionado Presidencial para Asuntos Indígenas [Presidential Commission for Indigenous Issues], 'Informe de la Comision Verdad Historica y Nuevo

Trato con Los Pueblos Indígenas' [Report of the Historical Trust Commission and New Agreement] (First Edition, October 2008)

Commonwealth of Australia, 'A Review of Indigenous Involvement in Water Planning' (National Water Commission, April 2013) http://webarchive .nla.gov.au/gov/20160615062953/www.nwc.gov.au/publications/topic/wat er-planning/indigenous-involvement-in-water-planning

Commonwealth of Australia, 'Closing the Gap Prime Minister's Report 192014' (2014) www.dpmc.gov.au/publications/docs/closing_the_ gap_2014.doc

Conference of the Parties, United Nations Framework Convention on Climate Change, *Report of the Conference of the Parties on Its Sixteenth Session, Held in Cancun from 29 November to 10 December 2010 – Addendum – Part Two: Action Taken by the Conference of the Parties at Its Sixteenth Session* (UN Doc FCCC/CP/2010/7/Add.1 (15 March 2011) Decision 1/CP.16 ('The Cancun Agreements: Outcome of the Work of the Ad Hoc Working Group on Long-Term Cooperative Action under the Convention')

Contesse, Jorge, 'The Rebel Democracy: A Look into the Relationship between the Mapuche People and the Chilean State' (2006) 26 *Chicano-Latino Law Review* 131

Corporación Nacional de Desarollo Indígena [National Indigenous Development Corporation] and Dirección General de Aguas [General Water Directorate], 'Convenio Dirección General de Aguas y Corporacion Nacional de Desarollo Indígena' [Convention between the National Indigenous Development Corporation and General Water Directorate] (Interdepartmental Convention, on file with the author 2000)

Corporación Nacional de Desarollo Indígena [National Indigenous Development Corporation] and Dirección General de Aguas [General Water Directorate], 'Convenio Marco Para La Protección, Constitución y Reestablecimiento de los Derechos de Agua de Propiedad Ancestral de las Comunidades Aymaras y Atacamenas' [Convention for the Protection, Constitution and Reestablishment of the Ancestral Water Property Rights of the Aymara and Atacamena Communities] (on file with the author) (Gobierno de Chile (Intergovernmental Agreement), 1997)

Cortez, Héctor González and Hans Gundermann Kröll, 'Acceso a la Propiedad de la Tierra, Comunidad e Identidades Colectivas Entre los Aymaras del Norte de Chile (1821–1930)' [Land Property Right Access, Community and Collective Identities among Aymara Communities in Northern Chile (1821–1930)] (2009) 41(1) *Chungara: Revista de Antropología Chilena* 51

Cotterrell, Roger, 'Subverting Orthodoxy, Making Law Central: A View of Sociolegal Studies' in Roger Cotterrell (ed.), *Living Law Studies in Legal and Social Theory* (Ashgate, 2008)

Coulthard, Glen Sean, *Red Skin, White Masks: Rejecting the Colonial Politics of Recognition* (University of Minnesota Press, 2014)

Coulthard, Glen, 'Subjects of Empire: Indigenous Peoples and the Politics of Recognition' in Canada' (2007) 6(4) *Contemporary Political Theory* 437

Council of Australian Governments, 'National Indigenous Reform Agreement (Closing the Gap)' (2012) www.federalfinancialrelations.gov.au/content/np a/health_indigenous/indigenous-reform/national-agreement_sept_12.pdf

Cranney, Kate and Poh-Ling Tan, 'Old Knowledge in Freshwater: Why Traditional Ecological Knowledge Is Essential for Determining Environmental Flows in Water Plans' (2011) 14(2) *The Australasian Journal of Natural Resources Law and Policy* 71

Crase, Lin, *Water Policy in Australia: The Impact of Change and Uncertainty, Issues in Water* (Resources for the Future, 2011)

Crown Law, 'Closing Submissions on Behalf of the Crown in the National Freshwater and Geothermal Resources Inquiry' (20 November 2018) https:// forms.justice.govt.nz

Cuadra, Manuel, 'Teoría Práctica de los Derechos Ancestrales de Agua de las Comunidades Atacameñas' [Practical Theory of the Ancestral Water Rights of the Atacamena Communties] [2000] (19) *Estudios Atacameños* 93

Davidson, BR, *Australia: Wet or Dry? The Physical and Economic Limits to the Expansion of Irrigation* (Melbourne University Press, 1969)

Davies, M and N Naffine, *Are Persons Property? Legal Debates about Property and Personality* (Ashgate Publishing, 2001)

De Alessi, Michael, 'The Political Economy of Fishing Rights and Claims: The Maori Experience in New Zealand' (2012) 12(2/3) *Journal of Agrarian Change* 390

De Lucia, Vito, 'Towards an Ecological Philosophy of Law: A Comparative Discussion' (2013) 4(2) *Journal of Human Rights & the Environment* 167

de Soto, Hernando, *The Mystery of Capital: Why Capitalism Triumphs in the West and Fails Everywhere Else* (London: Bantam, 2000)

De Stefano, Lucia, 'International Initiatives for Water Policy Assessment: A Review' (2010) 24(11) *Water Resources Management* 2449

Debelo, Asebe Regassa, 'Contrast in the Politics of Recognition and Indigenous People's Rights' 7(3) *AlterNative: An International Journal of Indigenous Peoples* [258]

Defensoría del Pueblo de Colombia, *Avance del Derecho Humano al Agua en la Constitución, La Jurisprudencia y los Instrumentos Internacionales 2005–2011* [The Advance of the Human Right to Water in the Constitution, Jurisprudence and International Instruments 2005–2011] (2012)

Defensoría del Pueblo, 'Crisis Humanitaria en el Chocó: Diagnóstico, Valoración y Acciones de la Defensoría del Pueblo' [The Humanitarian Crisis in Chocó: Diagnosis, Analysis and Action by the Public Defender's Office] (2014) www.defensoria.gov.co/public/pdf/crisishumanitariachoco .pdf

Departamento Nacional de Estadística, 'Censo General 2005: "Proyecciones Nacionales y Departamentales de Población 2005–2020"' (2010)

Diccionario de la Lengua Espanola [Spanish Language Dictionary] (Real Academia Espanola, 2011)

Dodds, Susan, 'Justice and Indigenous Land Rights' (1998) 41(2) *Inquiry. An Interdisciplinary Journal of Philosophy* 187

Dodson, Mick, 'Mabo Lecture: Asserting Our Sovereignty' in *Dialogue about Land Justice: Papers from the National Native Title Conferences* (Aboriginal Studies Press, 2010) 13 http://search.informit.com.au/documentsummary;d n=008566462202658;res=ielind

Donnelly, B and P Bishop, 'Natural Law and Ecocentrism' (2006) 19(1) *Journal of Environmental Law* 89

Dorsett, Shaunnagh and Shaun McVeigh, *Jurisdiction* (Routledge, 2012)

Durette, M et al., *Māori Perspectives on Water Allocation* (Wellington: Ministry for the Environment, 2009)

Durette, Melanie, 'A Comparative Approach to Indigenous Legal Rights to Freshwater: Key Lessons for Australia from the United States, Canada and New Zealand' (2010) 27(4) *Environmental and Planning Law Journal* 296

Durie, Taihakurei, 'Ngā Wai o Te Māori: Ngā Tikanga Me Ngā Ture Roia' [The Waters of the Māori: Māori Law and State Law] (Paper Prepared for the New Zealand Māori Council) (2017)

Dworkin, Ronald, *Law's Empire* (Cambridge: Belknap Press, 1986)

Easter, K William, Mark W Rosegrant and Ariel Dinar, 'Formal and Informal Markets for Water: Institutions, Performance, and Constraints' (1999) 14 (1) *The World Bank Research Observer* 99

Encuentro Nacional de Pueblos Indígenas [National Alliance of Indigenous Peoples] and Don Patricio Aylwin Azocar, 'Acta de Compromiso [Agreement], Nueva Imperial' (1 December 1989)

Engle, Karen, *The Elusive Promise of Indigenous Development: Rights, Culture, Strategy* (Duke University Press, 2010)

Environment Canterbury, 'Report of the Hearing Committee on an Application to Vary a National Water Conservation Order for Lake Ellesmere/Te Waihora in Canterbury' (2011)

Erueti, Andrew, 'Translating Maori Customary Title into a Common Law Title' [2003] *New Zealand Law Journal* 421(3)

Evans, B and P Howsam, 'A Critical Analysis of the Riparian Rights of Water Abstractors in England and Wales' (2005) 16(3) *Journal of Water Law* 90

Fisher, DE, 'Markets, Water Rights and Sustainable Development' (4) 23(2) *Environmental and Planning Law Journal* 100

Fisher, DE, *Water Law* (LBC Information Services, 2000)

Fitzpatrick, Daniel, '"Best Practice" Options for the Legal Recognition of Customary Tenure' (2005) 36(3) *Development and Change* 449–475

Fraser, Nancy, 'From Redistribution to Recognition? Dilemmas of Justice in a "Post-Socialist" Age' (1995) 21(2) *New Left Review* 68

García Márquez, Gabriel, *Love in the Time of Cholera* (Camberwell, Victoria Penguin, 2008)

García Pachón, María del Pilar, *Régimen Jurídico de los Vertimientos en Colombia: Análisis Desde el Derecho Ambiental y el Derecho de Aguas* [The Legal Regime for Wastewater in Colombia: An Environmental and Water Law Analysis] (Universidad Externado de Colombia, 2017)

Gardner, Alexander Walter et al., *Water Resources Law* (LexisNexis Butterworths, 2009)

Gluckman, Sir Peter, 'New Zealand's Fresh Waters: Values, State, Trends and Human Impacts' (Office of the Prime Minister's Chief Advisor, 12 April 2017)

Gobierno de Colombia, 'Una Nación Multicultural – Su Diversidad Étnica' (DANE – Departamento Administrativo Nacional de Estadísticas, May 2007) www.dane.gov.co/files/censo2005/etnia/sys/colombia_nacion.pdf

Godden, Lee et al., 'Accommodating Interests in Resource Extraction: Indigenous Peoples, Local Communities and the Role of Law in Economic and Social Sustainability' (2008) 26(1) *Journal of Energy & Natural Resources Law* 1

Godden, Lee, 'Governing Common Resources: Environmental Markets and Property in Water' in *Property and the Law in Energy and Natural Resources* (Oxford University Press, 2010) 413

Godden, Lee, 'Water Law Reform in Australia and South Africa: Sustainability, Efficiency and Social Justice' (2005) 17(2) *Journal of Environmental Law* 181

Godden, Lee, Raymond L Ison and Philip J Wallis, 'Water Governance in a Climate Change World: Appraising Systemic and Adaptive Effectiveness' (2011) 25(15) *Water Resources Management* 3971

Good, Meg, 'The River as a Legal Person: Evaluating Nature Rights-Based Approaches to Environmental Protection in Australia' [2013] (1) *National Environmental Law Review* 34

Gover, Kirsty, 'Legal Pluralism and State-Indigenous Relations in Western Settler Societies' (International Council on Human Rights Policy, 2009)

Gover, Kirsty, *Indigenous Rights and Governance in Canada, Australia, and New Zealand* (Electronic Resource, Oxford University Press, 2012)

Gover, Kirsty, *Tribal Constitutionalism: States, Tribes, and the Governance of Membership* (Oxford University Press, 2010)

Government, Northern Territory, 'Draft Water Allocation Plan: Oolloo Aquifer' (2012)

Grafton, R Quentin et al., 'An Integrated Assessment of Water Markets: A Cross-Country Comparison' (2011) 5(2) *Review of Environmental Economics and Policy* 219

Grafton, R Quentin and W Nikolakis, 'Assessment of the Potential Costs and Benefits of Water Trading Across Northern Australia' (Report, Tropical Rivers and Coastal Knowledge, March 2011) 217

Grainger, Corbett A and Christopher Costello, 'The Value of Secure Property Rights: Evidence from Global Fisheries' [2011] (No. w17019) *NBER Working Paper*

Grammond, Sebastien, 'The Reception of Indigenous Legal Systems in Canada' in Albert Breton et al. (eds.), *Multijuralism: Manifestations, Causes, and Consequences* (Ashgate Publishing Limited, 2009)

Grandón, Javier Barrientos, 'Juan Sala Bañuls (1731–1806) y el "Código Civil" de Chile (1855)' [Juan Sala Bañuls (1731–1806) and the Chilean 'Civil Code' (1855)] (2009) 31 *Revista de Estudios Historico-Juridicos* 351

Harvey, David, *The New Imperialism* (Oxford University Press, 2003)

Havemann, Paul (ed.), *Indigenous Peoples' Rights: In Australia, Canada and New Zealand* (Oxford University Press, 1999)

Heise, Wolfram, 'Indigenous Rights in Chile: Elaboration and Application of the New Indigenous Law (Ley No. 19.253) of 1993' in Rene Kuppe and Richard Potz (eds.), *Law and Anthropology: International Yearbook for Legal Anthropology* (Kluwer Law International, 2001)

Hendrix, Burke A, 'Context, Equality, and Aboriginal Compensation Claims' (2011) 50 *Dialogue: Canadian Philosophical Review* 669

Hepburn, Samantha, 'Native Title Rights in the Territorial Sea and Beyond: Exclusivity and Commerce in the Akiba Decision' (2011) 34(1) *University of New South Wales Law Journal* 159

Hepburn, Samantha, *Principles of Property Law* (Cavendish Publishing, 1998)

Hepi, Maria et al., 'Enabling Mātauranga-Informed Management of the Kaipara Harbour, Aotearoa New Zealand' (2018) 52(4) *New Zealand Journal of Marine and Freshwater Research* 497

Hernandez, Camilo Antonio, *Ideas y Practicas Ambientales del Pueblo Embera del Choco* (Cerec, 1995)

Hewitt, Anne, 'Commercial Exploitation of Native Title Rights – a Possible Tool in the Quest for Substantive Equality for Indigenous Australians?' (February 12) 32(2) *Adelaide Law Review* 227

Hitchcock, Joshua, 'Financing Maori Land Development: The Difficulties Faced by Owners of Maori Land in Accessing Finance for Development and a Framework for the Solution' (2008) 14 *Te Mata Koi: Auckland University Law Review* 217

Hoeke, Mark Van, 'Deep Level Comparative Law' in Mark Van Hoeke (ed.), *Epistemology and Methodology of Comparative Law* (Hart Publishing, 2004) 165

Hogue, Emily J and Pilar Rau, 'Troubled Water: Ethnodevelopment, Natural Resource Commodification, and Neoliberalism in Andean Peru' [2008] (3/4) *Urban Anthropology & Studies of Cultural Systems & World Economic Development* 283

Holley, Cameron and Darren Sinclair (eds.), *Reforming Water Law and Governance: From Stagnation to Innovation in Australia* (Springer, 2018)

Hook, Gary Raumati and Lynne Parehaereone Raumati, 'Cultural Perspectives of Fresh Water' (2011) 2 *MAI Review* 1

Hutchison, Abigail, 'The Whanganui River as a Legal Person' (2014) 39(3) *Alternative Law Journal (Gaunt)* 179

Imran, Sophia, Khorshed Alam and Narelle Beaumont, 'Reinterpreting the Definition of Sustainable Development for a More Ecocentric Reorientation: Reinterpreting the Definition of Sustainable Development' (2014) 22(2) *Sustainable Development* 134

Instituto de Estudios Indígenas, 'Los Derechos de los Pueblos Indígenas en Chile' [Indigenous Peoples' Rights in Chile] (Programa de Derechos Indígenas, Universidad de la Frontera, 2003)

Instituto Nacional de Estadísticas Chile, 'Síntesis de Resultados Censo 2017' [Synthesis of Census Results 2017] www.censo2017.cl/descargas/home/sintesis-de-resultados-censo2017.pdf

International Labour Organization, *Convention Concerning Indigenous and Tribal Peoples in Independent Countries* (opened for signature 27 June 1989, 1650 UNTS 383) (entered into force 5 September 1991) ('ILO Convention 169')

Iorns Magallanes, Catherine J, 'Maori Cultural Rights in Aotearoa New Zealand: Protecting the Cosmology That Protects the Environment' (2015) 21(2) *Widener Law Review* 273

Ivison, Duncan, 'The Logic of Aboriginal Rights' (2003) 3(3) *Ethnicities* 321

Ivison, Duncan, Paul Patton and Will Sanders, 'Introduction' in Duncan Ivison, Paul Patton and Will Sanders (eds.), *Political Theory and the Rights of Indigenous Peoples* (Cambridge University Press, 2000)

Jackson, Sue and Jon Altman, 'Indigenous Rights and Water Policy: Perspectives from Tropical Northern Australia' (2009) 13(1) *Australian Indigenous Law Review* 27

Jackson, Sue and Marcia Langton, 'Trends in the Recognition of Indigenous Water Needs in Australian Water Reform: The Limitations of "Cultural" Entitlements in Achieving Water Equity' (2012) 22(2/3) *Journal of Water Law* 110

Jackson, Sue and Marcus Barber, 'Recognition of Indigenous Water Values in Australia's Northern Territory: Current Progress and Ongoing Challenges for Social Justice in Water Planning' (2013) 14(4) *Planning Theory & Practice* 435

Jackson, Sue, 'Aboriginal Access to Water in Australia: Opportunities and Constraints' in R Quentin Grafton and Karen Hussey (eds.), *Water Resources Planning and Management* (Cambridge University Press, 2011)

Jackson, Sue, 'Background Paper on Indigenous Participation in Water Planning and Access to Water – Report Prepared for the National Water Commission' in CSIRO (ed.) (2009)

Jackson, Sue, 'Enduring and Persistent Injustices in Water Access in Australia' in *Natural Resources and Environmental Justice: Australian Perspectives* (CSIRO Publishing, 2017)

Jackson, Sue, 'National Indigenous Water Planning Forum – Background Paper on Indigenous Participation in Water Planning and Access to Water' (National Water Commission, CSIRO, 2009)

Jackson, Sue, 'Recognition of Indigenous Interests in Australian Water Resource Management, with Particular Reference to Environmental Flow Assessment' (2008) 2(3) *Geography Compass* 874

Johnson, Miranda, 'Reconciliation, Indigeneity, and Postcolonial Nationhood in Settler States' (6)14(2) *Postcolonial Studies* 187

Jones, Carwyn, *New Treaty, New Tradition: Reconciling New Zealand and Māori Law* (New Zealand Victoria University Press, 2016)

Kahui Legal, 'Closing Submissions on Behalf of the Freshwater [Te Pou Taiao] Iwi Leaders Group in the National Freshwater and Geothermal Resources Inquiry' (14 November 2018) https://forms.justice.govt.nz

Kahui, Viktoria and Amanda Richards, 'Lessons from Resource Management by Indigenous Māori in New Zealand: Governing the Ecosystems as a Commons' (2014) 102 *Ecological Economics* 1

Kauffman, Craig M and Pamela L Martin, 'Can Rights of Nature Make Development More Sustainable? Why Some Ecuadorian Lawsuits Succeed and Others Fail' (2017) 92 *World Development* 130

Keal, Paul, 'Indigenous Self-Determination and the Legitimacy of Sovereign States' (2007) 44(2/3) *International Politics* 287

Kerins, Seán, *Social Enterprise as a Model for Developing Aboriginal Lands* (Australian National University, 2013)

Kukathas, Chandran, 'Are There Any Cultural Rights?' [1992] (1) *Political Theory* 105

Kuppe, René, 'The Three Dimensions of the Rights of Indigenous Peoples' (2009) 11(1) *International Community Law Review* 103

Kymlicka, Will, *Multicultural Citizenship: A Liberal Theory of Minority Rights* (Oxford University Press, 1996)

La Follette, Cameron and Chris Maser, *Sustainability and the Rights of Nature: An Introduction* (CRC Press, 2017)

Lane, Patricia, 'Native Title and Inland Waters' (2000) 4(29) *Indigenous Law Bulletin*

Langford, Malcolm and Anna FS Russell, *The Human Right to Water: Theory, Practice and Prospects* (Cambridge University Press, 2017)

Langford-Smith, T and J Rutherford, *Water and Land: Two Case Studies* (Australia National University Press, 1966)

Larrain, Sara and Pamela Poo, 'Conflictos por el agua en Chile: entre los derechos humanos y las reglas del mercado' [Water Conflicts in Chile: Between Human Rights and Market Rules] (Programa Chile Sustentable, 2010)

Lasser, Mitchel De S-O-L'E, 'The Question of Understanding' in Pierre Legrand and Roderick Munday (eds.), *Comparative Legal Studies: Traditions and Transitions* (Cambridge University Press, 2003)

Lawlab, 'Options Paper for the First Peoples' Water Engagement Council (FPWEC): Options for an Indigenous Economic Water Fund (IEWF)' www.nwc.gov.au

Lennox, James, Wendy Proctor and Shona Russell, 'Structuring Stakeholder Participation in New Zealand's Water Resource Governance' (2011) 70(7) *Ecological Economics* 1381

Luis Felipe Guzmán Jiménez, *Las Aguas Residuales en la Jurisprudencia del Consejo de Estado: Periodo 2003–2014* [Wastewater in the Jurisprudence of the Administrative Court of Colombia: 2003–2014] (Universidad Externado de Colombia, 2015)

MacDonnell, Lawrence J and Neil S Grigg, 'Establishing a Water Law Framework: The Colombia Example' (2007) 32(4) *Water International* 662

Macpherson, Elizabeth and Clavijo Ospina, Felipe, 'The Pluralism of River Rights in Aotearoa New Zealand and Colombia' (2018) 25 *Journal of Water Law* 283

Macpherson, Elizabeth and Erin O'Donnell, '¿Necesitan Derechos los Ríos? Comparando Estructuras Legales Para Regulación de Ríos en Nueva Zelanda, Australia y Chile' [Do Rivers Need Rights? Comparing Legal Structures for River Regulation in New Zealand, Australia and Chile] (Pontificia Universidad Catolica de Chile, 2017)

Macpherson, Elizabeth and Erin O'Donnell, '¿Necesitan Derechos Los Ríos? Comparando Estructuras Legales para la Regulación de los Ríos en Nueva Zelanda, Australia y Chile' [Do Rivers Need Rights? Comparing Legal Structures for River Regulation in New Zealand, Australia and Chile] (2017) 25 *Revista de Derecho Administrativo Económico*

Macpherson, Elizabeth et al., 'Lessons from Australian Water Reforms: Indigenous and Environmental Values in Market-Based Water Regulation' in *Reforming Water Law and Governance: From Stagnation to Innovation in Australia* (Springer, 2018)

Macpherson, Elizabeth, 'Beyond Recognition: Lessons from Chile for Allocating Indigenous Water Rights in Australia' (2017) 40(3) *University of New South Wales Law Journal* 1130

Macquarie Dictionary [Electronic Resource]: Australia's National Dictionary Online (Macquarie Library, 2003)

Maloney, Michelle and Peter Burdon, *Wild Law – In Practice* (Taylor and Francis, 2014)

Maloney, Michelle, 'Building an Alternative Jurisprudence for the Earth: The International Rights of Nature Tribunal' [2016] (1) *Vermont Law Review* 129

Margil, Mari, 'The Standing of Trees: Why Nature Needs Legal Rights' (2017) 34(2) *World Policy Journal* 8

Marsden, Māori, *The Woven Universe: Selected Writings of Rev. Māori Marsden* (Estate of Rev. Māori Marsden, 2003)

Marshall, Virginia, *Overturning Aqua Nullius: Securing Aboriginal Water Rights* (Aboriginal Studies Press, 2017)

McAvoy, Tony, 'The Human Right to Water and Aboriginal Water Rights in New South Wales' (2008) 17(1) *Human Rights Defender* 6

McFarlane, Bardy, 'The National Water Initiative and Acknowledging Indigenous Interests in Planning' (2004)

McHugh, Paul G, *Aboriginal Title [Electronic Resource]: The Modern Jurisprudence of Tribal Land Rights* (Oxford University Press, 2011)

McKay, Jennifer, 'The Legal Frameworks of Australian Water: Progression from Common Law Rights to Sustainable Shares' in L Crase (ed.), *Water Policy in Australia: The Impact of Change and Uncertainty* (Resources for the Future, 2008) 46

McNeil, Kent, *Common Law Aboriginal Title* (Clarendon Press, 1989)

McRae, Heather, *Indigenous Legal Issues: Commentary and Materials* (Thomson Reuters (Professional) Australia, 4th ed, 2009)

Memon, P and N Kirk, 'Role of Indigenous Māori People in Collaborative Water Governance in Aotearoa/New Zealand' (2012) 55(7) *Journal of Environmental Planning and Management* 941

Miller, Robert J, 'The International Law of Colonialism: A Comparative Analysis' (2011) 15(4) *Lewis & Clark Law Review* 847

Miller, Robert J, 'The International Law of Discovery, Indigenous Peoples, and Chile' (2010) 89 *Nebraska Law Review* 819

Ministry for the Environment and Māori Crown Relations Unit, 'Shared Interests in Freshwater: A New Approach to the Crown/Māori Relationship for Freshwater' (2018)

Ministry for the Environment, 'Briefing to the Incoming Minister for the Environment' (December 2017) www.beehive.govt.nz/sites/default/files/20 17–12/water.pdf

Molano Bravo, Alfredo, *De Río en Río: Vistazo a los Territorios Negros* (Editorial Aguilar, 2017)

Morgan, B, *The Intersection of Rights and Regulation: New Directions in Sociolegal Scholarship* (Aldershot: Ashgate, 2007)

Morgan, Bronwen, *Water on Tap: Rights and Regulation in the Transnational Governance of Urban Water Services* (Cambridge University Press, 2011)

Morgan, Monica, Lisa Strelein and Jessica Weir, *Indigenous Rights to Water in the Murray Darling Basin: In Support of the Indigenous Final Report to the Living Murray Initiative* (Native Title Research Unit, Australian Institute of Aboriginal and Torres Strait Islander Studies, 2004)

Morgera, Elisa and Kati Kulovesi, *Research Handbook on International Law and Natural Resources* (Edward Elgar Publishing, 2016)

Morris, James and Jacinta Ruru, 'Giving Voice to Rivers: Legal Personality as a Vehicle for Recognising Indigenous Peoples' Relationships to Water' (2010) 14(2) *Australian Indigenous Law Review* 49

Muru-Lanning, Marama, *Tupuna Awa: People and Politics of the Waikato River* (Auckland University Press, 2016)

Mutu, Margaret, 'Māori Issues' (2017) 29(1) *The Contemporary Pacific* 144

Naffine, Ngaire, *Law's Meaning of Life: Philosophy, Religion, Darwin and the Legal Person* (Hart Publishing, 2009)

National Water Commission, 'National Water Initiative: Securing Australia's Water Future: 2011 Assessment' (Commonwealth of Australia, 2011)

New South Wales Office of Water, 'Our Water Our Country: An Information Manual for Aboriginal People and Communities about the Water Reform Process' (2012) www.water.nsw.gov.au

New Zealand and Ministry for the Environment, *Next Steps for Fresh Water: Consultation Document* (2016)

New Zealand and Office of Treaty Settlements, *Ka Tika ā Muri, Ka Tika ā Mua: He Tohutohu Whakamārama i Ngā Whakataunga Kerēme e Pā Ana Ki Te Tiriti o Waitangi Me Ngā Whakaritenga Ki Te Karauna = Healing the Past, Building a Future: A Guide to Treaty of Waitangi Claims and Negotiations with the Crown* (2015)

Nicholson, Penelope (Pip) and Sarah Biddulph (eds.), *Examining Practice, Interrogating Theory: Comparative Legal Studies in Asia* (Nijhoff, 2008)

Niezen, Ronald, *The Origins of Indigenism: Human Rights and the Politics of Identity* (University of California Press, 2002)

Nikolakis, William and R Quentin Grafton, 'Fairness and Justice in Indigenous Water Allocations: Insights from Northern Australia' (2014) 16 *Water Policy* 19

Nikolakis, William D, R Quentin Grafton and Hang to, 'Indigenous Values and Water Markets: Survey Insights from Northern Australia' (2013) 500 (12) *Journal of Hydrology* 12

Nikolakis, William, 'Providing for Social Equity in Water Markets: The Case for an Indigenous Reserve in Northern Australia' in R Quentin Grafton and

Karen Hussey (eds.), *Water Resources Planning and Management* (Cambridge University Press, 2011)

NIWA, '2016 Pilot Waikato River Report Card: Methods and Technical Summary – Prepared for Waikato River Authority' (NIWA, March 2016)

Northern Australian Land and Sea Management Alliance, 'Knowledge Series: Indigenous People's Right to the Commercial Use and Management of Water on Their Traditional Territories' (2013) www.nailsma.org.au

Northern Australian Land and Sea Management Alliance, 'Knowledge Series: Indigenous People's Right to the Commercial Use and Management of Water on Their Traditional Territories' (2013) www.nailsma.org.au/hub/resources/publication/indigenous-peoples-right-commercial-use-and-management-water-policy

Nozick, Robert, *Anarchy, State, and Utopia* (New York: Basic Books, 1974)

O'Bryan, Katie, 'New Law Finally Gives Voice to the Yarra River's Traditional Owners' [2017] *The Conversation* http://theconversation.com/new-law-finally-gives-voice-to-the-yarra-rivers-traditional-owners-83307

O'Bryan, Katie, 'The National Water Initiative and Victoria's Legislative Implementation of Indigenous Water Rights' (2012) 7(29) *Indigenous Law Bulletin*

O'Bryan, Katie, *Indigenous Rights and Water Resource Management: Not Just Another Stakeholder* (Routledge, 2018)

O'Connor, Pamela, 'Aboriginal Land Rights at Common Law: Mabo v Queensland' (1992) 18(2) *Monash University Law Review* 251

O'Donnell, Erin, *Constructing the Aquatic Environment as a Legal Subject: Legal Rights, Market Participation, and the Power of Narrative* (PhD Thesis, University of Melbourne, 2017)

O'Donnell, Erin and Elizabeth Macpherson, 'Voice, Power and Legitimacy: The Role of the Legal Person in River Management in New Zealand, Chile and Australia' [2018] *Australasian Journal of Water Resources* 1

O'Donnell, Erin and J Talbot-Jones, 'Creating Legal Rights for Rivers: Lessons from Australia, New Zealand, and India' (2018) 23(1) *Ecology and Society*

O'Donnell, Erin and Macpherson, Elizabeth, 'Challenges and Opportunities for Environmental Water Management in Chile: An Australian Perspective' (2012) 23(1) *Journal of Water Law* 24

O'Donnell, Erin, 'At the Intersection of the Sacred and the Legal: Rights for Nature in Uttarakhand, India' [2017] *Journal of Environmental Law* 1

O'Donnell, Erin, 'Competition or Collaboration? Using Legal Persons to Manage Water for the Environment in Australia and the United States' (2017) 34 *Environmental and Planning Law Journal* 503

O'Donnell, Erin, 'Institutional Reform and the Victorian Environmental Water Holder' (2011) 22(2/3) *Journal of Water Law* 78

O'Donnell, Erin, *Legal Rights for Rivers: Competition, Collaboration and Water Governance* (Routledge, 2018)

O'Donnell, Michael, 'Briefing Paper for the Water Rights Project by the Lingiari Foundation and ATSIC' in Lingiari Foundation (ed.), *Background Briefing Papers* (2002)

O'Donnell, Michael, '*Indigenous Rights in Water in Northern Australia*' (NAILSMA – TRaCK, 2011)

O'Neill, Lily Maire, *A Tale of Two Agreements: Negotiating Aboriginal Land Access Agreements in Australia's Natural Gas Industry* (PhD Thesis, University of Melbourne, 2016)

O'Neill, Lily, 'The Role of State Governments in Native Title Negotiations: A Tale of Two Agreements' (2014) 18(2) *Australian Indigenous Law Review* 29

OECD, *OECD Environmental Performance Reviews: New Zealand 2017* (OECD Publishing, 2017)

Orucu, A Esin, 'Methodology of Comparative Law' in JM Smits Elgar (ed.), *Encyclopedia of Comparative Law* (Edward Elgar Publishing, 2006) 442

Oscar Darío Amaya Navas, *La Constitución Ecológica de Colombia: Análisis Comparativo Con el Sistema Constitucional Latinoamericano* (Universidad Externado de Colombia, 2002)

Pearson, Noel, 'Land Is Susceptible of Ownership' in Lisa Palmer, Maureen Tehan and Kathryn Shain (eds.), *Honour among Nations?: Treaties and Agreements with Indigenous People* (Melbourne University Press, 2004)

Pearson, Noel, 'Properties of Integration' *The Australian* (Australia), 14 October 2006

Pearson, Noel, 'The Concept of Native Title at Common Law' [1997] (5) *Australian Humanities Review*

Perreault, Tom, 'Dispossession by Accumulation? Mining, Water and the Nature of Enclosure on the Bolivian Altiplano' (2013) 45(5) *Antipode* 1050

Perreault, Tom, 'Tendencies in Tension: Resource Governance and Social Contradictions in Contemporary Bolivia' in *Governing Resource Extraction* (2017)

Pohle, Brady, 'Possessory Title in the Context of Aboriginal Claimants' [1995] *Queensland University of Technology Law Journal*

Popic, Linda, 'Sovereignty in Law: The Justiciability of Indigenous Sovereignty in Australia, the United States and Canada' (2005) 4 *Indigenous Law Journal* 117

Pratt, Angela, 'Treaties vs. Terra Nullius: Reconciliation, Treaty-Making and Indigenous Sovereignty in Australia and Canada' (2004) 3 *Indigenous Law Journal* 43

Prieto, Manuel, *Privatizing Water and Articulating Indigeneity: The Chilean Water Reforms and the Atacameño People (Likan Antai)* (PhD Thesis, The University of Arizona, 2014)

Radonic, Lucero, 'Through the Aqueduct and the Courts: An Analysis of the Human Right to Water and Indigenous Water Rights in Northwestern Mexico' (2017) 84 *Geoforum* 151

Rawls, John, *A Theory of Justice* (Revised Edition) (Oxford University Press, 1999)

Rendic Veliz, Dinko Tomislav, *Derechos de Agua y Pueblos Indígenas* [Water Rights and Indigenous Peoples] (Librotecnia, 2009)

Report of the United Nations Conference on Environment and Development (UN Doc A/CONF.151/26 (Vol. I) (12 August 1992) annex I ('Rio Declaration on Environment and Development')

Reynolds, Henry, 'New Frontiers' in Paul Havemann (ed.), *Indigenous Peoples' Rights: In Australia, Canada & New Zealand* (Oxford University Press, 1999)

Richards, Patricia, 'Of Indians and Terrorists: How the State and Local Elites Construct the Mapuche in Neoliberal and Multicultural Chile' (2010) 42 *Journal of Latin American Studies* 59

Ritter, David Laurence, *Contesting Native Title: From Controversy to Consensus in the Struggle over Indigenous Land Rights* (Allen & Unwin, 2009)

Riquelme Salazar, Carolina de Lourdes, *El Derecho al Uso Privativo de las Aguas en España y Chile: Un Estudio de Derecho Comparado* [Exclusive Water Rights in Spain and Chile: A Comparative Law Study] (PhD Thesis, Universitat Rovira i Virgili, 2013)

Roa-García, María Cecilia, Patricia Urteaga-Crovetto and Rocío Bustamante-Zenteno, 'Water Laws in the Andes: A Promising Precedent for Challenging Neoliberalism' (2015) 64 *Geoforum* 270

Rodríguez, Gloria Amparo, Carlos Lozano Acosta and Andrés Gómez Rey, *Protección Jurídica del Agua En Colombia* (Grupo Editorial Ibáñez, 2011)

Rogers, Nicole and Michelle M Maloney, *Law as If Earth Really Mattered: The Wild Law Judgement Project* (Routledge, 2017)

Rose, Carol M, 'Expanding the Choices for the Global Commons: Comparing Newfangled Tradeable Allowance Schemes to Old Fashioned Common Property Regimes' (1999) 10 *Duke Environmental Law & Policy Forum*

Ruckert, Arne, 'Towards an Inclusive-Neoliberal Regime of Development: From the Washington to the Post-Washington Consensus' (2006) 39(1) *International Development Studies* 34

Ruru, Jacinta, 'Indigenous Restitution in Settling Water Claims: The Developing Cultural and Commercial Redress Opportunities in Aotearoa, New Zealand' (2013) 22(2) *Pacific Rim Law & Policy Journal* 311

Ruru, Jacinta, 'Māori Legal Rights to Water: Ownership, Management, or Just Consultation?' [2011] *Resource Management Theory & Practice*

Ruru, Jacinta, 'Māori Rights in Water – the Waitangi Tribunal's Interim Report' [2012] *Māori Law Review* http://maorilawreview.co.nz/2012/09/maori-rights-in-water-the-waitangi-tribunals-interim-report/

Ruru, Jacinta, 'Undefined and Unresolved: Exploring Indigenous Rights in Aotearoa New Zealand's Freshwater Legal Regime' (2009) 20(5–6) *Journal of Water Law* 236

Salmond, Anne, 'Tears of Rangi: Water, Power, and People in New Zealand' (2014) 4(3) HAU: *Journal of Ethnographic Theory* 285

Sanders, Katherine, '"Beyond Human Ownership?" Property, Power and Legal Personality for Nature in Aotearoa New Zealand' (2018) 30(2) *Journal of Environmental Law* 1 207–234

Schlager, Edella and Elinor Ostrom, 'Property-Rights Regimes and Natural Resources: A Conceptual Analysis' (1992) 68(3) *Land Economics* 249

Schönsteiner, Judith and Javier Couso, 'La Implementación de las Decisiones de los Órganos del Sistema Interamericano de Derechos Humanos en Chile: Ensayo de un Balance' [The Implementation of Decisions of the Interamerican Human Rights System in Chile: The Study of a Balance] (2015) 22(2) *Revista de Derecho Universidad Católica del Norte Sección: Estudios* 315

Seeman, Miriam, *Water Security, Justice and the Politics of Water Rights in Peru and Bolivia* (Palgrave Macmillan, 2016)

Short, Damien, *Reconciliation and Colonial Power: Indigenous Rights in Australia* (Ashgate, 2008)

Silva, Hernando, 'El Derecho Humano al Agua de los Pueblos Indígenas y el Proyecto de Reforma al Código de Aguas' [Indigenous Peoples' Human Right to Water and the Reform of the Water Code] in *Derechos humanos y pueblos indígenas en Chile hoy: las amenazas al agua, a la biodiversidad y a la protesta social* [Human Rights and Indigenous Peoples in Chile today: Threats to Water, Biodiversity and Social Protest] (Observatorio Ciudadano, 2017)

Simmons, A John, 'Historical Rights and Fair Shares' (1995) 14(2) *Law and Philosophy* 149

Sin, Marcus, Kieran Murray and Sally Wyatt, *The Costs and Benefits of an Allocation of Freshwater to Iwi* (Sapere Research Group, 2014) www.iwichairs.maori.nz/kaupapa/fresh-water/

Solaiman, SM, 'Legal Personality of Robots, Corporations, Idols and Chimpanzees: A Quest for Legitimacy' [2017] (2) *Artificial Intelligence and Law* 155

Solís, D and A Luis, 'Memoria: Comisión Especial de Pueblos Indígenas' [Memoir: Special Commission for Indigenous Peoples] (Comisión Especial de Pueblos Indígenas, 1993)

Sorrenson, MPK, 'Folkland to Bookland: F.D. Fenton and the Enclosure of the Māori "Commons"' (2011) 45 *New Zealand Journal of History* 149

State of Queensland, 'Cape York Draft Water Plan—Statement of Intent' (June 2018) 41 www.dnrme.qld.gov.au

State of Victoria, 'Water for Victoria: Securing Victoria's Future' (2016) www.water.vic.gov.au/water-for-victoria

Stewart, Phoebe, 'Indigenous Water Reserve Policy Tap Turned Off ABC *News (online)*, 10 October 2013

Stone, Christopher, 'Should Trees Have Standing? Towards Legal Rights for Natural Objects' (1972) 45 *Southern California Law Review* 450

Stone, Christopher, *Should Trees Have Standing? Law, Morality, and the Environment* (Oxford University Press, 2010)

Strack, Mick, 'Land and Rivers Can Own Themselves' (2017) 9(1) *International Journal of Law in the Built Environment* 4

Strelein, Lisa and Tran Tran, 'Building Indigenous Governance from Native Title: Moving Away from "Fitting in" to Creating a Decolonized Space' (2013) 18(1) *Review of Constitutional Studies* 19

Strelein, Lisa, 'Conceptualising Native Title' (2001) 23 *The Sydney Law Review* 95

Suárez Gómez, Gabriel Andrés and Giovanni José Herra Carrascal, 'El Agua Como Sujeto de Derechos' [Water as the Subject of Rights] in *Tratado de Derecho de Aguas [Water Law Treatise]* (Universidad Externado de Colombia, 2018) Vol. I

Sunstein, Cass R, 'Incompletely Theorized Agreements in Constitutional Law' (2007) 74(1) *Social Research* 1

Taiepa, Todd et al., 'Co-Management of New Zealand's Conservation Estate by Māori and Pakeha: A Review' (1997) 24(3) *Environmental Conservation* 236

Tan, Poh-Ling, 'National Indigenous Water Planning Forum – A Review of the Legal Basis for Indigenous Access to Water' (Report prepared for the National Water Commission, Griffith Law School, February 2009)

Tan, Poh-Ling, *Legal Issues Relating to Water Use, in Property: Rights and Responsibilities, Current Australian Thinking* (Land and Water Australia, 2002)

Tanasescu, Mihnea, 'The Rights of Nature in Ecuador: The Making of an Idea' (2013) 70(6) *International Journal of Environmental Studies* 846

Tanasescu, Mihnea, *Environment, Political Representation, and the Challenge of Rights: Speaking for Nature* (Houndmills, Basingstoke, Hampshire: Palgrave Macmillan, 2016)

Taylor, Charles, 'Multiculturalism: Examining the Politics of Recognition' in Amy Gutmann (ed.), *The Politics of Recognition* (Princeton University Press, 1994)

Te Aho, Linda, 'Corporate Governance: Balancing Tikanga Maori with Commercial Objectives' [2005] (2) *Yearbook of New Zealand Jurisprudence* 300

Te Aho, Linda, 'Indigenous Challenges to Enhance Freshwater Governance and Management in Aotearoa New Zealand – The Waikato River Settlement' (2009) 20(5–6) *Journal of Water Law* 285

Te Aho, Linda, 'Tikanga Maori, Historical Context and the Interface with Pakeha Law in Aotearoa/New Zealand' [2007] *Yearbook of New Zealand Jurisprudence* 10

Te Puni Kokiri, *He Tirohanga o Kawa Ki Te Tiriti o Waitangi = A Guide to the Principles of the Treaty of Waitangi as Expressed by the Courts and the Waitangi Tribunal* (Te Puni Kokiri, 2002)

Tehan, Maureen, 'A Hope Disillusioned, and Opportunity Lost? Reflections on Common Law Native Title and Ten Years of the Native Title Act' (2003) 27(2) *Melbourne University Law Review* 523

Tehan, Maureen, 'Customary Land Tenure, Communal Titles and Sustainability: The Allure of Individual Title and Property Rights in Australia' in *Comparative Perspectives on Communal Lands and Individual Ownership: Sustainable Futures* (Routledge, 2010)

The Land and Water Forum, 'The Fourth Report of the Land and Water Forum' (2015) www.landandwater.org.nz/site/about_us/default.aspx

The New Zealand Labour Party, 'Manifesto – Water Policy' (2017)

The Waitangi Tribunal, *Report on the Kaituna River Claim (WAI 4)* (Department of Justice, 1984)

The Waitangi Tribunal, *Report on the Manukau Claim (WAI 8)* (Department of Justice, 1985)

The Waitangi Tribunal, *Te Ika Whenua Rivers Report (WAI 212)* (GP Publications, 1998)

The Waitangi Tribunal, *The Mohaka River Report (WAI 119)* (Department of Justice, 1992)

Tobin, Brendan, *Indigenous Peoples, Customary Law and Human Rights: Why Living Law Matters* (Routledge, Taylor & Francis Group, 2014)

Toki, Valmaine, 'Rights to Water an Indigenous Right?' (2012) 20 *Waikato Law Review: Taumauri* 107

Tomas, Nin, 'Maori Concepts of Rangatiratanga, Kaitiakitanga, The Environment, and Property Rights' in David Grinlinton and Prue Taylor (eds.), *Property Rights and Sustainability* (BRILL, Martinus Nijhoff Publishers, 2011) Vol. 11

Tribe, Laurence H, 'Ways Not to Think about Plastic Trees: New Foundations for Environmental Law' (1874) 83 *Yale Law Journal* 1315

Tully, James, *Strange Multiplicity* (Cambridge University Press, 1995)

Urueña, Rene, 'The Rise of the Constitutional Regulatory State in Colombia: The Case of Water Governance' (2012) 6(3) *Regulation & Governance* 282

Valenzuela Reyes, Mylene and Sergio Oliva Fuentealba, *Recopilación de Legislación del Estado Chileno para los Pueblos Indígenas, 1813–2006* [Compilation of Legislation of the State of Chile for Indigenous Peoples 1813–2006] (Librotecnia, 2007)

van der Hammen, Maria Clara, 'The Indigenous Resguardos of Colombia: Their Contribution to Conservation and Sustainable Forest Use' (Netherlands Committee for IUCN The World Conservation Union, NC-IUCN / GSI Series 1) https://cmsdata.iucn.org/downloads/the_indigenous_resguardos_of_colombia.pdf

Van Koppen, Barbara, 'Water Allocation, Customary Practice and the Right to Water' in *The Human Right to Water: Theory, Practice and Prospects* (Cambridge University Press, 2017)

Van Meijl, Toon, 'Changing Property Regimes in Maori Society: A Critical Assessment of the Settlement Process in New Zealand' (2012) 121(2) *The Journal of Polynesian Society* 181

Varsi Rospigliosi, Enrique, *Tratado de Derecho de las Personas* (Coedición Universidad de Lima – Gaceta Jurídica, 2014)

Venn, Tyron J, 'Economic Implications of Inalienable and Communal Native Title: The Case of Wik Forestry in Australia' (2007) 64(1) *Ecological Economics* 131

Vergara Blanco, Alejandro, 'Comentario: Regularización de Derechos de Aguas y Publicidad en el Uso de las Mismas' [Commentary: Regularisation of Water Rights and Publicity of Their Use] (1996) VII *Revista de Derecho de Aguas* 254

Vergara Blanco, Alejandro, 'Contribución a la Historia del Derecho de Aguas: Fuentes y Principios del Derecho de Aguas Chileno Contemporáneo (1818–1981)' [Contribution to the History of Water Law: Sources and Principles of Contemporary Chilean Water Law (1818–1981)] (1989) 1 *Revista de Derecho de Minas y Aguas* 118

Vergara Blanco, Alejandro, 'Las Aguas Como Bien Público (No Estatal) y lo Privado en el Derecho Chileno: Evolución Legislativa y su Proyecto de Reforma' [Water as a Public (Non-State) and Private Good in Chilean Law: Legislative Evolution and Reform] (2002) 1 *Revista De Derecho Administrativo Económico*

Vergara Blanco, Alejandro, *Derecho de Aguas* [Water Law] (Editorial Jurídica de Chile, 1998)

Vergara, Jorge Iván, Hans Gundermann and Rolf Foerster, 'Legalidad y Legitimidad: Ley Indígena, Estado Chileno y Pueblos Originarios (1989–2004)' [Legality and Legitimacy: Indigenous Law, the State, and Native Peoples in Chile (1989–2004)] (2006) 24(71) *Estudios Sociologicos* 331

von der Porten, Suzanne, Rob E de Loë and Deb McGregor, 'Incorporating Indigenous Knowledge Systems into Collaborative Governance for Water: Challenges and Opportunities' (2016) 50(1) *Journal of Canadian Studies/Revue d'études canadiennes* 214

Waitangi Tribunal, *Ko Aotearoa Tēnei: A Report into Claims Concerning New Zealand Law and Policy Affecting Māori Culture and Identity, Te Taumata Tuatahi (WAI 262 Volume 1)* (Legislation Direct, 2011)

Waitangi Tribunal, *The Ngawha Geothermal Resource Report 1993, WAI 304* (Legislation Direct, 2006)

Waitangi Tribunal, *The Stage 1 Report on the National Freshwater and Geothermal Resources Claim: WAI 2358* (Legislation Direct, 2012)

Waitangi Tribunal, *The Whanganui River Report (WAI 167)* (GP Publications, 1999)

Waldron, Jeremy, 'Superseding Historic Injustice' (1992) 103(1) *Ethics The University of Chicago Press* 4

Waldron, Jeremy, *Why Is Indigeneity Important?* (American Political Science Association, 2005) 1

Walker, Margaret Urban, *What Is Reparative Justice?* (Marquette University Press, 2010)

Ward, Alan, *An Unsettled History: Treaty Claims in New Zealand Today* (Bridget Williams Books, 1999)

Watson Hamilton, Jonnette and Nigel Banks, 'Different Views of the Cathedral: The Literature on Property Law Theory', in *Property and the Law in Energy and Natural Resources* (Oxford University Press, 2010) 19

Watson, Susan et al., *Corporate Law in New Zealand* (Thomson Reuters, 2018)

Webber, Jeremy, 'Beyond Regret: Mabo's Implications for Australian Constitutionalism' in Duncan Ivison, Paul Patton and Will Sanders (eds.), *Political Theory and the Rights of Indigenous Peoples* (Cambridge University Press, 2000) 60

West, Robert C, *La Minería de Aluvión en Colombia Durante el Período Colonial* [Alluvial Mining in Colombia during the Colonial Period] (Imprenta Nacional de Colombia, 1972)

Whittemore, Mary Elizabeth, 'Problem of Enforcing Nature's Rights under Ecuador's Constitution: Why the 2008 Environmental Amendments Have No Bite' [2011] (3) *Pacific Rim Law & Policy Journal* 659

Williams, Caroline, 1962-, *Between Resistance and Adaptation: Indigenous Peoples and the Colonisation of the Chocó, 1510–1753* (Liverpool University Press, 2004)

Williams, Jim, 'Resource Management and Māori Attitudes to Water in Southern New Zealand' (2006) 62(1) *New Zealand Geographer* 73

Williams, Joe, Australian Institute of Aboriginal and Torres Strait Islander Studies and Native Title Research Unit, *Confessions of a Native Judge: Reflections on the Role of Transitional Justice in the Transformation of Indigeneity* (Native Title Research Unit, Australian Institute of Aboriginal and Torres Strait Islander Studies, 2008)

Williams, Joseph, 'Lex Aotearoa: An Heroic Attempt to Map the Māori Dimension in Modern New Zealand Law' (2013) 21 *Waikato Law Review* 1

Wilson, Zaryd, 'Whanganui River Representatives Appointed' *New Zealand Herald*, 5 September 2017 www.nzherald.co.nz/nz/news/article.cfm?c_id=1&objectid=11916893

Woodward Law, Capital Chambers and Thorndon Chambers, 'Closing Submissions on Behalf of the Claimant (New Zealand Māori Council) in the National Freshwater and Geothermal Resources Inquiry' (26 October 2018) https://forms.justice.govt.nz

Woodward, AES, 'Aboriginal Land Rights Commission: Second Report' (April 1974)

Yañez, Nancy and Raul Molina, *Las Aguas Indígenas en Chile* [Indigenous Waters in Chile] (LOM Ediciones, 2011)

Young, Simon, *The Trouble with Tradition: Native Title and Cultural Change* (Federation Press, 2008)

Zorzi, C A, 'The *"Irrecognition" of Aboriginal Customary Law* (Brisbane Institute Seminar, 2000)

Zweigert, Konrad and Hein Kötz, *Introduction to Comparative Law* (Oxford University Press, 1998)

B   CASES

1   Australia

Akiba on behalf of the Torres Straight Regional Seas Claim Group v. Commonwealth of Australia [2013] HCA 33

BP (Deceased) on behalf of the Birriliburu People v. State of Western Australia [2014] FCA 715

Brooks on behalf of the Mamu People v. State of Queensland (No. 4) [2013] FCA 1453

Commonwealth v. Yarmirr (2001) 208 CLR 49

De Rose v. South Australia (No. 2) (2005) 145 FCR 290

Harris v. Great Barrier Reef Marine Park Authority [2000] FCA 603

ICM Agriculture v. The Commonwealth [2009] 240 CLR 140

Japalyi v. Northern Territory of Australia [2014] FCA 421

Kaurareg People v. Queensland [2001] FCA 657

Lampton on behalf of the Juru People v. State of Queensland [2014] FCA 736

Mabo and Others v. The State of Queensland [No. 2] (1992) 175 CLR 1

Mabo v. Queensland [No. 1] (1988) 166 CLR 186

Members of the Yorta Yorta Aboriginal Community v. State of Victoria and Others (2002) 214 CLR 422

Milirrpum and Others v. Nabalco Pty Ltd and the Commonwealth of Australia (1971) 17 FLR 141

Northern Territory of Australia v. Alyawarr, Kaytetye, Warumungu, Wakaya Native Title Claim Group (2005) 145 145 FCR 442

Northern Territory v. Arnhem Land Aboriginal Land Trust (2008) 236 CLR 24

Rrumburriya Borroloola Claim Group v. Northern Territory of Australia [2016] FCA 776

Western Australia v. Ward (2002) 213 CLR 1

Wik Peoples v. Queensland (1996) 187 CLR 1

Willis on behalf of the Pilki People v. State of Western Australia [2014] FCA 714

Wurridjal v. The Commonwealth 237 CLR 309

Yanner v. Eaton (1999) 201 CLR 351

2 New Zealand

Aoraki Water Trust v. Meridian Energy Ltd (2005) 2 NZLR 268

Attorney-General v. Ngati Apa (2003) 3 NZLR 643

Beadle v. Minister of Corrections [2002] BCL 701 BC200269088 EC

Bleakley v. Environmental Risk Management Authority [2003] 3 NZLR 213 HC

Fleetwing Farms v. Marlborough [1997] 3 NZLR 257 CA

Huakina Development Trust v. Waikato Valley Authority [1989] 3 NZLR 257

In re the Bed of the Wanganui River [1962] NZLR 600 CA

Lake Omapere [1929] 11 MB BI 253 MLC

McGuire v. Hastings District Council [2000] UKPC 43

McGuire v. Hastings District Court [2002] NZLR 577 PC

Mueller v. Taupiri Coal Mines [1990] 20 NZLR 89 CA

New Zealand Māori Council v. Attorney-General [1987] 1 NZLR 641

New Zealand Māori Council v. Attorney-General [2013] NZSC

Ngati Apa v. Attorney-General (2003) 3 NZLR 643

Outstanding Landscape Protection Society v. Hastings District Council [2008] NZRMA 8 EC

Paki v. Attorney General (No. 1) [2012] NZSC 50

Paki v. Attorney General (No. 2) [2015] 1 NZLR 67 SC

R v. Symonds [1847] NZPCC 387

Takamore v. Clarke [2012] 2 NZLR 733 SC

Tamihana Korokai v. Solicitor General [1912] 32 NZLR 321 CA

Watercare Services Ltd v. Minhinnick (1998) 1 NZLR 63

Wi Parata v. The Bishop of Wellington [1877] 3 NZLR 72

3 Colombia

Andrea Lozano Barragán, Victoria Alexandra Arenas Sánchez, Jose Daniel y Felix Jeffry Rodríguez peña and Others v. The President of the Republic and Others, No STC4360-2018, Corte Suprema de Justicia [Supreme Court of Justice], Sala de Casación Civil [Appeals Chamber] *(Colombia)* (4 April 2018)

Centro de Estudios para la Justicia Social 'Tierra Digna' and Others v. the President of the Republic and Others [2016] Corte Constitucional [Constitutional Court], Sala Sexta de Revision [Sixth Chamber] (Colombia) No T-622 of 2016 (10 November 2016)

Consejo Comunitario Mayor Cuenca Río Cacarica v. the Ministry of Environment and Others, No T-955 of 2003, Corte Constitucional [Constitutional Court], *Sala Octava de Revision [Eighth Chamber] (Colombia)* (17 October 2003)

Fundación Botánica y Zoológica de Barranquilla (Fundazoo) and Others v. Ministeria de Ambiente y Desarollo Sostenible and Others (Unreported, Corte Suprema de Justicia [Supreme Court of Justice], 16 August 2017)

Gustavo Moya Ángel y Otros v. Empresa de Energia de Bogota y Otros [Gustavo Moya Angel and Others v. The Bogotá Energy Company and Others] [2014] Consejo de Estado, Sala de Contencioso Administrativo, Sección Primera AP-25000–23-27–000-2001–90479-01 (28 March 2014)

José Manuel Rodríguez Rangel v. Enrique Chartuny González [1992] Corte Constitucional de Colombia [Constitutional Court of Colombia] T-406/92 (5 June 1992)

Marcos Arrepiche contra el Alcalde del Municipio de Puerto López y el Gobernador del Meta [Marcos Arrepiche v. The Mayor of Puerto López and the Governor of Meta] [2010] Corte Constitucional de Colombia [Colombian Constitutional Court] T-143/10 (26 February 2010)

4 Chile

Admissibility Aymara Indigenous Community of Chusmiza-Usmagama and its Members, Chile [2013] Inter American Commission on Human Rights 1288–06, 29/13 (2013)

Agua Mineral Chusmiza SAIC con Comunidad Indígena Aymara Chusmiza [2006] Juzgado de Letras de Pozo Almonte [Pozo Almonte Civil Court] No. 1194–1996 (Chile) (31 August 2006)

Alejandro Papic con Comunidad Indígena Aymara Chuzmira y Usmagama [2008] Corte de Apelaciones de Iquique [Iquique Court of Appeal] No. 817–2006 (9 April 2008) (Chile)

Alejandro Papic Dominguez con Comunidad Indígena Aymara Chusmiza y Usmagama [2009] Corte Suprema [Supreme Court] No. 2840–2008 (Chile) (25 November 2009)

Codelco Chile División Chuquicamata v. Dirección General de Aguas y otra (Unreported, Corte de Apelaciones de Antofagasta, Chile [Court of Appeal of Antofagasta, Chile], Rol: 14003–2013, 15 May 2014

Comunidad Atacamena Toconce con Essan SA [2004] Corte Suprema [Supreme Court] No. 4064–2004 (Chile) (22 March 2004)

Comunidad Indígena Atacameña con Sociedad Química y Minera SA y otros [2018] Corte Suprema de Chile [Supreme Court of Chile] No 44.255–2017 (Chile) (22 August 2018)

Corporación Movimiento Unitario Campesino y Etnias de Chile con Dirección General de Aguas [2014] Corte Suprema [Supreme Court] (Chile) No. 7899–2013 (Chile) (5 May 2014)

Inscripción Sentencia Derechos de Approvechamiento de Aguas Comunidad Atacamena de Cupo [1997] Segundo Juzgado Civil de Calama, Chile [Second Civil Court of Calama, Chile] NR-II-1387 (19 November 1997)

Inscripción Sentencia Derechos de Approvechamiento de Aguas Comunidad Atacamena de Peine [1997] Segundo Juzgado Civil de Calama, Chile [Second Civil Court of Calama, Chile], NR-II-1383 (19 November 1997)

Inscripción Sentencia Derechos de Approvechamiento de Aguas Associacion Atacamena de Regantes y Agricultores de Aguas Blancas [1997] Segundo Juzgado Civil de Calama, Chile [Second Civil Court of Calama, Chile] NR-II-1381 (19 November 1997)

No 13 Comunidad de Aguas 'Canal dos de Quillagua' (Regularisation decisión 619/ 155, Conservador de Bienes Raises y Comercio de Tocopilla [Real Estate and Business Office of Tocopilla], *10 December 1986)*

5 Other jurisdictions

Acción de hábeas corpus presentada por la Asociación de Funcionarios y Abogados por los Derechos de los Animales (AFADA) [2016] Tercer Juzgado de Garantías, Mendoza (Argentina) P-72.254/15 (3 November 2016)

Calder v. *Attorney-General of British Colombia* (1973) 34 DLR (3rd) 145 (1973) 34 DLR 3rd 145 (Canada)

Case of the Saramaka People v. *Suriname (Preliminary Objections, Merits, Reparations and Costs)* [2007] Inter-American Court of Human Rights Series C No. 172 (28 November 2007)

Case of the Saramaka People v. *Suriname* Inter-Am. Court Hum. Rights Ser. C No. 18 17 Sept. 2003

Chasemore v. *Richards* (1843) 77 ER 82

Colorado River System v. *State of Colorado, Case No. 1:17-cv-02316-RPM (D Colo)*

Coos Bay, Lower Umpqua and Siuslaw Indian Tribes v. *United States* (1938) 87 Ct Cl 143

Delgamuukw v. *British Colombia* (1997) 153 DLR 4th 193

John Young and Co. v. *Bankier Distillery and Co.* [1893] AC 691

Jones v. *Kingborough* (1950) 82 CLR 282

Mayagna (Sumo) Awas Tingni Community v. *Nicaragua (Judgment)* [2001] Inter-American Court of Human Rights (Ser C) Case No. 79 (31 August 2001)

Mohd Salim con State of Uttarakhand and Others: 20 March 2017, High Court of Uttarakhand (WPPIL 126/2014) (India)

Sawhoyamaxa v. Paraguay, IACHR Series C No. 146 (29 March 2006)
Sierra Club v. Morton (1972) 405 USSC 727
Tsilhqot'in Nation v. British Colombia [2014] SCC 44
United States v. Santa Fe Public Railroad Company (1941) 314 US 339 62 Ct 248

C LEGISLATION

1 Australia
Aboriginal and Torres Strait Islander Act 1995 (Cth)
Aboriginal and Torres Strait Islander Amendment (Indigenous Land Corporation) Act 2018 (Cth)
Aboriginal Land Act 1991 (Qld)
Aboriginal Land Fund Act 1974 (Cth)
Aboriginal Land Rights (Northern Territory) Act 1976 (Cth)
Aboriginal Land Rights Act 1983 (NSW)
Aboriginal Land Trusts Act 1966 (SA)
Aboriginal Lands Act 1970 (Vic)
Aboriginal Lands Act 1995 (Tas)
Advancing the Treaty Process with Aboriginal Victorians Act 2018 (Vic)
Anangu Pitjantjatjara Yankunytjatjara Land Rights Act 1981 (SA)
Basin Plan 2012 (Cth)
Mineral, Water and Other Legislation Amendment Bill 2017 (Qld)
Native Title Act 1993 (Cth)
Racial Discrimination Act 1975 (Cth)
Rights in Water and Irrigation Act 1914 (WA)
The Irrigation Act 1886 (Vic)
Traditional Owner Settlement Act 2010 (Vic)
Water Act 1989 (Vic)
Water Act 1992 (NT)
Water Act 2000 (Qld)
Water Act 2007 (Cth)
Water and Catchment Legislation Amendment Bill 2017 (Vic)
Water Management (General) Regulation 2011 (NSW)
Water Management Act 2000 (NSW)
Water Plan (Cape York) 2018 (Draft) (Qld)
Water Resource (Gulf) Plan 2007 (Qld)
Water Resource (Mitchell) Plan 2007 (Qld)
Water Resource (Wet Tropics) Plan 2013 (Qld)
Water Resources Act 2007 (ACT)
Water Sharing Plan for the Bellinger River Area Unregulated and Alluvial Water Sources 2008 (NSW)

Water Sharing Plan for the Coffs Harbour Area Unregulated and Alluvial Water Sources 2009 (NSW)

Water Sharing Plan for the Lower North Coast Unregulated and Alluvial Water Sources 2009 (NSW)

Water Sharing Plan for the Tweed River Area Unregulated and Alluvial Water Sources 2010 (NSW)

Yarra River Protection (Wilip-Gin Birrarung Murron) Act 2017 (Vic)

2   New Zealand

Central North Island Forests Land Collective Settlement Act 2008 (NZ)

Coal Mines Act 1979 (NZ)

Coal Mines Amendment Act 1903 (NZ)

Coal Mines Amendment Act 1903 (NZ)

Conservation Law Reform Act 1990 (NZ)

Crown Forest Assets Act 1989 (NZ)

Maori Fisheries Act 2004 (NZ)

Native Lands Act 1862 (NZ)

Native Lands Act 1865 (NZ)

New Zealand Bill of Rights Act 1990 (NZ)

New Zealand Settlements Act 1863 (NZ)

Ngaa Rauru Kiitahi Claims Settlement Act 2005 (NZ)

Ngāi Tahu Claims Settlement Act 1998 (NZ)

Ngāti Awa Claims Settlement Act 2005 (NZ)

Ngāti Rangi Claims Settlement Bill 2018 (NZ)

Resource Management Act 1991(NZ)

State-Owned Enterprises Act 1986 (NZ)

Tauranga Moana Iwi Collective Redress and Ngā Hapū o Ngāti Ranginui Claims Settlement Bill (NZ)

Te Arawa Lakes Settlement Act 2006

Te Awa Tupua (Whanganui River Claims Settlement) Act 2017 (NZ)

Te Ture Whenua Māori Act 1993 (NZ)

Te Urewera Act 2014 (NZ)

The Māori Fisheries Act 2004 (NZ)

Treaty of Waitangi (Fisheries Claims) Settlement Act 1992 (NZ)

Treaty of Waitangi Act 1975 (NZ)

Waikato-Tainui Raupatu Claims (Waikato River) Settlement Act 2010 (NZ)

Water and Soil Conservation Act 1967 (NZ)

Water Power Act 1903 (NZ)

3   Colombia

Acta de Confederación de la Provincias Unidas de la Nueva Granada [Act of Confederation of the United Provinces of New Granada] *1811* (Colombia)

Código Civil [Civil Code] *1887* (Colombia)

Código Nacional de Recursos Naturales Renovables y de Protección al Medio Ambiente (Decreto 2811 del 18 de Diciembre de 1974) [National Code of Renewable Natural Resources and Protection of the Environment (Decree 2811 of 18 December 1974] (Colombia)

Constitución de Rionegro 1863 [Constitution of Rionegro 1863] (Colombia)

Constitución Política de Colombia 1886 [Political Constitution of Colombia 1886] (Colombia)

Constitución Política de Colombia 1991 [Political Constitution of Colombia 1991] (Colombia)

Constitución Política de la Confederación Granadína 1858 [Political Constitution of the Granadine Federation 1858] (Colombia)

Constitución Política de la Nueva Granada 1853 [Political Constitution of New Granada 1853] (Colombia)

Decree 1541 1978 (Colombia)

Decree 2164 1995 (Colombia)

Ley de Protección Animal Número 1774 de 2016 [Animal Protection Act 2016] (Colombia)

4   Chile

Código Civil de la República de Chile [Civil Code of the Republic of Chile] *1855* (Chile)

Código de Aguas 1951 [Water Code 1951] (Chile)

Constitución Política de la República de Chile [Political Constitution of the Republic of Chile] *1980* (Chile)

Decreto 30 'Crea Comision Especial de Pueblos Indígenas' [Creating the Special Commission of Indigenous Peoples] *1990* (Chile)

Decreto 395 Que Aprueba el Reglamento Sobre el Fondo de Tierras y Aguas 1994 [Decree 395 Approving the Land and Water Fund Regulations 1994] (Chile)

Decreto Ley 2.603 Modifica y Complementa Acta Constitucional N° 3; y Establece Normas Sobre Derechos de Aprovechamiento de Aguas y Facultades para el Establecimiento del Regimen General de las Aguas [Decree Law 2.603 To Modify and Complement Constitutional Act 3; and Establish Rules about Water Rights and Arrangements for the Establishment of a General Water Regime] *1979* (Chile)

Decreto Ley No 2.568 1979 (Modifica Ley N° 17.729, Sobre Proteccion de Indígenas, y Radica Funciones del Instituto de Desarrollo Indígena en el Instituto de Desarrollo Agropecuario) [Decree Law No. 2.568 1979] (Modifies Law No. 17.729, about Protection of the Indigenous, and Transfers the Functions of the Institute of Indigenous Development to the Institute of Agricultural and Fishing Policy) (Chile)

Decreto No 60 1964 (Aprueba el Reglamento Organico de la Dirección de Asuntos Indígenas) [Decree No. 60/1964 (Approve the Regulation of the Directorate for Indigenous Issues)] (Chile)

Decreto Supremo 236 [Supreme Decree 236] *2 October 2008 (Diario Oficial* [Official Gazette] 14 October 2008) (Chile)

Decreto Supremo N°1.220 Que Aprueba el Reglamento del Catastro Público de Aguas [Supreme Decree No. 1.220 Approving the Public Water Cadastre Regulations] (Diario Oficial [Official Gazette] 25 July 1998) *1997* (Chile)

Ley 19.145 Modifica Articulos 58 and 63 Codigo de Aguas [Law 19.145 to Modify Articles 58 and 63 of the Water Code] *1992* (Chile)

Ley No 16.615 Modifica La Constitucion Politica del Estado 1967 [Law No. 16.615 to Modify the Political Constitution of State] *1967* (Chile)

Ley No 19.253 (Establece Normas Sobre Protección, Fomento y Desarrollo de Los Indígenas, y Crea La Corporación Nacional de Desarrollo Indígena) [Law No. 19.253 (Establish Norms for the Protection, Creation and Development of the Indigenous, and to Create the National Corporation of Indigenous Development)] *1993* (Chile)

Ley No. 17.729 1972 (Establece Normas Sobre Indígenas y Tierras de Indígenas, Transforma la Dirección de Asuntos Indígenas en Instituto de Desarrollo Indígena, Establece Disposiciones Judiciales, Administrativas y de Desarrollo Educacional en la Materia y Modifica o Deroga Los Textos Legales Que Señala) [Law No. 17.729 1971 (Establish Norms about Indigenous People and Lands, Transform the Directorate of Indigenous Issues into an Institute of Indigenous Development, Establish Legal, Administrative and Educational Development Dispositions on the Issue and Modify or Repeal Specified Legal Texts)] (Chile)

Modifica la Ley No 19.253, Relativa a la Proteccion, Fomento y Desarrollo de los Pueblos Indígenas, Estableciendo la Regularizacion de Derechos de Agua Potable Rural [Modification of Law No. 19.253, Relating to the Protection, Promotion and Development of Indigenous Peoples, Establishing the Regularisation of Rights for Rural Drinking Water] *2012* (Chile)

Proyecto de Reforma al Código de Aguas [Water Code Amendment Bill 2018] (Chile)

Reforma el Codigo de Aguas, Examinando del Pago de Patente a Pequenos Productores Agricolas y Campesinos, a Comunidades Agricolas y a Indígenas y Comunidades Indígenas Que se Senalan' [Reform of the Water Code, Examining the Payment of Tax by Small Agricultural Producers, to Agricultural and Indigenous Communities Included] (Boletin No 8315–01) (Chile)

5 Other jurisdictions

Constitución de la República del Ecuador 2008 (Ecuador)
Constitución del Estado Plurinacional de Bolivia 2009 (Bolivia)
Constitución Política de la Ciudad de México 2016 (Mexico)

Ley de Derechos de la Madre Tierra (Ley 071) 2010 (Bolivia)

D INTERNATIONAL TREATIES AND DOCUMENTS

Conference of the Parties, United Nations Framework Convention on Climate Change, Adoption of the Paris Agreement, UN Doc FCCC/CP/ 192015/L.9/ Rev.1 (12 December 2015) Annex ('Paris Agreement')
Convention Concerning Indigenous and Tribal Peoples in Independent Countries (No. 169) [1989] 28 *ILM* 1382 (Entered into Force 5 September 1991) ('Convention 169')
Report to the United Nations General Assembly on Harmony with Nature (UN Doc A/C.2/73/L.39/Rev.1) (21 November 2018)
UN General Assembly, The Human Right to Water and Sanitation (Sixty-Fourth Session, 2010) UN Doc A/64/L.63/Rev.1 2010
United Nations Declaration on the Rights of Indigenous Peoples, GA Res 61/295, UN GAOR, 61st Sess, 107th Plen Mtg, Agenda Item 68, Supp No. 49, UN Doc A/RES/61/295 (2 October 2007, Adopted 13 September 2007)
Declaration of the United Nations Conference on the Human Environment, Report of the United Nations Conference on the Human Environment, Stockholm (5–17 June 1972) A/CONF.48/14/Rev.1 (1972) www.un-documents.net/acon f48-14r1.pdf

E INTERVIEWS

Interview with Viviana González Moreno (Bogotá, 2 September 2017)
Interview with Maria Angelica Alegria (Santiago, 17 November 2011)
Interview with Milka Castro (Santiago, 9 November 2011)
Interview with Carlos Herrera Inzunza (Temuco, 11 November 2011)
Interview with Hernando Silva (Temuco, 11 November 2011)
Interview with Juan Carlos Araya (Santiago, 15 November 2011)
Interview with Waldo Contreras (Santiago, 16 November 2011)
Interview with Gonzalo Arevalo (Santiago, 18 November 2011)
Interview with Rodrigo Weisner (Santiago, 18 November 2011)
Interview with Daniela Rivera (Santiago, 22 November 2011)
Interview with Nancy Yanez (Santiago, 22 November 2011)
Interview with Manuel Cuadra (Antofagasta, 23 November 2011)
Interview with Manuel Prieto (Santiago, 4 September 2013)
Interview with Diego Sotomayor (Santiago, 23 December 2013)
Interview with Fransisco Huenchumilla (Temuco, 22 January 2014)
Interview with Alvaro Duran (Santiago, 29 August 2017)
Interview with Andres Gomez Rey (Bogotá, 30 August 2017)
Interview with Pilar Garcia (Bogotá, 30 August 2017)

Interview with Cristian Carabaly (Bogotá, 31 August 2017)
Interview with Eugenia Ponce (Bogotá, 1 September 2017)
Interview with Sara Larrain (Temuco, 6 September 2017)
Interview with Carlos Estevez (Santiago, 8 September 2017)
Interview with Sophia Angelis (Melbourne, 6 July 2018)
Interview with Bryony Grice (Melbourne, 9 July 2018)
Interview with Erin O'Donnell (Melbourne, 9 July 2018)
Interview with Reuben Berg (Melbourne, 10 July 2018)
Interview with Lauren Butterly (Sydney, 11 July 2018)
Interview with Fred Hooper (Queensland, 12 July 2018)
Interview with Rene Woods (New South Wales, 12 July 2018)
Interview with Poh-Ling Tan (Brisbane, 13 July 2018)
Interview with Felipe Clavijo (Bogotá, 12 August 2018)
Interview with Richard Fowler (Wellington, 17 August 2018)
Interview with Baden Vertongen (Wellington, 28 August 2018)
Interview with Tania Gerrard (Wellington, 28 August 2018)
Interview with Chris Finlayson (Wellington, 4 October 2018)
Interview with Jacinta Ruru (Christchurch, 7 December 2018)
Interview with Riki Ellison (Wellington) (5 February 2019)

F   OTHER

Australian Bureau of Statistics, *Census: Aboriginal and Torres Strait Islander Population* (27 June 2017) www.abs.gov.au
Caribe, Comisión Económica para América Latina y el, *Los pueblos indígenas en América Latina. Avances en el último decenio y retos pendientes para la garantía de sus derechos. Síntesis* [Indigenous Peoples in Latin America. Advances in the last decade and current objectives for the guarantee of their rights] (27 October 2014) www.cepal.org
Chapman Tripp, *Te Ao Māori: Trends and Insights* (Piripi 2017) www .chapmantripp.com
Colombia: A Country Study Library of Congress, Washington, D.C. 20540 USA www.loc.gov/item/2010009203/
Commonwealth of Australia and the Governments of New South Wales, Victoria, Queensland, South Australia, the Australian Capital Territory and the Northern Territory, 'Intergovernmental Agreement on a National Water Initiative'
Community Environmental Legal Defence Fund, *Press Release: Ho-Chunk Nation General Council Approves Rights of Nature Constitutional Amendment* (18 September 2016) https://celdf.org/2016/09/press-release-ho-chunk-nation-general-council-approves-rights-nature-constitutional-amendment

Council of Australian Governments, 'Water Reform Framework (Communiqué)' www.environment.gov.au/water/publications/action/pubs/policyframework.pdf

David Espinoza Quezada, 'Regularizaciones Remitidas por DGA Región de Antofagasta a Tribunales Competentes' [Regularisations Remitted by the DGA in the Region of Antofagasta to Competent Courts]

Earth Law Center, *Mexico on the Vanguard for Rights of Nature* (21 November 2017) Earth Law Centre www.earthlawcenter.org/blog-entries/2017/11/mexico-on-the-vanguard-for-rights-of-nature

El Gobierno Colombiano Recibe Propuestas Para el Plan de Descontaminación del Río Atrato www.iagua.es/noticias/minambiente/gobierno-colombiano-recibe-propuestas-plan-descontaminacion-rio-atrato

El Ministerio de Ambiente Ordena Suspender la Minería en el Río Quito, Chocó [The Minister for the Environment Orders the Suspension of Mining in the Quito River, Chocó] El Espectador www.elespectador.com/noticias/medio-ambiente/ordenan-suspender-toda-la-mineria-en-uno-de-los-rios-mas-importantes-de-choco-articulo-754898

Goodland, Marianne, *Lawsuit Seeking 'personhood' for Colorado River Dismissed* (5 December 2017) Colorado Springs Gazette http://gazette.com/lawsuit-seeking-personhood-for-colorado-river-dismissed/article/1616604

Guardianes del río Atrato: amenazados e ignorados [Guardians of the Atrato River: threatened and ignored] Colombia 2020 https://colombia2020.elespectador.com/territorio/guardianes-del-rio-atrato-amenazados-e-ignorados

Ministerial de Ambiente y Desarrollo Sostenible Resolución N. 0115 de 26 Enero 2018 Por Medio de la Cual Se Asignan Funciones al Interior del Ministerio de Ambiente y Desarollo Sostenible a Efectos de dar Cumplimiento a lo Dispuesto en la Sentencia T'622 de 2016 [Ministry of Environment and Sustainable Development Resolution No. 0115 of 26 January 2018 Which Assigns Functions Internal to the Ministry of Environment and Sustainable Development in Compliance with that Provided for in Sentence T-622 of 2016] *(Colombia)*

Iwi Chairs Forum – Fresh Water https://iwichairs.maori.nz/our-kaupapa/fresh-water/

Laura Villa, *The Importance of the Atrato River in Colombia Gaining Legal Rights* (5 May 2017) Earth Law Center www.earthlawcenter.org/new-blog-1/2017/5/the-importance-of-the-atrato-river-in-colombia-gaining-legal-rights

Ministerio de Ambiente y Desarollo Sostenible [Ministry for the Environment and Sustainable Development, 'Decreto No. 1148 por el Cual se Designa al Representante de los Derechos del Río Atrato En Cumplimiento de la Sentencia de T-622 de 2016 de la Corte Constitucional [Decree No. 1148 Designating the Representative of the Rights of the Atrato River Giving Effect to Sentence T-622 of 2016 of the Constitutional Court']

Ministry for the Environment, 'National Policy Statement for Freshwater Management (Amended 2017)' www.mfe.govt.nz/sites/default/files/media/Fresh%20water/nps-freshwater-ameneded-2017_0.pdf

Ministry for the Environment, *Kahui Wai Māori Group to Work on Freshwater Press Release* (3 August 2018) www.beehive.govt.nz/release/kahui-wai-m%C4%81ori-group-work-freshwater

Murray Lower Darling Rivers Indigenous Nations and Northern Murray–Darling Basin Aboriginal Nations, *Agreed Definition of Cultural Flows* www.mdba.gov.au/explore-the-basin/communities/indigenous-communities/

National Native Title Tribunal, *National Native Title Register* www.nntt.gov.au

New South Wales Government Department of Primary Industries, *How Water Sharing Plans Work* (2012) www.water.nsw.gov.au/Water-management/Water-sharing-plans/How-water-sharing-plans-work/how-water-sharing-plans-work/default.aspx

North Australian Indigenous Land and Sea Management Alliance, A *Policy Statement on North Australian Indigenous Water Rights* (November 2009) www.nailsma.org.au/nailsma/forum/downloads/Water-Policy-Statement-web-view.pdf

OAS, IACHR, *Progress on Friendly Settlements in Petitions and Ongoing Cases before the IACHR Concerning Chile* (23 June 2016) www.oas.org/en/iachr/media_center/PReleases/2016/085.asp

Sebastian Piñera, *en Nuestro Gobierno Vamos a Asegurar la Disponibilidad de Agua* [In Our Government we are Going to Secure Water Availability] (12 October 2018) www.sebastianpinera.cl/sebastian-pinera-en-nuestro-gobierno-vamos-a-asegurar-la-disponibilidad-de-agua

The Hague Principles for a Universal Declaration on Human Responsibilities and Earth Trusteeship | Earth Trusteeship www.earthtrusteeship.world/the-hague-principles-for-a-universal-declaration-on-human-responsibilities-and-earth-trusteeship/

The New Zealand Labour Party, *Clean Water for Future Generations* (2018) www.labour.org.nz/water

¡Todas y Todos Somos Guardianes del Atrato! [We Are All the Guardians of the Atrato!] (4 September 2017) https://co.boell.org/es/2017/09/04/todas-y-todos-somos-guardianes-del-atrato

Commonwealth of Australia, *Parliamentary Debates*, 16 November 1993

Australian Museum, *Introduction to Indigenous Australia* (15 May 2018) https://australianmuseum.net.au/indigenous-australia-introduction

Referendum Council, *Uluru Statement from the Heart* (26 May 2017) www.referendumcouncil.org.au

INDEX

CAMBRIDGE STUDIES IN LAW AND SOCIETY

Edited by John R. Bowen, Christophe Bertossi, Jan Willem Duyvendak, and Mona Lena Krook

Environmental Litigation in China: A Study in Political Ambivalence
Rachel E. Stern

Indigeneity and Legal Pluralism in India: Claims, Histories, Meanings
Pooja Parmar

Paper Tiger: Law, Bureaucracy and the Developmental State in Himalayan India
Nayanika Mathur

Religion, Law and Society
Russell Sandberg

The Experiences of Face Veil Wearers in Europe and the Law
Edited by Eva Brems

The Contentious History of the International Bill of Human Rights
Christopher N. J. Roberts

Transnational Legal Orders
Edited by Terence C. Halliday and Gregory Shaffer

Lost in China? Law, Culture and Society in Post-1997 Hong Kong
Carol A. G. Jones

Security Theology, Surveillance and the Politics of Fear
Nadera Shalhoub-Kevorkian

Opposing the Rule of Law: How Myanmar's Courts Make Law and Order
Nick Cheesman

The Ironies of Colonial Governance: Law, Custom and Justice in Colonial India
James Jaffe

The Clinic and the Court: Law, Medicine and Anthropology
Edited by Ian Harper, Tobias Kelly, and Akshay Khanna

A World of Indicators: The Making of Government Knowledge through Quantification
Edited by Richard Rottenburg, Sally Engle Merry, Sung-Joon Park, and Johanna Mugler

Contesting Immigration Policy in Court: Legal Activism and Its Radiating Effects in the United States and France
Leila Kawar

The Quiet Power of Indicators: Measuring Governance, Corruption, and Rule of Law
Edited by Sally Engle Merry, Kevin Davis, and Benedict Kingsbury

www.ingramcontent.com/pod-product-compliance
Ingram Content Group Australia Pty Ltd
76 Discovery Rd, Dandenong South VIC 3175, AU
AUHW010841240225
407430AU00009B/77